T 32

# CELL MOVEMENT
## AND CELL
# BEHAVIOUR

# CELL MOVEMENT
## AND CELL
# BEHAVIOUR

J. M. LACKIE

*Department of Cell Biology, University of Glasgow*

London
ALLEN & UNWIN
Boston          Sydney

**Allen & Unwin (Publishers) Ltd,**
**40 Museum Street, London WC1A 1LU, UK**

Allen & Unwin (Publishers) Ltd,
Park Lane, Hemel Hempstead, Herts HP2 4TE, UK

Allen & Unwin, Inc.,
8 Winchester Place, Winchester, Mass 01890, USA

Allen & Unwin (Australia) Ltd,
8 Napier Street, North Sydney, NSW 2060, Australia


First published in 1986

**British Library Cataloguing in Publication Data**

Lackie, J. M.
Cell movement and cell behaviour.
1. Cells – Motility
I. Title
574.87′64     QH647
ISBN 0-04-574034-8
0-04-574035-6 Pbk

**Library of Congress Cataloging in Publication Data**

Lackie, J. M.
Cell movement and cell behaviour.
Bibliography: p.
Includes index.
1. Cells – Motility. I. Title. [DNLM: 1. Cell
Movement.  2. Cells – physiology.   QH 647 L141c]
QH647.L33  1985     575.87′64     85-20041

Set in 9 on 10 point Melior by Columns, Caversham,
Reading
and printed in Great Britain by Mackays of Chatham

*In memory of my mother*

# PREFACE

Some years ago a book reviewer, perhaps with Freudian honesty, remarked that the book in question 'filled a much needed gap in the literature'. That phrase has haunted the writing of this gap-filler and this preface may be considered an apologia.

For a number of years I have found myself teaching various groups of students about cell locomotion and cell behaviour; sometimes science students specializing in cell or molecular biology, sometimes immunologists or pathologists who only wanted a broad background introduction. Those students who were enthusiastic, or who wished to appear so, asked for a general background text (to explain my lectures perhaps), and that is what I hope this book will provide. With luck, other scientists who have only a peripheral interest in cell movement will also find this a useful overview. The more proximate origin of the book was a special 'option' subject which I taught for two years to our Senior Honours Cell Biology students in Glasgow, an option cunningly juxtaposed against a very non-cellular topic so that I had a reasonably sized group of test consumers. Since more students chose the option the second time round I felt slightly encouraged. If writing the book does nothing else, it will make it easier to teach the option!

Although I knew from the outset that cell movement was a very broad topic I soon came to realize how small an area I really knew well and how superficial my treatment of many topics was going to be. Inevitably I have relied on other people's reviews and books. The virtue of a monograph is that it is a single person's view of things and is in theory more integrated than a multi-author text; its weakness is that one head contains more ignorance than two, surprising though that might seem. My approach has been to try to extract those things which I myself have found interesting or illuminating and to concentrate more on those areas, sacrificing completeness in so doing. The bibliography is by no means exhaustive; the papers I have quoted are the ones which I have found useful and which are reasonably accessible (not just the ones I have reprints of!), but many excellent papers have been omitted. Too many references make the text unreadable and my intention was to put in only enough to lead the reader into the right part of the primary literature in a fairly direct manner, and I have not attempted to be comprehensive. At the end of each chapter I have listed a few reviews, with cryptic comments on their relevance. The main reference list is at the end of the book. Jargon is always a problem to the uninitiated which is why we use it (it keeps out those who might otherwise see things too clearly) so I have traitorously put in a glossary, which may demystify some terms.

One major problem in dealing with the topic of cell movement is in knowing where to start; should a description of the cellular phenomena precede the detailed discussion of the molecular machinery, or should we start with the motor system and come to the more complicated controls later? Although I have chosen to put the horse (the motor) before the cart, this is not necessarily the way in which the book has to be read. Those who know the general biological background will find the order of the chapters straight-forward; others may prefer to read Chapters 5 and 6 first. There are three major sections to the book, Chapters 2 4 which deal with motors, Chapters 5 and 6 which deal with the mechanism of movement, and Chapters 7, 8, 9, and 10 which deal with factors controlling and directing movement. The sections are more-or-less independent, but I have referred back more than I have referred forward, for obvious reasons. In teaching I have invariably followed the sequence of the chapters.

Several of the discussions draw on an analogy between the behaviour of cells and the behaviour of traffic seen from a vantage point; the analogy is given in detail in Chapter 1, and it is deliberately far-fetched because so many students confuse the use of analogy and homology. The discerning reader may recognize the origin of some passages as verbal harangues aimed at students on sleepy afternoons; perhaps that is no bad thing, since it may also stir the somnolent reader. Not everybody cares for analogy of course, and there are traps in anthropomorphic thinking; in fairness to my colleagues I should point out that they did try hard to curb my excesses.

Although my fondness for analogy has not been cured, the comments of many colleagues have been very helpful, and I am grateful to all those who have listened, commented, argued or even remained indifferent – conspic-uous boredom in the face of an argument is very instructive. I would like to record my gratitude to those who read drafts of chapters for me: John Edwards, John Freer, Bob Hard, Joan Heaysman, Ann Lackie, Doug Neil, Peter Sheterline, Mike Vicker and Peter Wilkinson. Their comments were a great help, and they cannot be blamed for the mistakes and the infelicities of style which are mine alone. Many people have helped by lending material for figures; they are acknowledged in the legends. The first draft of the book was written while visiting Oregon State University at Corvallis, and I am grateful to all those people who made the summer there such an enjoyable experience. Finally, I would like to thank my colleagues in the Cell Biology Department, who put up with my constant talk of writing; Unity Miller who helped transform manuscript to disc; the students who make it all necessary and last, but by no means least, my family who helped, encouraged, interrupted rarely, and constantly brought me back to reality.

J. M. Lackie
*Glasgow*

# ACKNOWLEDGEMENTS

The following organizations and individuals are thanked for permission to reproduce illustrative material (figure numbers in parentheses):

Figure 2.1 reproduced from *Mechanisms of cell motility: molecular aspects of contractility* (P. Sheterline) by permission of the author (who also gave permission for 2.14b) and Academic Press; Figure 2.5 reproduced from *The Journal of Cell Biology* 1978, **79**, 846–52 by copyright permission of The Rockefeller University Press; P. J. Knight (2.6); H. Elder (2.10d & e, 2.11); Figure 2.13d reproduced from *The Journal of Cell Biology* 1970, **44**, 192–209 by copyright permission of The Rockefeller University Press; D. Szollosi (2.13d); G. Campbell (2.14a); P. Wilkinson (2.16b, 8.8, 9.12c); Scothorne (2.19); Figures 3.1b & c reproduced from *The Journal of Cell Biology* 1979, **80**, 266–76 by copyright permission of The Rockefeller University Press; Figure 3.5 reproduced from *Cell motility* (H. Stebbings & J. S. Hyams) by permission of H. Stebbings and Longman; Figure 3.8 reproduced from *The Journal of Cell Biology* 1980, **87**, 509–15 by copyright permission of The Rockefeller University Press; U. Euteneuer (3.8); K. Vickerman (3.10b); L. Tetley (3.10b, 5.4, 6.14); Figures 3.14 & 5.5 reproduced from M. A. Sleigh and D. I. Barlow, *Symp. Soc. Exp. Biol.* **35**, 139–58 by permission of M. A. Sleigh and Cambridge University Press; Figure 3.15 reproduced from K. Takahashi *et al.*, *Symp. Soc. Exp. Biol.* **35**, 159–78 by permission of K. Takahashi and Cambridge University Press; Figure 3.16 reproduced from P. Satir, *Symp. Soc. Exp. Biol.* **35**, 179–202 by permission of the author and Cambridge University Press; Figure 3.18 reproduced from *The Journal of Cell Biology* 1977, **74**, 377–88 by copyright permission of The Rockefeller University Press; K. McDonald (3.18); C. D. Ockleford (3.21a); Figures 3.21b & c reproduced from C. D. Ockleford and J. B. Tucker, *J. Ultrastruct. Res.* 44, 369–87 by permission of C. D. Ockleford and Academic Press; Figure 3.22 reproduced from *The Journal of Cell Biology* 1974, **61**, 757–79 by copyright permission of The Rockefeller University Press; D. B. Murphy (3.22); L. G. Tilney (3.22, 4.5–8); Figure 4.1 reproduced from R. Kamiya *et al.*, *Symp. Soc. Exp. Biol.* **35**, 53–76 by permission of R. Kamiya and Cambridge University Press; W. B. Amos (4.4); Figure 4.4c reprinted by permission from *Nature* **236**, 301–4, © 1972 Macmillan Journals; Figures 4.5 and 4.6 reproduced from *The Journal of Cell Biology* 1975, **64**, 289–310 by copyright permission of The Rockefeller University Press; Figure 4.7 reproduced from *The Journal of Cell Biology* 1982, **93**, 812–19 by copyright permission of The Rockefeller University Press; S. Inoué (4.7); Figure 4.8a reproduced from *The Journal of Cell Biology* 1978, 77, 551–64 by copyright permission of The Rockefeller University Press; Figure 4.8b reproduced from *The Journal of Cell Biology* 1979, **81**, 608–23 by copyright permission of The Rockefeller University Press; Figure 5.1a reproduced from *Cell biology in medicine* (M. A. Sleigh) by permission of the author and John Wiley & Sons; The Company of

Biologists (5.1b, 6.12); C. J. Brokaw (5.1b); B. Nisbet (5.3); Figures 5.7 and 9.6 reproduced from *Protozoology* (K. G. Grell) by kind permission of the author and Springer Verlag; Figure 5.9 reproduced from *The Journal of Cell Biology* 1981, **89**, 495–509 by copyright permission of The Rockefeller University Press; S. L. Tamm (5.9); Figure 5.10 reproduced from M. E. J. Holwill, *Symp. Soc. Exp. Biol.* **35**, 289–312 by permission of the author and Cambridge University Press; D. L. Taylor (6.5–6); Figure 6.6 reproduced from *The Journal of Cell Biology* 1981, **91**, 26–44 by copyright permission of The Rockefeller University Press; Figures 6.11 and 10.1 reproduced from M. Abercrombie, *Proc. R. Soc. Lond.* **207**, 129–47 by kind permission of the publisher; J. P. Heath (6.11–12, 6.15b, 6.17, 8.5, 10.1); Figure 6.15a reproduced from Abercrombie *et al.*, *Exp. Cell Res.* **62**, 389–98 by permission of Academic Press; T. M. Preston and D. H. Davies (6.16); Figures 6.17b & c reprinted by permission from *Nature* **302**, 532–4, © 1983 Macmillan Journals; M. Lydon (6.18, 7.9b, 8.4b); D. Bray (6.20); Figure 6.21 reproduced from *The Journal of Cell Biology* 1981, **91**, 26–44 by copyright permission of The Rockefeller University Press; S. Ward (6.21); Anatomy Dept, Glasgow University (7.6); A. F. Brown (7.8); Figures 7.9a & 10.4 reproduced from *Cell Behaviour* (R. Bellairs *et al.*, eds) by permission of the authors and Cambridge University Press; Figures 8.2 & 9.8–9 reproduced from *Biology of the chemotactic response* (J. M. Lackie & P. C. Wilkinson, eds) by permission of Cambridge University Press; G. A. Dunn (8.2, 8.5–6); Figure 8.3 reproduced from *The Journal of Cell Biology* 1984, **98**, 2204–14 by copyright permission of The Rockefeller University Press; M. G. Vicker (8.3); Figure 8.5 reproduced from G. A. Dunn & J. P. Heath, *Exp. Cell Res.* **101**, 1–14 by permission of the authors and Academic Press; Figure 8.7 reproduced from P. Weiss, *Int. Rev. Cytol.* **7**, 391–423 by permission of Academic Press; Figure 8.9 reproduced from P. C. Wilkinson & J. M. Lackie, *Exp. Cell Res.* **145**, 255–64 by permission of Academic Press; *Journal of General Microbiology* (9.7); P. C. Newell (9.7–9); Figure 10.2 taken from J. E. M. Heaysman & S. M. Pegrum, *Exp. Cell Res.* **78**, 71–8 & 479–81 by permission of Academic Press and J. Heaysman; D. E. Sims (10.3); Figure 10.6b reproduced from P. B. Armstrong, *Bioscience* **27**, 803–9, © 1977 by the American Institute of Biological Sciences; C. Kerr (10.7).

# CONTENTS

CONTENTS

# 1
# INTRODUCTION

## 1.1 Why is the movement of cells interesting?

'It's alive – it's moving!'

All the movements of which living systems are capable derive from movement within cells or groups of cells, and the study of cellular movement forms a bridge between the disciplines of biochemistry and of whole-organism biology or physiology. A lot of cellular activities can be considered under the general heading of movement, and a range of different cellular systems must operate to bring about movement; this means that we need a multidisciplinary approach, and this is not an area solely for the narrow specialist. The phenomena of cell movement have an appeal at a primitive or childlike level. Movement catches the eye and begs the questions: 'how?, why?, what for?'. The apparent naivety of the questions does not mean that they will be answered easily.

## 1.2 What do we mean by movement?

One of the characteristics of living things is their ability to move. The movement may be conspicuous, as with animals, or slow and rather restricted, as with plants: we would expect to find movement even at the cellular level. Movement is, however, a very broad term and we must distinguish between movements which are brought about by the activities of the organism and movements imposed by external forces, as well as distinguishing the various categories of active movement which occur. An analytical approach to the question of what we mean by movement will assist us in subdividing the topic of cell movement and will form the basis of the subdivisions of the book.

We will not concern ourselves with the movements of small molecules, although the movement of ions across membranes is crucial to the normal function of most cells, and the movement of specific molecules from one compartment to another plays an important part in the internal economy of the cell. With the exception of active transport, these movements are imposed by external forces, the kinetic energy of the molecular species involved, and are regulated by the manipulation of permeability. Arbitrarily we will limit our discussion to the movement of very large molecules which are in multimolecular arrays, although in many cases the movement of sub-units will be important and the boundary between this level and that of small molecules is somewhat imprecise. The movement of organelles within the cell is a category of movement that we will need to consider, and it is clear that cells have had to tackle this problem as their size increased. Within a bacterium the diffusion of molecules suffices to keep things going, bringing substrates into contact with enzymes and permitting all of the cytoplasm to be maintained at an homogeneous level of ions, oxygen and so on. Once organelles become 'necessary' to increase the local concentrations of enzymes, substrates, or information, then inequality of production, inhomog-eneity of the cytoplasm and the separation of nucleus from cytoplasm means that some distributive mechanism is needed. Once the cell becomes very large, as for example the giant amoeba *Chaos chaos*, the acellular slime mould *Physarum polycephalum*, or the internodal cells of characean algae such as *Nitella*, then the 'soup' must be stirred by active cytoplasmic streaming lest it 'burn onto the pan'. Such streaming probably occurs on a small scale in most cells but its conspicuous manifestations in these giant cells capture our attention and provide models with which to play.

No student of cell biology seriously thinks of cells as smooth spheres, although sometimes we pretend that they are, and appendages dangle from many cells, often with dimensions at around the limit of resolution of the light microscope. These appendages include those designed for specific purposes: for moving the cell around, for increasing the surface area of the cell, or serving as a reservoir of membrane to permit rapid expansion of the cell volume or to allow a marked change in geometry. In fact appendages for a variety of functions exist and we will need to consider them in some detail since they are not static projections but move, either in space or time. The cell, while remaining stationary, may gesture wildly with its appendages, reaching out to sample or manipulate the environment.

The cell may change its shape conspicuously: in cytokinesis where the change is irreversible, in phagocytosis when objects are engulfed within a pseudopodial cup, in contraction, as in a muscle cell, and during the movement of the whole cell. Less conspicuous shape changes may cause an epithelial sheet to curl up into a tube, cause a tube of cells such as a capillary blood vessel to change its internal diameter, or may cause a multicellular array to move, as with the shape changes of cells within the pulvinus which move the whole leaf.

Last, but by no means least, the cell may move position, moving from one location to another, either freely within its environment as in the case of unicellular organisms, or within a multicellular array as, for example, leucocytes during inflammation, tumour cells in malignant invasion, epithelial sheets in wound healing and embryonic cells in **morphogenesis**. The locomotory activity of cells is of particular interest because it is perhaps the most sophisticated level of cellular activity, certainly the most complex of cellular movements. Some of the movements mentioned previously will play a part in cell locomotion, but the activity is coordinated to achieve an overall movement which may be influenced in a variety of ways by the nature of the environment: the behaviour of a cell moving as a unit offers a variety of problems which will occupy much of the last part of this book.

We may, therefore, subdivide movement into the movement of:

(a)    molecules,
(b)    parts within the cell,
(c)    appendages,
(d)    the whole cell in one place,
(e)    the whole cell from place to place.

We should then begin to ask another sort of question, 'what causes the various kinds of movement?'

## 1.3    What causes movement?

Having in the previous section subdivided 'movement' according to what is happening, we can go further and ask how the movement comes about, and this in turn will generate a lot of other questions. Within the general topic of cell movement we can recognize certain levels or areas that are of interest. The way in which the motor system is organized and controlled within the cell has an appeal, particularly to the more biochemically and ultrastructurally minded. Problems concerning the nature of the components, the way in which they interact to generate and control movement, to circulate the cytoplasm, to redistribute membrane proteins or to distribute the products of chromosomal division are among the other questions that come to mind. The contribution of the motor system to the mechanical properties of the cell as a whole, and the influence of the motor system on the shape of the whole cell are related questions.

The movement of appendages on a cell may influence the local environment: ciliary tracts on the gills of bivalve molluscs beat in a coordinated way to sort and to transport food particles; **microvilli** on intestinal epithelial cells may stir the local nutritive soup, facilitating absorption; coordinated shape changes in groups of cells may generate waves of peristalsis or, in the most dramatic of examples, drive whole animals by the operation of skeletal (and cardiac) musculature. Movements are involved in phagocytosis, and phagocytes within Metazoa are important elements of the body's defence system. Small free-living organisms are motile by means of the activity of cellular appendages that enable the organism to swim, or by the ability of the cell to crawl. In the examples above the motile activity is being considered in a more complex way, the activity of the motor is taken for granted, the question of interest is the use to which the motor is being put. Then we can ask questions about cell behaviour, much as animal behaviourists and plant physiologists have done with whole multicellular organisms. Why, for example, do cells accumulate in a particular place, how do they respond to directional cues, and what happens when they collide with one another?

A tertiary level of interest in cell movement is in the consequences of movement and the way in which cellular movements influence the spatial organization of tissues and of the embryo – the processes of **histogenesis** and morphogenesis. Pathologists also have an interest in movement: wound-healing involves movement of cells, as does **invasion** of tissues by neoplastic cells. The movements of leucocytes in inflammation and the coming together of cells cooperating in immune responses are motile phenomena, and their perturbation is of clinical significance.

Not only is movement complex but in all of the examples mentioned, even malignant invasiveness, the outcome of movement is a precise ordering of the population. This is particularly true in morphogenetic processes and we will be interested, particularly in the later chapters, in trying to determine the environmental factors that influence movement and the cessation of movement.

As indicated above, cell movement is a broad topic and it has a further interest for the cell biologist: it is one of the most complex of cellular activities and involves the cooperative interaction of many cellular processes. The ability of a cell to move depends on the normal functioning of a range of cellular systems associated with energy redistribution, contractile events and environmental sensing; the challenge of trying to understand this activity lies partly in its very complexity.

## 1.4  An analogy

The aim of this book is to cover the general topic of cell movement, a fairly ambitious task. It may be useful to develop an analogy to which we can refer and which will to some extent dictate the structure of the succeeding chapters.

Imagine that you stand or hover at some vantage point overlooking the streets of a busy town and its surrounding countryside. That the town is inhabited is clear from the movement of small objects – people, dogs, vehicles etc. The larger moving objects are motor vehicles, and as an enquiring person you wonder how they move . . . what sort of engine do they have, is it reciprocating or rotary, is it petrol, diesel or coal that provides the fuel? Is the engine mounted in front or behind? How does the engine make the car move? A wheeled vehicle may be front- or rear-wheel drive (or both), and the careful observer may notice that the vehicles driving on smooth city streets are less likely to have four-wheel drive than those on rough country roads. The tracked vehicles are either tanks or bulldozers, and stop for nothing. Cars and other vehicles, being larger, attracted our attention first but there are also people moving around clearly using a very different mechanism, and not everything is restricted to moving over the ground surface (moles, underground trains, birds, helicopters and aeroplanes!).

If we concentrate again on the cars: they do not, despite first impressions, move completely randomly, they are being steered and their speed can vary: some element of control exists. More complex rules also seem to govern movement, cars move only in one direction on one side of the road, stop when lights shine red or when waved at by the arms of policemen, and do not collide with one another very often (altering the frictional interaction with the road surface can dramatically alter collision frequency). Some vehicles are restricted to roads whilst others roam freely over marshy ground and up steep slopes. People and cars seem to move independently though collision may lead to cessation of movement. Some vehicles pick up passengers, others release them in larger numbers at particular sites.

Let us attempt an experiment from our vantage point: arrange that the surfaces of roads are covered by snow. What effect will it have on the movement and what does it tell us? Our questions fall into four major categories, concerning:

(a)    engines,
(b)    traction methods,
(c)    steering and control devices,
(d)    traffic rules.

These are the headings we will use in the following chapters.

## 1.5    Motor design – an abstract exercise

Can we design movement systems in abstract, can we play at inventing ways of causing the movements mentioned above? This sort of exercise, of trying to design systems, might help us to understand what is actually going on and provide a framework of hypothetical possibilities which can be explored in seeking to find the basis of movements we discover when we look at real cells. Having tackled the problem from a position of ignorance we can go on, in later chapters, to see how cells have adapted particular mechanisms to achieve a desired end, although Table 1.1 gives a sneak preview. In fact there seems to be a real example of almost all the possible solutions to the problem of generating movement. So, given a basic construction kit of molecules, how could we cause a movement to occur?

Consider first an 'open' system, an unbounded compartment in which our components are unrestrained. This is useless, although we can get the molecules themselves to change their shape (Fig. 1.1a) we cannot make use of these shape changes (unless we want to control the rate of a reaction through an allosteric change in an enzyme for example); we must put some constraints on the system. If we link molecules together we could achieve

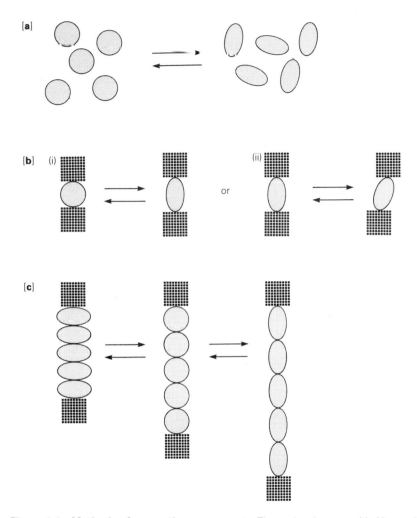

**Figure 1.1  Methods of generating movement**  The molecular assembly kit consists of motor components which play an active rôle (circles), anchorage sites (squares or rectangles), and receptors – molecules that can bind to other molecules. In (a) molecules change shape in an open, unconstrained system, whereas in (b) they are attached to anchorages and in changing shape they cause a change in the relative positions of the anchorage points: a movement. By linking such molecules in series (c) or parallel (d) the movement or the force can be amplified. If to one anchorage we fasten receptors which can bind ligands on the other anchorage then exchanging receptor–ligand pairings may cause (random) movement (e); if the rule is to form new pairings only to the right of the previous ones then the two anchorages will slide (f). The relative movement of the anchorages might come about as a result of electrostatic repulsion or attraction if the charge on the components was varied (g), and this could be used in a manner analogous to that of (f), as shown in (h). Assembly of protomers as in (i) might push the terminal anchorages apart and disassembly might permit their closer approach. Linked assembly and disassembly, as in (j) might be used to push/pull with the cursor (▲) being sent hand-to-hand, or the whole complex might treadmill (see later chapters). The expansion of components, perhaps by altering the hydration, could also push things apart (k).

[d]

[e]

[f]

[g]

[h]

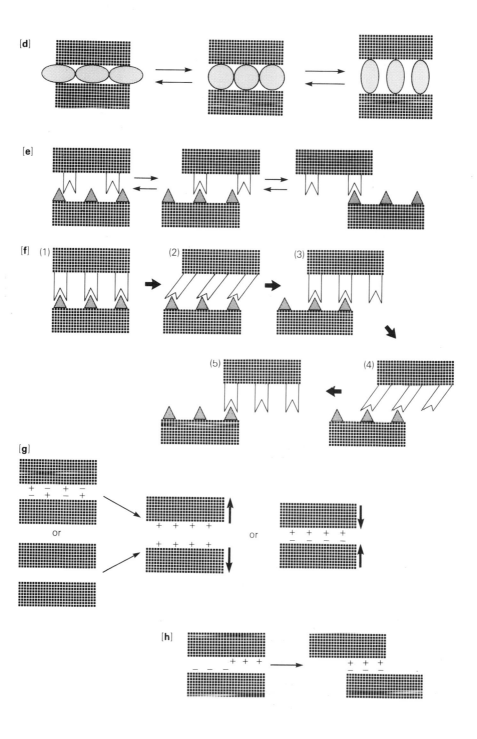

[i]

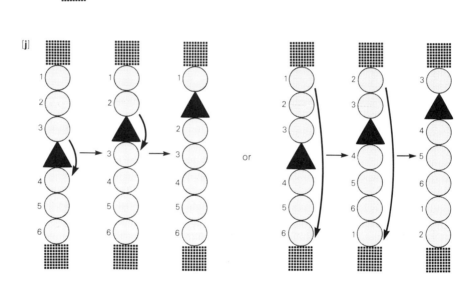

[j]

[k]

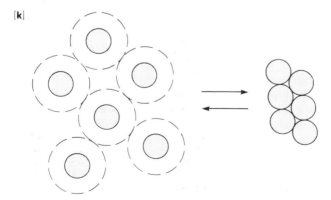

**Figure 1.1** – *continued*

MOTOR DESIGN – AN ABSTRACT EXERCISE

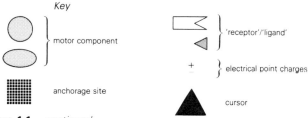

Key

motor component

'receptor'/'ligand'

electrical point charges

anchorage site

cursor

**Figure 1.1** – *continued*

**Table 1.1** A summary of possible motor mechanisms and systems in which they are thought to operate.

| Mechanism | | Movement system | Details |
|---|---|---|---|
| (a) | shape change | allosteric effects on enzymes | |
| (b) | shape change with anchorage | control in striated muscle | Chapter 2 |
| (c) | shape change with anchorage: series linked | spasmoneme of vorticellids | Chapter 4 |
| (d) | shape change with anchorage: parallel linked | acrosomal process of *Limulus* | Chapter 4 |
| (e) | oscillating plates | ? | |
| (f) | sliding filaments | striated muscle | Chapter 2 |
| | | sliding microtubules | Chapter 3 |
| (g) | electrostatic | bacterial flagella | Chapter 4 |
| (h) | electrically driven sliding | some bacteria? | |
| (i) | assembly | separation of chromosomes in bacteria | |
| | | acrosomal process of *Thyone* | Chapter 4 |
| | disassembly | nowhere proven | |
| (j) | treadmilling | chromatid movement? | Chapter 3 |
| (k) | volume change | nematocyst discharge | Chapter 4 |
| | | stomata, leaf movement etc. | Chapter 4 |
| (h) | gel formation | protrusion of pseudopodia | Chapter 6 |

some useful degree of movement, the shape change of one molecule might bring two others closer together or separate them (Fig. 1.1b): we could link our shape-changing molecules together in series (Fig. 1.1c) or in parallel (Fig. 1.1d) to increase the magnitude of the movement or the force exerted. The only clear example of this simple sort of mechanism is the spasmoneme of vorticellids, discussed in Chapter 4. Notice that we have now differentiated active and passive components, the active ones being the molecules which are changing shape, the passive ones being anchorages. For a motor to be of use it must act on something; we do not need to specify much about the passive components except that the linkage to the active components should exist.

A further possibility exists: the linkage can be a temporary one which can be made and broken. The formation of irreversible linkages will tend to cause clustering and will accumulate linked molecules in a particular area, a purely passive phenomenon which does not constitute a motor system (as such, it is a flocculation or coagulation process). If, however, we couple linkage control with shape change we can, with some ingenuity, devise a more sophisticated motor (Fig. 1.1e), although not a very useful one because it oscillates randomly. We need to impose some polarity as in Figure 1.1f, in which we have reinvented the sliding filament system – or a zip fastener.

9

This game gets quite complicated as we introduce more rules, but the basic idea of shape change, reversible linkage and polarity of the components is a powerful method of generating movement and one which we will obviously consider in more detail when we look at the two major motor systems, actomyosin and tubulin–dynein, which use this method. The polarity used above is a simple linear one, but the linear element could be curved and rejoined to make a rotatory system. The movement in the system above consists of a shape change which exerts a force, either pushing or pulling, but the force could equally well be exerted by changing the charge on the components (perhaps by manipulating the local pH above or below the pI of the molecule using a proton pump) and using electrostatic repulsion or attraction to serve as the force generator; self-electrophoresis has been proposed as a mechanism for organelle movement and the movement of some bacteria. By arranging the components in a suitable array the **filaments** could slide past one another (Fig. 1.1h), plates or filaments could be brought together or pushed apart (Fig. 1.1g), or plates or rings could rotate relative to one another (the latter seems to be the basis of movement in bacterial **flagella**, see Ch. 4).

Reversible linkage does allow another set of possibilities, the controlled assembly or disassembly of sub-units to bring about gross movement. Passive elements could be separated by assembling material between them (Fig. 1.1i) or allowed to come into contact by disassembly of the spacer. A neater way of moving a passive component from place to place might be to have a push–pull system whereby the assembly on one side is associated with disassembly on the other (Fig. 1.1j). Whether chromosomes in the mitotic spindle are moved in this way is still contentious (§ 3.6). Growth is, of course, a process where irreversible assembly can be used, as in the production of the cell wall of bacteria and plants. The assembly–disassembly process might be of the pile-of-bricks type illustrated, or might be a more complex polymerization or gelation. The formation of a cross-linked filamentous network from more mobile components could push things around as could, perhaps, the controlled hydration of a large tangled molecule such as hyaluronic acid to form a highly viscous pseudogel (Fig. 1.1k).

So far we have been working with an open, unbounded system, although the constraint of linkage was necessary. Within closed systems we have another possibility, i.e. shape change by altering the volume of the compartment. If a flat (non-spherical) sac has air or water pumped into it, it will expand (at the expense of the external compartment) and can force adjacent objects apart. The usefulness of pneumatic or hydraulic mechanisms has been much exploited in macroscopic engineering and by multicellular organisms, especially those which burrow through semi-solid media (nematodes, annelids, molluscs, etc.). It is less commonly encountered at the cellular level, although the inflation and deflation of plant cells plays a major part in movements of the whole plant and of leaves.

## 1.6   How are movements controlled?

From the above discussion we can extract certain basic themes: that movement might arise by shape change of permanently linked elements, by reversible interactions causing movement of elements relative to one another, by reversible assembly and disassembly, or by changing the volume of a closed compartment. In devising these mechanisms we have made certain assumptions, notably that it is possible to induce changes in shape, charge or interaction of the components. That such changes are possible is implicit in the design of the hypothetical construction kit and we know that such changes can occur in proteins, the most likely molecules serving these functions in real movement systems. But what drives and controls such changes? Unless work is done in altering the composition of the environment of the components then we are getting movement for nothing, which would contravene the laws of thermodynamics.

As we look at particular systems in detail we will see how these changes are brought about in reality: in principle the problem is not a difficult one. The shape of the sub-unit, the tertiary structure of a protein, can be altered (with greater or lesser ease depending on the primary amino acid sequence) by altering the ionic milieu or by making other molecules available. Changes in the local ionic environment could lead to the binding of ions (such as calcium), or to change in charge on the protein (if the local proton concentration changes): these changes might in turn lead to changes in the level of hydration thereby altering the volume-fraction of the molecule. Indirect changes might also play a rôle, phosphorylation (a common control mechanism) can be brought about by activating a kinase, increased by inhibiting a phosphatase or facilitated by raising the concentration of a nucleotide triphosphate. Control systems are not difficult to invent although cells have been frugal (or conservative) in their use of a limited range of control systems: the implications of this frugality are a topic in themselves. The control of assembly–disassembly is likely to involve changes in the properties of the sub-units, and the control systems turn out to be similar to those involved in altering the properties of shape-changing systems.

When we are dealing with the closed-compartment systems we need to hypothesize the control of movement of small molecules, we need a pump and valve system for our hydraulics. Again the control of protein conformation is likely to play a part; the opening of channels through which ions will pass is almost certainly due to changes in membrane-embedded proteins. Alternatively, changes in osmotic pressure (resulting in movement of water to inflate the compartment) might be brought about by switching on enzymes which cleave polymers to yield large numbers of monomers, by altering the solubility of inorganic crystals by changing the pH, or by actively pumping ions or other small molecules into the compartment. Further schemes could undoubtedly be devised, but we should be looking for a common theme in the design of control systems.

In most, if not all, of the control systems a change in the environment brings about a change in the properties of the motor, acting either directly or indirectly on the components of the motor. The ability or property of cellular membranes to regulate the composition of one compartment relative to another is well recognized as essential to the maintenance of the thermo-dynamically improbable order which exists within cells: it should not surprise us that motor systems are likely to be sensitive to local environmental changes which follow from changes in membrane permeability. Perhaps the most basic biological switching mechanism, the regulated passage of material from one compartment to another, is going to be as important in motile systems as in others.

## 1.7 Which motor for the task?

Given the diversity of motor systems we have been able to invent, how should our customer, the cell, decide which of these is most suitable for a particular task? In other words, are some of our motors better for some jobs than for others? Without labouring the point unduly it seems very likely that this will be the case. Intuitively it seems probable that conformational changes will be faster and more easily reversible than assembly–disassembly or hydraulic mechanisms. Equally, having once assembled a supporting column or spacer it can be left in place for relatively little extra cost provided the disassembly process is blocked (if the disassembly process cannot be blocked then severe constraints are placed upon subsequent changes in the local environment, which must be maintained in the appropriate disassembly-inhibiting condition, almost certainly at the cost of metabolic energy). There are also simple restraints upon motor choice associated with the way in which the motor must act. Tension can be transmitted through flexible linkages, rope-like elements, whereas resistance to compression requires more rigid structures especially if lateral bending stresses are encountered. So, depending upon the type of movement, we might expect to find different motor systems being used.

In the chapters which follow we will look at motor systems, devoting a chapter to each of the well-known sliding filament systems and one to miscellaneous motors; traction, under the headings of swimming and crawling; and traffic rules. The latter topic will be subdivided into sections on uniform and non-uniform environments, and the final chapter will be concerned with interactions between cells.

# References

The following books are important sources, although none covers all of the topics. Some, particularly the multi-author volumes, are aimed at those working on cell motility, rather than at a general audience.

Stebbings, H. and J. S. Hyams 1979. *Cell motility*. London: Longman. A good general book but with little on tissue cells, bacteria or cell behaviour.
Trinkaus, J. P. 1984. *Cells into organs. The forces that shape the embryo*, 2nd edn. Englewood Cliffs, NJ: Prentice-Hall. Written far more for the developmental biologist than this book; covers many aspects of cell behaviour in more detail.
Inoué, S. and R. E. Stephens 1975. *Molecules and cell movement*. New York: Raven. A multi-author volume. Some topics have moved forward considerably since these semi-review chapters were written.
Goldman, R., T. Pollard and J. Rosenbaum 1976. *Cell motility*. Cold Spring Harbor Conferences on Cell Proliferation. Vol. 3. A three-volume set covering (a) *Motility, muscle and non-muscle cells*, (b) *Actin, myosin and associated proteins*, and (c) *Microtubules and related proteins*. Very much devoted to the internal machinery, and ten years old in a fast-moving field.
*Prokaryotic and eukaryotic flagella* 1982. Symposium of the Society for Experimental Biology **35**. A valuable collection of review-type articles with some primary data.

There are other books which cover aspects of cell movement; the ones mentioned are ones I have found particularly useful. Other general sources will be quoted at the end of each chapter.

# 2
# MOTORS BASED ON ACTOMYOSIN

## 2.1   Introduction

Engines or motors based upon **actin** and **myosin** are probably the most common amongst eukaryotic cells; the components required are found in almost all the cell types that have been investigated although the activity of the motor has only been demonstrated in a few cases. The way in which the components of the motor are arranged does vary, according to the task for which it is used, but the basic principle of its operation is assumed to remain more-or-less constant. The mechanism by which the motor operates is normally introduced by discussing the mechanism of force generation in vertebrate striated muscle, partly because this is the best understood and best studied example, and partly because historically it was the first such system to be worked out. In some ways this is unfortunate because it has led to the feeling, frequently not recognized or acknowledged explicitly, that all other forms of the motor are degenerate examples of striated muscle. Nothing could be further from the truth; striated muscle is merely a very specialized form, a highly-tuned engine using the motor for a simple and clearly defined task, of generating tension in a linear fashion.

In this chapter we will look first of all at the molecular basis for movement, fairly briefly because this topic is well discussed in innumerable texts, then go on to look at the way in which the system is organized in increasingly complicated ways in order to deal with the problems of generating linear, planar- and solid-shape change. We will then turn to the problems of controlling the motor.

## 2.2   Components of the motor

The two basic components from which the engine is constructed are the proteins actin and myosin. All the other so-called 'muscle' proteins are bolt-on extras required to connect the actomyosin system to the plasma membrane, and to regulate the organization and rate of operation of the motor; and we can ignore these for the time being.

### 2.2.1   Actin

Actin is probably one of the most abundant of proteins, present in all eukaryotic cells and constituting 5–15 per cent of the total protein content of the cell. Since actin is part of the motor then movement must be very important for the cells to devote so much of their protein-synthesizing capacity to this one species although not all actin is necessarily involved in the actomyosin motor as we will see later. Not only is actin an abundant protein but its amino acid sequence is very highly conserved, so that the differences between actin extracted from plant cells, protozoa, and vertebrates are trivial and do not appear to affect its functional capacity when we test for its ability to promote the ATPase activity of myosin from another species, or its ability to assemble under defined conditions. Minor differences do, however, occur and it is becoming clear that some organisms have multiple copies of the actin gene that differ slightly in nucleotide sequence, not all of which may be fully expressed. There are two possible reasons for the occurrence of multiple copies of the actin gene: one that it happens to be convenient for the way in which different parts of the genome are activated, the other that the apparent identity is spurious and represents simply our inability to distinguish between the rôles of different **isoforms** of actin which

14

perhaps serve slightly different functions. In *Dictyostelium discoideum*, the cellular slime mould (see Ch. 9), there are 17 actin genes, of which approximately six (but not always the same six) are actively expressed to a greater or lesser extent (McKeown & Firtel 1982). In chick **myoblasts** there are three distinguishable actin gene sets, the β- and γ-gene products being conspicuous whilst the cells are individually motile, the α-gene(s) being predominant in the post-fusion differentiated myotube (Storti *et al.* 1976). No doubt an increasing number of examples will be documented as genomes become more exhaustively dissected, but the crucial point is that all the actin types or isoforms have the basic properties needed for the motor system. Multiplicity of actin genes probably just represents a convenient way of switching genome expression: actin is one of the 'standard' components and as such must be available no matter which route of differentiation the cell follows. Another sign of the importance of actin is the paucity of examples of mutations in actin-gene expression, presumably because deletion of the gene or a non-functional gene product would be lethal to a developing organism at a very early stage. It is of course possible that conditional mutations in the actin gene could occur, but it is difficult to visualize a protocol by which such mutants could be selected experimentally.

Actin is a protein of 43 kilodalton (kd) that can self-assemble into a helical filamentous polymer (**F-actin**). The amino acid sequence is known in full and there are a few unusual amino acids present (3-methylhistidine, dimethyllysine). Although F-actin has usually been considered a double-start (double-stranded) helix, more recent image reconstruction studies suggest that it is better thought of as a single-start (single stranded) helix with the sub-units of globular (G-) actin staggered in such a way as to give the impression of a twin-stranded rope (Fig. 2.1). The filament, when assembled, has a polarity determined by the sub-units and has a diameter of 5–7 nm and a major repeat distance of 38 nm; it is the **microfilament** (mf) of non-muscle cells and the **thin filament** of striated muscle although other proteins may be associated with the actin, particularly in the latter example. Assembly of the **G-actin** into the polymeric F-actin form depends upon $Mg^{2+}$ and ATP both of which must be bound to the monomer if polymerization is to occur. Addition of the G-actin to a growing filament is associated with the hydrolysis of bound ATP to ADP so that within the filament each actin molecule is associated with an ADP molecule. The assembly process itself is polarized and, although monomer can be added at either end, there are preferred assembly and disassembly ends (Fig. 2.2). The length of a filament is indeterminate; there is no reason, in principle, why the filament should not grow infinitely long, provided of course that conditions remain favourable to the assembly rate exceeding the disassembly rate. The factors limiting the monomer–polymer equilibrium seem to be within a range which is physiologically realistic and could be controlled by the cell.

As we shall see later (§§ 2.6.1, 2.7.3, 2.7.5, Table 2.1) actin interacts with many other proteins as well as with other actin molecules and the multiplicity of these interactions probably accounts for the highly conserved nature of the primary sequence.

## 2.2.2 Myosin

Since actin is highly conserved and since the **actomyosin** motor is variable in its properties we might expect that the other component, myosin, would account for most of the variability. This expectation is amply fulfilled. Myosin is not a single protein but is a hexamer (480 kd) composed of two heavy

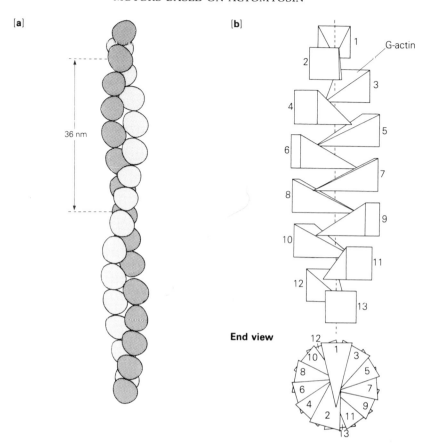

**Figure 2.1 Structure of F-actin** Although normally drawn as though two strings of beads were intertwined (as in (a)), each molecule of G-actin probably forms only two links, and the structure is a single twisted rope and not two ropes. The structure shown in (b) illustrates the proposed relationship between monomers and is redrawn from Sheterline (1983), with permission from the author and publisher.

chains (200 kd each) and two pairs of light chains (16–22 kd each), the light-chain pairs being different (Fig. 2.3). From our point of view, myosin has three important properties: it is an ATPase (of a somewhat unusual type), it will assemble into multimolecular bipolar filaments of determinate length, and, most importantly, part of the molecule can undergo the conformational change which is used to generate movement. Before discussing the way in which myosin and actin together produce a movement let us dispose of the properties of myosin, or rather the properties of myosins as a class since the properties of myosin from different sources vary considerably.

The myosin heavy chain is highly asymmetric with a linear rod-like segment at the carboxy-terminal end and a more globular head region. Proteolytic cleavage of the molecule with trypsin will separate the rod-like segment (light **meromyosin**, or **LMM**) from the head (heavy meromyosin, **HMM**), the latter region having the ATPase activity. Cleavage with papain splits the head region into two portions, HMM-subfragments 1 and 2 or **S1**

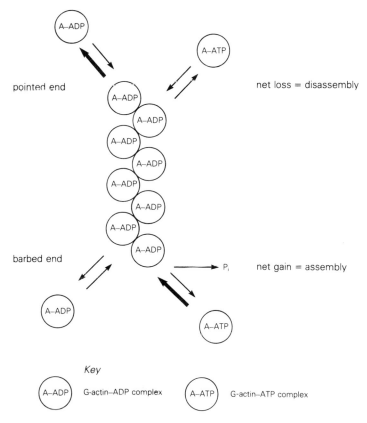

**Figure 2.2  Assembly of F-actin**   The F-actin polymer has distinct polarity, as can be shown by decorating with heavy meromyosin, and the kinetics of assembly and disassembly are different at the 'barbed' and 'pointed' ends. The addition of G-actin–ATP to the barbed end is faster than addition to the pointed end and the disassembly (of G-actin–ADP) is faster at the pointed end, although all the interactions are reversible. Under steady-state conditions addition at the barbed (preferred assembly) end exactly balances losses at the pointed (preferred disassembly) end and so that the filament remains of constant length, although the position of individual molecules within the polymer changes constantly in the process known as treadmilling. Under physiological conditions the rate of addition to the barbed end is probably limited by collision probability.

and **S2** regions respectively. The most distal region from the rod, S1, retains the ATPase activity even after cleavage, and the S2 region, which is predominantly alpha-helical, is thought to act as a flexible linkage between head and tail. These digestion products are more than mere curiosities, the three regions play different parts in the function of the molecule. The linear LMM region is important in assembly of myosin molecules to form a **thick filament** structure; the ability to assemble resides in and is retained by the rod-like fragment after cleavage. The enzymatically active portion of HMM, the globular S1 fragment, also has F-actin-binding capacity which it retains. The latter property has been utilized extensively by electron-microscopists who find in HMM (or S1 alone) a specific ultrastructural 'label': actin microfilaments in permeabilized cells can be 'decorated' with HMM and the

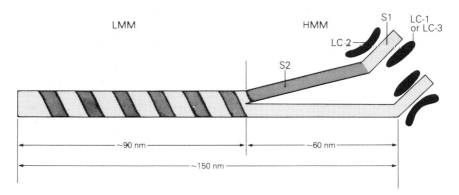

**Figure 2.3  The myosin molecule**  Six components make up the myosin complex: two heavy chains and four light chains. Two kinds of light chain are found per hexamer. The heavy chain can be split into light meromyosin (LMM), which is predominantly alpha-helical, and heavy meromyosin (HMM). HMM can be split further into S1 and S2 fragments, the S1 fragment having actin-binding and ATPase activity. The angle made by the head region with the LMM tail is variable.

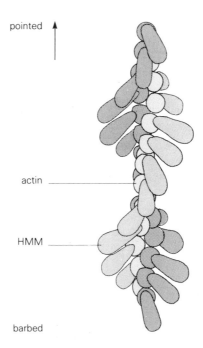

**Figure 2.4  The decoration of F-actin with HMM**  Heavy meromyosin (or S1 alone) will bind to F-actin. The interaction is specific and polarized, and F-actin is helical; thus the effect (provided there is no ATP available) is to 'decorate' the filament with 'arrowheads': the arrows point in the direction in which a thin filament from striated muscle would slide. Microfilaments, which can be decorated just like F-actin, are always anchored to membranes by their barbed ends (see also Fig. 2.5).

decoration can be visualized in the electron microscope (Figs 2.4 & 2.5). The binding is reversible on the addition of $Mg^{2+}$-ATP, which further suggests that the interaction is similar to the functional interaction in muscle. Not only is the 'stain' specific for actin and similar or identical to the interaction involved in force generation (see below), but it also indicates the polarity of the microfilament because of a tilt in the binding of the HMM which gives rise to the characteristic 'arrowhead' pattern. This reagent therefore enables us to determine the nature of filaments seen within the cell, and to predict that they could, in principle if not in practice, be used in the motor system. So extensively has this decoration with HMM been used that the two ends of the actin filament are now just referred to as the 'barbed' and 'pointed' ends respectively; attachment of the microfilament to other structures is almost always by the barbed end.

## MYOSIN LIGHT CHAINS

Much of the diversity in the properties of myosin from different tissues probably resides in the **light chains** which are very variable, although their functional properties in general have been conserved. An exhaustive discussion of this topic is inappropriate (for more detail see Sheterline 1983) but an outline will be given here. Two classes of light chain exist, the regulatory light chains (LC-2 in more recent literature, DTNB-light chains in the older literature) and so-called essential light chains (LC-1, LC-3 or alkali light chains); one pair of each of these classes being found in each myosin hexamer. The light chains are all related to divalent-cation-binding proteins and have amino acid sequence **homology** with **parvalbumins** (from teleost and amphibian muscle), **troponin C** (from vertebrate skeletal muscle), and **calmodulin** – the ubiquitous $Ca^{2+}$-dependent regulatory protein. However, only the regulatory light chains seem to have a functional divalent-cation binding site. The loss of this property might imply that a second function has been selected for, but the nature of this function is not obvious.

(a) Regulatory light chains, as their name implies, probably play a rôle in controlling the activity of the HMM-S1 ATPase; either as a direct consequence of calcium binding by the light chain, as in molluscan muscle, or indirectly following their phosphorylation by myosin light chain kinase (**MLCK**), a calmodulin-regulated kinase the activity of which may be important in controlling the motor in smooth muscle and in non-muscle cells. Their derivation from $Ca^{2+}$-binding proteins suggests, perhaps, that their original mode of action involved direct cation-binding, and that other regulatory mechanisms have been superimposed.

(b) Essential light chains: although these light chains clearly play some rôle in regulating the characteristics of the myosin ATPase their name is misleading; they are not essential to function, although they may well be important for physiologically normal activity of the myosin complex. Different forms of these light chains (LC-1 and LC-3) are found in **fast-twitch skeletal muscle**, one of the few muscle types to have been studied in this sort of detail. Since the two isoforms apparently coexist within a single cell the implications of their occurrence are unclear. Their only known functional rôle is in regulating the characteristics of actin-activation of the myosin-ATPase activity of the heavy chain. They are not substrates for the myosin light chain kinase, nor do they bind calcium

19

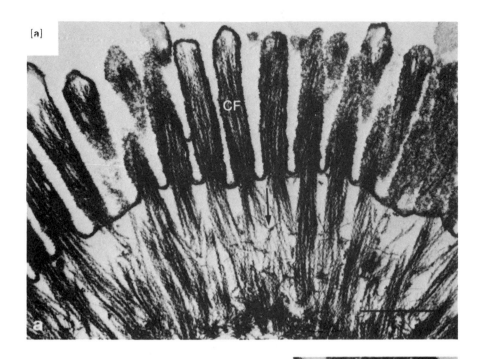

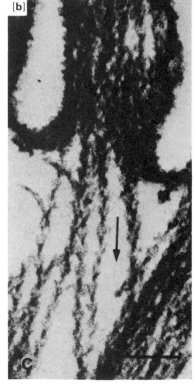

**Figure 2.5 Decorated microfilaments**
(a) Microfilaments in the microvilli and terminal web region of intestinal epithelial cells (brush border). The section was decorated with S1 and fixed in the presence of tannic acid: even the microfilaments of the core (CF) are decorated; bar represents 500 nm. (b) Higher-power view of portion of (a) in which the basally-directed polarity of the core filaments (arrow) can be seen; bar represents 100 nm. (c) Antiparallel microfilaments in a stress fibre. The cells were glycerinated before decoration, but otherwise treated in the same way as in (a) and (b); bar represents 200 nm (all from Begg *et al.* 1978, with permission).

ions – an indication perhaps that there is a third regulatory mechanism for the motor about which we are, as yet, ignorant.

MYOSIN FILAMENT ASSEMBLY

Although the polymerization of myosin into multimeric arrays may not be strictly necessary for the operation of the motor, it does seem to be the general rule (we may well be able to justify this on mechanical grounds somewhat later), and we can conveniently consider the formation of myosin filament structures at this stage. Myosin extracted from vertebrate muscle will self-assemble *in vitro* and does so to produce filaments which are similar to the filaments seen in muscle. As mentioned already, the linear tail of the molecule is largely responsible for the assembly behaviour although the HMM region may have some influence.

(a) Vertebrate striated muscle myosin. The naturally occurring form of myosin in striated muscle is in the 'thick' bipolar filaments which characterize the **A-band**. These filaments are approximately 1.6 μm long with a diameter of 15 nm and have a central 'bare' region with lateral projections, the HMM heads, at the ends (Fig. 2.6). Two features of the assembly process are of particular importance. First, the filament is determinate in length with a fixed number (approximately 300) of

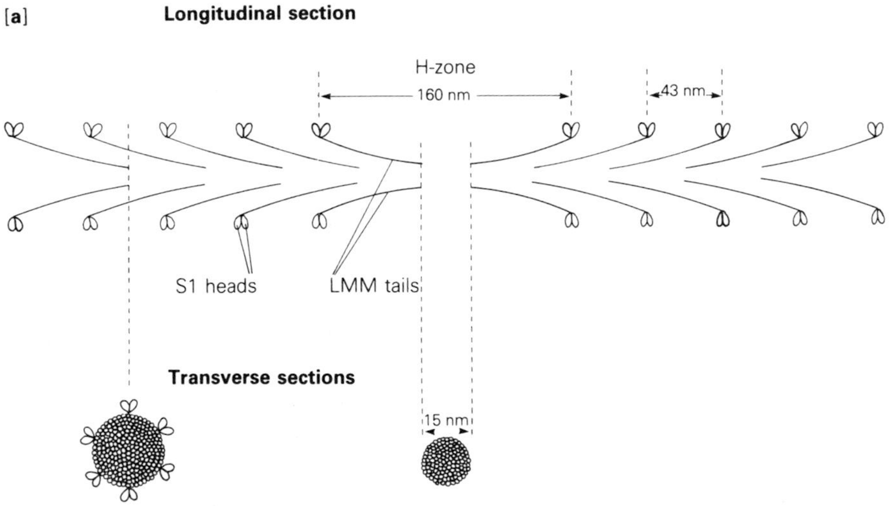

[a]    **Longitudinal section**

**Figure 2.6  Structure of the thick filament from striated muscle**  Unlike the F-actin polymer, the filament which is assembled from striated muscle myosin is bipolar, having a central bare zone devoid of head groups. In section the appearance is thought to be as shown in the diagram, with S1 heads repeating every 43 nm. The heads repeat axially every 14.3 nm; there are two sets above and below the plane of the longitudinal section, offset from each other and from the ones shown (one may alternatively describe this as a three-start helical arrangement). In a muscle each thick filament would normally be surrounded by six thin filaments, and the central bare zone would be located in the centre of the sarcomere (the H-zone). The micrographs show thick filaments from rabbit psoas muscle which have been negatively stained. Remnants of the M-line can be seen in the central bare region. Myosin heads are arrowed in the high-power view; scale represents 100 nm. (Micrographs courtesy of Dr P. J. Knight, Meat Research Institute; first published in *ARC Meat Res. Inst. Biennial Rep.* 1981–3.)

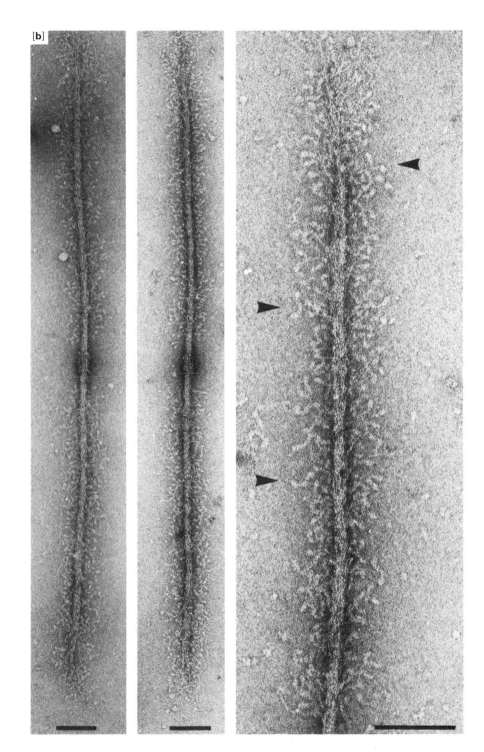

hexamers (although the filaments formed *in vitro* are rather more variable than those found *in vivo*). Secondly, the two ends of the filament have opposite polarity, the molecules have their heads distally disposed, the tails lying along (and comprising) the shaft of the filament – which explains the central bare region. Although the filaments formed *in vitro* resemble those found *in vivo*, their length is rather greater *in vitro* (up to 5 μm), implying that although the myosin filament is more determinate in length than the actin filament there is some metrical constraint imposed by other components of the muscle cell.

The packing of myosin molecules within the filament is regular and there is a helical repeat of 43 nm and an axial periodicity of 14.3 nm. There are probably three myosin heads projecting in each 14.3 nm repeat, an interesting figure only in that each thick filament *in situ* is normally surrounded by six thin filaments and there appears to be a discrepancy between the repeat distances between the myosin heads and the axial periodicity of actin filaments.

(b) Smooth muscle and non-muscle myosin. The structure of the thick (myosin) filaments in smooth muscle and cells derived from other tissues has for long been an area of dispute; the very existence of thick filaments in smooth muscle was doubtful for some time. The problems arose from two sources: the variability in appearance of the filamentous components depending upon the state of relaxation or otherwise of the cell when fixed for electron microscopy, and the much lower level of myosin within these tissues compared with striated muscle making thick filaments much less conspicuous, especially since they are not regularly disposed. A further complication may have contributed to some of the uncertainty, i.e. the instability of the myosin assemblies under certain 'physiological' conditions. It now appears that the assembly of myosin into filaments in these cell types depends not only upon $Mg^{2+}$-ATP but also upon the presence of $Ca^{2+}$-calmodulin and myosin light chain kinase. Phosphorylation of the regulatory (LC-2) light chain seems to be important in assembly as well as in control of the interaction with actin (Scholey *et al.*, 1980, see also § 2.7.5). The phosphorylation of the myosin heavy chain in *Dictyostelium discoideum* inhibits assembly (Kuczmarski & Spudich 1980), which introduces a further complication since the opposite is true for *Physarum* (Ogihara *et al.* 1983).

Filaments, once assembled, are of more variable length than those from striated muscle; ranging from 2–8 μm *in vitro* when the myosin is from vertebrate smooth muscle (Hinnsen *et al.* 1978), and being much smaller (0.2 μm) when the myosin is from non-muscle cells (Goldman, Yearna & Schloss 1976, Kuczmarski & Rosenbaum 1979). There is another more interesting difference, the filaments do not have a central bare zone but oppositely-polarized myosins seem to be arranged in longitudinal arrays (Fig. 2.7). The axial periodicity of 14.3 nm is, however, retained. Despite the differences, two essential points of similarity with striated muscle thick filaments remain, their determinate length and the disposition of myosin molecules with opposite polarity.

## 2.2.3 *Tropomyosin*

As with the other components of our motor system, **tropomyosin** is a generic name for a group of closely related proteins which bind to F-actin and which may stabilize the polymer. In addition, tropomyosin provides a switching system which will operate on several actin molecules in the thin filaments of

[a] **Longitudinal section**

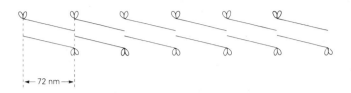

$\leftarrow$ 72 nm $\rightarrow$

[b] **Transverse section**

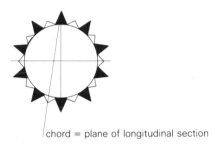

chord = plane of longitudinal section

**Figure 2.7 The thick filament from smooth muscle** The arrangement which is thought to exist is of alternate rows of myosins with opposite polarity, arranged in a slightly twisted fashion and repeating every 72 nm. The longitudinal section does not pass through the centre of the filament – if it did the upper and lower rows of myosin would have identical and not opposite polarity – but is a chord as shown in (b). In the cross section (b) the myosin heads alternate polarity around the circumference with some pointing towards you and some away (they are also at different levels, of course). Notice that the whole of the length of the filament is useful, there is no bare zone.

striated muscle. Tropomyosins from vertebrate skeletal muscle are found as dimers (monomers of 33 kd) which are long and thin (41 nm × 2 nm) and which will associate end-to-end at low ionic strength *in vitro*, with some overlap of the molecules. The linear structures generated in this way lie along the grooves of the F-actin helix. There is a seven-fold periodicity both of the primary sequence and of the secondary coiling of tropomyosin which would permit each molecule to interact with seven actin monomers. The end-to-end association of tropomyosin is such that this interaction will remain in register in adjacent molecules. In principle an indefinite length of F-actin can therefore be stabilized. In vertebrate striated muscle at least four genes for tropomyosin have been identified and two major isoforms, α and β, are found (Mak *et al.* 1980). The dimer of tropomyosin may not be a random association of α- and β-chains, but since the ratio of α to β varies both between muscle types and during development of a single type of muscle the situation is not as simple as with the tubulin heterodimer (§ 3.2.1). The implications of the differences in tropomyosin for the properties of the composite actin–tropomyosin filaments will, no doubt, become clear in time. Tropomyosin from tissues other than skeletal and cardiac muscle is clearly different, and platelet tropomyosin spans only six actin monomers (Cote & Smillie 1981).

Not only does tropomyosin interact with actin but, in skeletal muscle at least, it has a binding site for troponin T, which in turn interacts with the other troponins (I and C), and so forms part of the calcium-dependent regulatory system (see § 2.7.3).

## 2.3   The basic motor

We can now combine our two principal components, actin and myosin, and see how they might be used to achieve a movement. Our understanding of the machine comes from consideration of vertebrate striated muscle, where the filaments are laid out very neatly for our inspection. However, the molecular basis for movement does not depend on such an elaborate organization.

We need only two components to make the simplest motor, a single molecule of actin and one of myosin (Fig. 2.8). The transformation of chemical energy into mechanical work is thought to depend upon a conformational change in the myosin molecule which occurs when myosin binds to actin. 'Chemical energy' comes from breakdown of ATP to yield ADP and inorganic phosphate.

This single movement does not, however, get us very far – only about 7.5 nm in fact (and myosin is only activated by F-actin). If we make use of the fact that actin will polymerize we can string actin molecules together into a

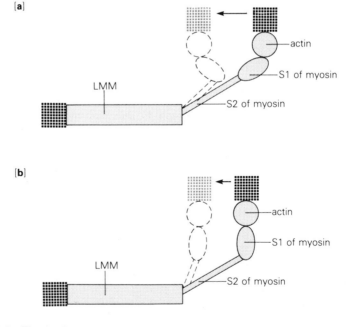

**Figure 2.8   The basic motor movement**   Two possibilities exist; though most people favour the one shown in (a), that the S1 region of myosin undergoes a conformational change and rotates about the actin molecule to which it is bound. On this model the S2 region acts as a hinged joint and transmutes the conformational change into a tension along the axis of the thick filament which then slides relative to the thin filament (or *vice versa* if you stand elsewhere). The alternative, shown in (b), is that the S2 region shortens. Distinguishing between the models requires the unequivocal demonstration of a change in the angle of the S1 region relative to actin.

linear polymer which can be attached to a more substantial structure some distance away. Provided the myosin can be persuaded to repeat the conformational change upon each actin in turn the 'boat' (myosin) can be hauled 'hand-over-hand' towards the distant anchorage. We have linked our small movements in a temporal series to amplify the amount of movement that can be achieved. We require that the attachment to the first actin be relinquished and that the motor be recharged. (There are other problems which should be obvious and to which we will return.) The biochemical basis for this is shown in Figure 2.9; each movement or step 'costs' an ATP molecule per myosin head. In this simple model we have used a single actin molecule (linked to some sort of anchor) and a single myosin (attached to a 'boat') and used the conformational change in the myosin, either in S1 (as in Fig. 2.8a) or in S2 (as in Fig. 2.8b) to move the boat nearer the shore.

An obvious problem in this scheme is to ensure that the myosin does not reattach to the actin that was used in the previous cycle but moves onto the next actin in the chain. If there is no resistance to the movement of the whole myosin molecule then the geometry of the system might suffice to guarantee progressive movement; as a consequence of the first conformational change the 'reactivated' myosin head will be nearer to the second actin than to the first. However, if we wish the motor to do useful work then the movement of the myosin must be resisted by the object which is being moved; if there is no resistance then no work is needed. The simplest solution, and the most obvious one, is to have two (or more) myosins and have them working out-of-phase, being careful not to let go before the new attachment is formed. Once we have two hands to pull on the rope we can indeed move hand-over-hand; the one-armed man can do no more than pull once.

Another point should be made about this motor system. The direction of movement is determined by the orientation of the myosin relative to the long axis of the actin filament and by reversing the myosin it should work in the opposite direction – but our analogy of pulling hand-over-hand turns out to be a good one: the actin rope is attached only at one end (defined by the polarity of the F-actin polymer which is attached only at the preferred-assembly or 'barbed' end) and, despite romantic notions to the contrary, the Indian rope trick is impossible, a flexible polymer-rope is a poor barge-pole and does not resist lateral stress. So, in this particular case, and for these particular molecules, the relative orientation of actin filament and myosin molecule is important if the motor is to work. In principle, however, there seems no reason why rigid elements should not be substituted for the flexible rope, or that the rope could not be fastened at both ends producing a system which could both push and pull, that has both forward and reverse gears. In practice the myosin will only operate if it is attached to the actin with the correct polarity. A 'winch' system such as the actomyosin motor can, nevertheless, be used in various ingenious ways.

Before going on to practical examples of the application of the actomyosin motor let us just briefly consider how it could be regulated. Several possibilities exist of which the most trivial is the scarcity of fuel. Without ATP the motor will stop working, but in a 'fail-safe' manner: movement will terminate (go into rigor) at the position reached after the last cross-bridge cycle was completed. If the absence of ATP led to relaxation then any work already done would be wasted, especially if there was an elastic series-resistance, and this may be a design feature of selective importance. Another possibility for control is that the cross-bridge cycle could be blocked, either by interfering with the thin filament or by regulating the conformational change in the myosin. The former might be achieved by making the actin

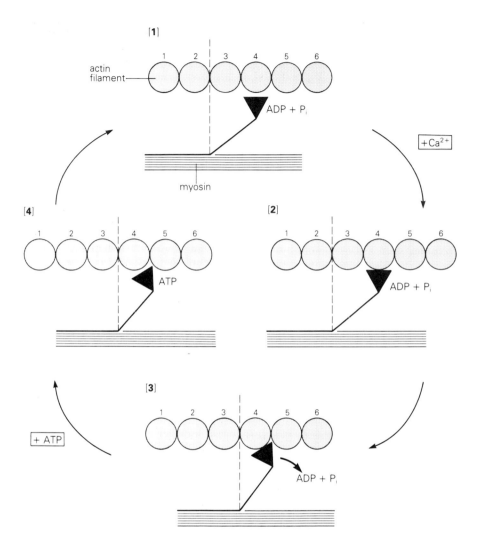

**Figure 2.9 The actomyosin mechanochemical cycle** Remembering that this is a continuous process we divide it up into four phases: in (**1**), the resting configuration, binding of myosin to the actin of the thin filament is blocked by the troponin–tropomyosin system (not shown) and the ATPase activity is minimal. Influx of calcium ions permits binding of the S1 of myosin (**2**) and the sarcomere would now resist passive extension. Release of ADP from the S1 is accompanied by a change in the relationship of the myosin to the actin (**3**) (shown here as a conformational change at the S1-S2 region): notice that one filament has moved relative to the other. Binding of ATP allows the myosin to release its binding to the actin (in the absence of ATP the muscle remains in rigor) and the system could now be passively extended easily (**4**). Hydrolysis of the ATP allows the myosin to regain its resting conformation and, unless the filaments have been dragged in the opposite direction by external forces, the myosin is in a position to reattach to an actin molecule (number 5 not number 4 this time) which is nearer to the Z-disc. Obviously the attachment of myosin to the thin filament will only occur at regular intervals along its length, the distance being set by the repeat distance of the F-actin helix.

unavailable for assembly into polymer, by destroying the integrity of the rope through disassembly of the filamentous actin polymer, or by masking or otherwise altering the binding site for myosin on the F-actin. Conformational change in the myosin could be blocked and the light chains are spatially well-placed to act upon the myosin head.

## 2.4   Linear contractile systems

### 2.4.1   Striated muscle

Striated muscle represents the most sophisticated development of the actomyosin motor. It has proved a fruitful system in which to study this interaction and has been the source of most of our information. Striated muscle is not restricted to vertebrates; the arthropods have also developed striated muscle, as seen, for example, in the flight muscle of insects. It is in fact an obvious solution to a simple problem, that of generating a powerful tension. Since the contraction required is linear, all the elements can be organized in a linear fashion and the individual elements can be linked both in series and in parallel. By increasing the series component the *extent* and speed of contraction can be increased; by increasing the parallel component the *force* can be increased. The work of which the muscle is capable remains the same, directly proportional to the number of conformational changes which occur in a given time, and can only be increased by increasing the rate at which the cycle of events occurs.

The variations in structure found in striated muscles from different tissues reflect the local requirements. There are advantages in using the system **isometrically**, or as nearly isometrically as possible, near to the sarcomere length which gives maximum filament overlap, since at this position the greatest number of cross-bridges can be used. Allowing the muscle to change length means inevitably that some cross-bridges are idle for part of the time and work must also be done to overcome the viscous drag of the surrounding **sarcoplasm**. This poses a major constraint in design of the musculature of animals, and most striated muscles normally only contract to a small extent (to around 80 per cent of the resting length). A further constraint that must not be forgotten is that the motor operates only as a winch and once contracted it will not extend unless work is done on the relaxed muscle, hence the antagonistic muscles familiar to the anatomist. A rich variety of alternative ways of restoring the muscle to its starting length have been devised for students of comparative morphology to memorize, but further discussion is beyond the scope of this volume: those interested will find the classic work of D'Arcy Thomson (1942) and more recently McNeill Alexander (1968) well worth reading.

In striated muscle the actin and myosin components are arranged in a very regular pattern (which was a great help in elucidating the mechanism), the myofibrils which lie within the syncytial muscle fibres being composed of repeated sarcomeres as shown in Figure 2.10. In the sarcomere the basic motor system discussed in Section 2.3 is arranged with mirror-image symmetry about the central **H-zone** which coincides with the central bare zone of the bipolar thick filaments. Opposite ends of the thick filament pull upon thin filaments composed of actin (with associated proteins) which are of opposite polarity. The thin filaments are inserted at their 'barbed' ends into the Z-line or **Z-disc**. When working, the myosin–actin cross-bridges operate in such a way as to bring the Z-lines closer together, each end of the bipolar myosin filament hauls thin filaments hand-over-hand. The effect is of filaments sliding past one another, rather than of filaments shortening. The evidence for the sliding-filament model (as it has become known) provides such an excellent illustration of the processes of scientific deductive logic that it is very frequently discussed in great detail and will probably be familiar to most readers. In barest outline the evidence may be summarized as follows:

(a)   constancy of thick filament length; the A-band (**anisotropic** when viewed in polarization optics) remains of constant length, whether the muscle is contracted or expanded;
(b)   constancy of thin-filament length, based upon measurements in electronmicrographs;
(c)   volume-constancy; shortening of the sarcomere is accompanied by an increase in diameter such that the volume remains constant (it cannot therefore be an hydraulic system);
(d)   tension depends upon filament overlap and the isometric tension which is generated is directly proportional to the extent of overlap (and to the number of cross-bridges which can be made);
(e)   the constancy of the axial periodicity of the thick and thin filaments, based on X-ray diffraction studies done on live muscle, shows very convincingly that the filaments themselves do not change in length.

The regularity of the structure and alignment of adjacent sarcomeres in register is perhaps one of the most surprising features of the system. Several things contribute to this regularity, and a variety of proteins give lateral stabilization both in the Z-discs, which are shared by linearly adjacent sarcomeres, and in the H-band, where myosin filaments are interlinked. The Z-discs are linked to one another laterally by intermediate filaments of **desmin,** and these tension-resisting elements are also involved in linking the **myofibril** to the cell boundary, where there may be linkage to another cell at the **intercalated disc** of cardiac muscle or to extracellular matrix in skeletal muscle. On a smaller scale the structural regularity is maintained in the constancy of thick-filament length (in a particular muscle), the constancy of thin-filament length (although there is considerable variation in some muscles, as shown by Traeger & Goldstein 1983), and the constancy of filament separation (which is of course essential to the operation of cross-bridges).

An aspect of the regularity of structure, which is often overlooked by those who delight in the opportunities such a structure offers for elegant image analysis, is the origin and maintenance of such a structure. Sarcomeres must presumably self-assemble and, as such, must be among the more complex end products of self-assembly; also, like any highly-tuned engine, main-tenance must be a problem. Reconstruction of a whole myofibril, to replace a single damaged sarcomere, would add greatly to the 'cost' of having such a sophisticated structure, but it seems that repair of sarcomeres within a myofibril may be possible.

We should not leave striated muscle without summarizing its design. The system is one in which bipolar myosin assemblies act upon antiparallel actin polymers to cause filaments to slide past one another. By linking the actin filaments to other structures, the two ends of the cell in most cases, the action of the motor is to bring the two anchorage points closer together, although the extent of contraction is generally limited to a shortening of some 20 per cent of the resting length of the contractile assembly. The structure is highly regular and the filaments are stable once assembled. The molecular basis for

30

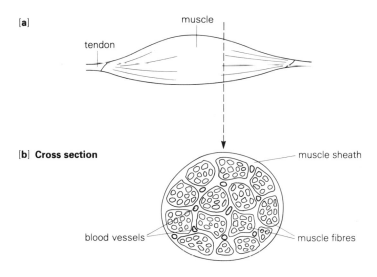

**[a]**

tendon

muscle

**[b] Cross section**

muscle sheath

blood vessels

muscle fibres

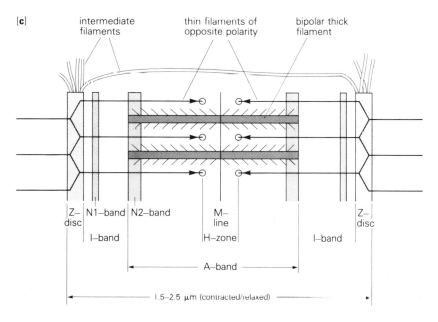

**[c]**

intermediate
filaments

thin filaments of
opposite polarity

bipolar thick
filament

Z–
disc

N1–band

N2–band

I–band

M–
line

H–zone

I–band

Z–
disc

A–band

1.5–2.5 μm (contracted/relaxed)

**Figure 2.10 Vertebrate striated muscle** The single muscle (a) of the gross anatomist (as flexed on beaches etc.) consists of numerous muscle fibres which are shown diagrammatically in transverse section in (b) and in detail in (c) and in the transmission micrographs (d) & (e). Within the syncytial fibre are numerous myofibrils, each composed of sarcomeres in series. The sarcomere is the basic unit and is shown in some detail in (c). Each sarcomere is bounded by Z-discs at the ends and sarcoplasmic reticulum around its circumference. The **I-band** is divided into two parts and may be very short in a contracted muscle. The A-band remains of constant length and remarkably constant position relative to the A-bands of parallel myofibrils. The central **M-line**, which contains **myomesin**, may serve to stabilize the structure of the A-band, in the same way as the lattice of the Z-disc ▶

[d]

◀ determines the pattern and spacing of the thin filaments. Further structural stability may be contributed by the **N-lines** (whose precise rôle is unclear), the end-capping of the thin filaments by β-actinin, and the linkage of the Z-discs to other cytoskeletal structures through intermediate filaments. Other stabilizing and structural components are known or believed to exist. (d) Transmission electron micrograph of frog sartorius muscle in longitudinal section. The various bands can easily be recognized. Parts of the T-tubule

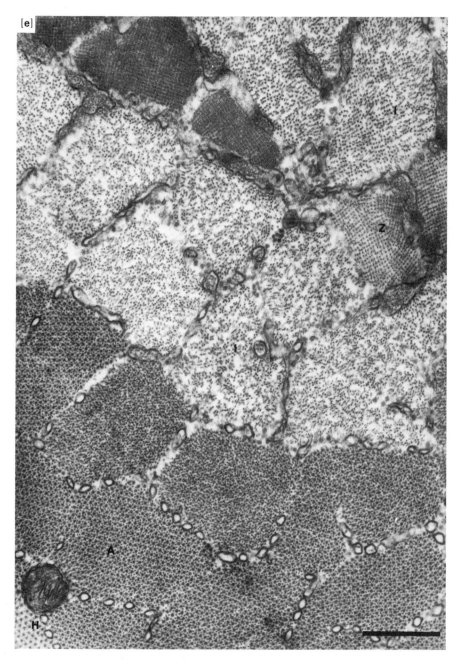

system can also be seen; bar represents 0.5 μm. (e) The same muscle in cross-section. Because the section is slightly oblique, myofibrils have been caught at different levels in the sarcomere. Notice the hexagonal packing of the H- and A-regions, the more random packing of the I-band and the rectangular symmetry of the Z-disc in section. The tubule system is also conspicuous. (d & e courtesy of Dr H. Elder, Physiology Dept, Glasgow University.)

the contraction is moderately well understood, certainly far more so than for any of the other systems we will encounter. Control of the system has been extensively studied but discussion of this will be reserved for a later section of this chapter (§ 2.7.3).

## 2.4.2 Smooth muscle

One of the problems with striated muscle is that it will contract at most to only a half of its maximum working length and for some tasks a greater range of movement is required. If the movement can be amplified with rigid levers then striated muscle will serve; the alternative is to redesign the contractile element. The limit on contraction in striated muscle is set by the organization of the sarcomere; if the thin filaments begin to overlap into the other half-sarcomere the interaction with the thick filament has the wrong polarity, and in general such an overlap does not seem to occur. Smooth muscle, which will contract to a much greater extent, down to 20 per cent of its resting length, is organized in a different way (Fig. 2.11). The basic mechanism is assumed to remain the same but the amount of myosin is relatively much lower.

Smooth muscle, which is found in various sites such as the walls of vertebrate gut, blood vessels and uterus, is composed of single cells which, in the relaxed state, are long and spindle-shaped (approximately 2–8 μm diameter, and anything from 50–400 μm long). The tension generating units, the cells, are arranged in an appropriate manner for the tissue in question. Thus the cells are arranged circumferentially in the blood-vessel wall, and in a criss-cross fashion in the wall of the uterus and of the bladder. Actin filaments are inserted into intermediate-type junctions (zonula adhaerens type) at the cell surface, with the filaments aligned approximately in the long axis of the cell (Schollmeyer et al. 1976). The integrity of the tissue may depend upon the mechanical linkage of filaments from cell to cell by means of these junctional specializations. Within the cell, contractile units – bundles of antiparallel actin filaments with associated myosin – may interlink by insertion into cytoplasmic dense-bodies, which are themselves constrained by intermediate filaments (although the evidence for microfilament insertion into the cytoplasmic dense-bodies is weak). The way in which the contractile elements within the cell are arranged may depend upon the exact shape change required of the cell, and one arrangement is to have them helically disposed, a sort of 'lazy tongs' system twisted upon itself (Fig. 2.12). The disposition of these elements is a concern of the ultrastructuralist; the important point is that each element should be capable of shortening markedly, which may involve the disassembly of component fibres either into sub-units or into a less organized tangle of unravelled fibres. Little is known about the mechanisms of contraction, and although the apparent instability of the thick filaments *in vitro* would be consistent with the idea that the structure is somewhat labile, this is by no means proven. There have been suggestions that the myosin is arranged as ribbons, which have opposite polarity on each face, rather than as bipolar filaments, but this is controversial.

The contractile elements are known to contain antiparallel actin filaments, myosin (which is distinct both antigenically and functionally from skeletal muscle myosin), tropomyosin, α-**actinin** and a $Ca^{2+}$-sensitive protein, **caldesmon**. Closely associated with the site of insertion of the actin filaments into the adhaerens junctions are two proteins, **vinculin** and α-actinin. The details of molecular organization, both of contractile elements and of the

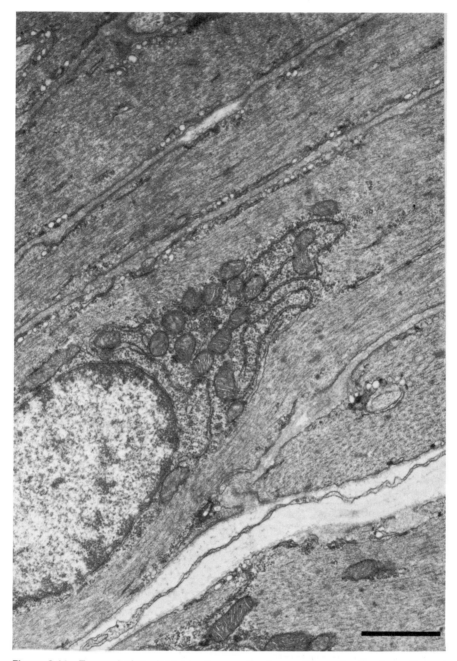

**Figure 2.11   Transmission electron micrograph of smooth muscle structure**   Several smooth muscle cells from the vas deferens of the mouse are seen in approximately longitudinal section. Dense-bodies and fibrils can be seen but the structure is much less ordered than striated muscle (see Fig. 2.10d); bar represents 1 μm. (Photograph courtesy of Dr H. Elder, Physiology Dept, Glasgow University.)

[a]

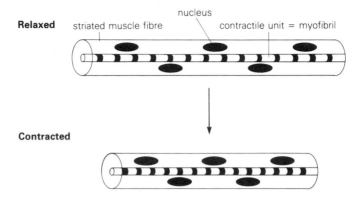

**Relaxed**   striated muscle fibre   nucleus   contractile unit = myofibril

**Contracted**

[b]

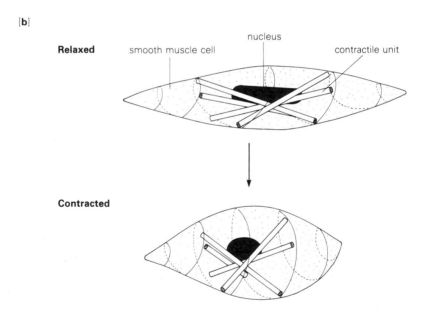

**Relaxed**   smooth muscle cell   nucleus   contractile unit

**Contracted**

**Figure 2.12  Arrangement of contractile units in striated and smooth muscle**   The myofibrils of striated muscle are arranged strictly parallel to the axis of the muscle fibre (a), whereas those of smooth muscle are probably arranged more as in (b), possibly with a helical pattern. The contractile elements of smooth muscle, which have been left deliberately rather imprecise, are inserted into dense-bodies on the cytoplasmic face of the plasma membrane, structures rather reminiscent of the adhaerens junctions into which microfilaments are inserted in non-muscle cells.

insertion sites remains unclear at present. If we wished to design a sliding contractile element capable of generating similar tensions at very different degrees of contraction, we might well want to have a collapsible thin-filament arrangement with the possibility of locking the filaments at a particular length; all the components required to build such a system are available in smooth muscle, but the evidence to confirm this model is not.

In smooth muscle we are dealing with a much less organized actomyosin system than in striated muscle, but this is a system specialized for contraction, and although it may be a better model for non-muscle contractile systems it cannot represent the most basic design.

The control of contraction in smooth muscle differs from that in striated muscle and we will return to this topic in Section 2.7.5.

## 2.4.3  Cytokinesis

An extreme example of a linear contractile element that must contract very considerably is the 'constriction ring' of microfilaments which is located equatorially and acts as a purse-string suture separating the two daughter cells in the final stages of cell division. This process of cytokinesis is distinct from that of nuclear division. (A similar process on a smaller scale occurs in **apocrine** secretion when membrane-bounded fragments are budded from the apical region of secretory cells.) The linear element is recurved upon itself to form a toroidal band (Fig. 2.13) which, particularly in the penultimate stages, must shorten to almost zero length. Whilst the mechanism of shortening is probably the same as in the two muscle systems discussed above, the arrangement of the filaments must be very different. The contractile ring contains antiparallel microfilaments (Schroeder 1973, 1976); it is assembled only at the appropriate stage in the cell cycle, and its disassembly probably plays an essential part in permitting closure of the ring (this is somewhat controversial). Myosin is also known to be present in the contractile ring region (Fujiwara & Pollard 1976), but the observation that drugs like the **cytochalasins** block this contractile process (in some cells at least) whilst having no effect upon muscle systems, indicates that the mechanism is indeed different. Actin polymerization is blocked by the cytochalasins, and therefore the assembly of the ring will not take place; in muscle the filaments are stable and assembly is not required.

## 2.4.4  Stress fibres

Some of the more conspicuous microfilaments in non-muscle cells are arranged in bundles which are anchored distally to the plasma membrane at sites of adhesion, and either have a less obvious proximal anchorage in a perinuclear zone or stretch from adhesion point to adhesion point (Fig. 2.14). These microfilament bundles are well developed in many cells grown *in vitro* where they can be demonstrated by staining with fluorescently-tagged antibodies directed against actin (Goldman *et al.* 1975), by careful observation with interference-contrast (Normarski) optics, or by electron microscopy using either thin sections or high-voltage electron microscopy (HVEM) (Goldman *et al.* 1979, Heath & Dunn 1978). Associated with the bundles are a variety of proteins such as myosin, tropomyosin and α-actinin; and the whole bundle has been shown to be contractile (Isenberg *et al.* 1976; Kreis & Birchmeier 1980). Unfortunately in some ways, these bundles are no more than linear contractile elements and they are most conspicuous in cells which are stationary; they are probably not involved in the movement of cells from

[a]

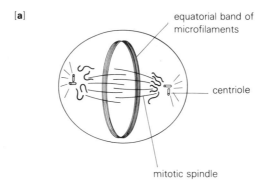

equatorial band of
microfilaments

centriole

mitotic spindle

[b]

daughter nucleus

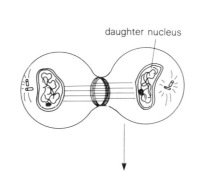

[c] **Transverse section through contraction band**

microfilament bundle

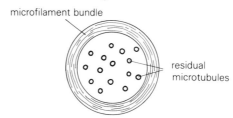

residual
microtubules

place to place, and may in many cases be artefacts of the two-dimensional rigid substratum of conventional tissue culture (Burridge 1981). That the microfilaments can become organized in this fashion, even if it is under rather unusual circumstances, tells us something about the properties of cellular filament systems.

The bundles contain several hundred antiparallel actin filaments and the myosin, tropomyosin and α-actinin may be distributed in a periodic fashion (Fig. 2.15). Lateral stability of the bundles probably comes partly from actin-binding proteins (see § 2.6.1) and partly from the discrete nature of the membrane insertion. Alpha-actinin and possibly vinculin appear to be involved in the anchorage of the filaments, by their barbed ends (Begg *et al.* 1978), in a manner similar to that of filament anchorage at the intermediate-type junctions of smooth muscle (Geiger *et al.* 1981). The periodicity of distribution of the various proteins demonstrated by immunocytochemistry

38

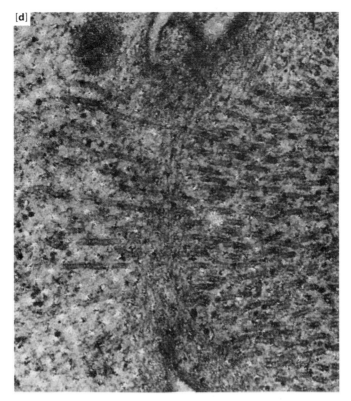

**Figure 2.13  The contractile ring of cytokinesis**   (a) The circumferential microfilament ring is shown in relation to the mitotic spindle   whose poles specify its equatorial position. In (b) the reduced diameter of the ring has constricted the cell – with a daughter nucleus in each half. The ultrastructure of the cleavage furrow between forming blastomeres in a rat egg is illustrated in (d) (from Szollosi 1970) in which both microfilaments of the contractile ring and residual microtubules of the mitotic spindle can be seen.

and the periodicity of electron density, gives the bundles an appearance reminiscent of striated muscle, although there are numerous points of difference (compare Figs 2.10 and 2.15). One interesting aspect of the periodicity is that the 'unit' seems to be of approximately the same length as the length of filaments self-assembled from myosin extracted from non-muscle cells. This may be purely fortuitous, but periodicity does demand some metric device and the length of the myosin-containing bands could certainly be explained on this basis. The cross-banding pattern may be in register with that of adjacent parallel bundles, implying some further cross-linking element between stress fibres, a rôle played in muscle by **intermediate filaments**.

The organization of filaments in striated muscle was justified on the grounds that a strong linear contraction was required, and on the same basis we might suppose that the arrangement of microfilaments in bundles produces a strong though nearly isometric tension and thus considerable stress on the attachment point. This is borne out by the tension which bundle-containing cells can exert upon a flexible substratum (see § 7.5), the

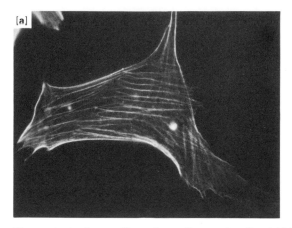

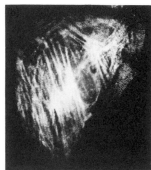

**Figure 2.14 Stress fibres in well-spread cells** (a) Microfilament bundles in a well-spread fibroblast stained with fluorescent phallocidin (photograph courtesy of G. Campbell, Cell Biology Dept, Glasgow University). (b) The periodic arrangement of myosin in stress fibres can be seen in this pig ovarian granulosa cell which has been stained with fluoresceinated antibody to smooth-muscle myosin (photograph courtesy of Peter Sheterline, Dept of Medical Cell Biology, Liverpool University).

tension that must be required to distort a spherical cell to the extremely flattened shape characteristic of fibroblasts in sparse culture, and the tension that is associated with wound healing *in vivo*. Under normal circumstances *in vivo* the extracellular matrix is unlikely to be able to resist such tension and the cell will not adopt a flattened morphology. Thus, fibroblasts in collagen matrices have a cylindrical rather than lamellar appearance and stress fibres are not present (Tomasek *et al.* 1982). It may be that because the rigid substratum provided *in vitro* resists deformation, the bundle maintains its isometric tension, whereas if the substratum is deformable the tension is dissipated and the bundle, being redundant, is disassembled.

## 2.5 Non-linear motors: planar actomyosin motors

Muscle, the contractile ring of cytokinesis, and stress fibres, are all straightforward applications of the actomyosin motor in situations where tension is required. The most organized structure, striated muscle, is both the most specialized and the most stable of the systems, and the degree of disorder increases as we move to systems which operate for a shorter period. Not all the applications of the motor have such simple geometry, and in the next sections we will consider the problems of planar and solid shape change.

Relatively few examples of true planar shape change are known but this is a convenient heading under which to cover two actomyosin-associated motile events which do occur in a restricted two-dimensional geometry, the phenomena of capping and of **cyclosis**. Also under this heading, although it is not clear that it is a system for generating movement, we will describe the organization of filaments in microvilli. The phenomenon of cell spreading, already alluded to in the previous section, is more aptly discussed in Chapter 6, in association with cell locomotion.

[a] **Myofibril**

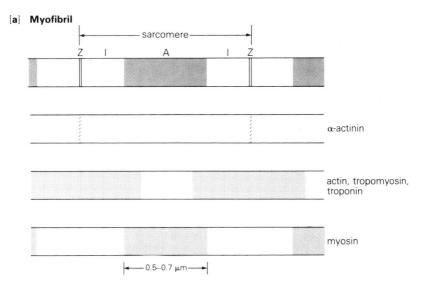

α-actinin

actin, tropomyosin, troponin

myosin

[b] **Stress fibre**

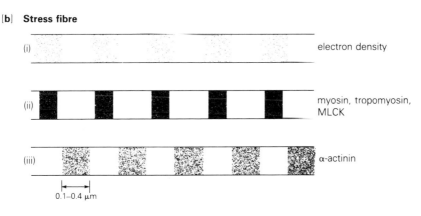

(i) electron density

(ii) myosin, tropomyosin, MLCK

(iii) α-actinin

**Figure 2.15 Periodicities in the sarcomere and the stress fibre** The periodicity of the sarcomere (a) is well understood; the various periodicities in the stress fibres (b) are less easily explained although increasingly thought to represent pseudo-sarcomeres. One important difference is that the myosin-containing bands shown in (b(ii)) shorten when the stress fibre contracts whereas the A-bands of the myofibril remain of constant length (as do all the filaments, although their overlapping alters).

## 2.5.1 Capping

If cells are labelled with a fluorescently-tagged polyvalent reagent (such as immunoglobulin directed against a cell surface antigen, or a **lectin** which binds to the carbohydrate moiety of integral membrane glycoproteins) the label, uniformly distributed at first, becomes patchy, then comes to lie as a cap over one part of the cell, and finally is internalized. The process can be considered in two phases, the aggregation of label to form patches (**patching**),

[a]

**Uniform**

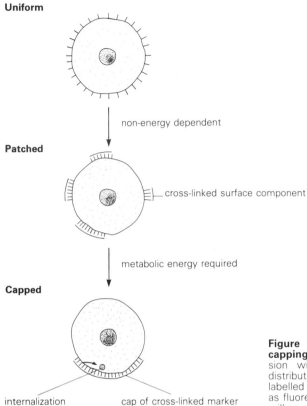

non-energy dependent

**Patched**

cross-linked surface component

metabolic energy required

**Capped**

internalization            cap of cross-linked marker

**Figure 2.16 Patching and capping** A fixed cell in suspension will exhibit an uniformly distributed ring of fluorescently-labelled cross-linking ligand, such as fluorescein-conjugated IgG, as will an unfixed cell for a minute or two before the label becomes discontinuously distributed — patchy. The ring-pattern is of course an artifact: only at the edges does the fluorescence become bright because here the membrane is viewed obliquely and many molecules contribute. The later stage of label redistribution, the capping process, requires that the cell be active: by the time all the label has reached the cap some will already have been internalized. Although patching does not require metabolic activity by the cell, it does depend upon the fluidity of the plasma membrane and may be reduced by cooling. All stages can be seen in the photograph, which is of lymphocytes treated with fluorescently-conjugated lectin (courtesy of Peter Wilkinson, Dept of Bacteriology and Immunology, Glasgow University).

[b]

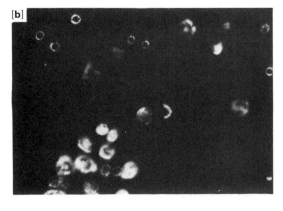

and the relocation of these patches to form a cap (**capping**) (Fig. 2.16). These two processes can be distinguished in several important respects. Patching is probably no more than an aggregation process restricted to a plane; the formation of the patch depends solely upon the random movement of the label in the plane of the membrane and stabilization of aggregates by the polyvalent cross-linking ligand. It is a passive process which does not require metabolic activity on the part of the cell, and it can only be stopped by restricting the lateral mobility of the label. The mobility of integral membrane components could be modified by altering the fluidity of the membrane, or by tethering the labelled membrane components to some form of cytoplasmic cytoskeletal system. Whether these methods are actually used by cells is not clear. In erythrocytes a sub-**plasmalemmal** lattice of **spectrin**, actin and **ankyrin** restricts the mobility of many integral membrane proteins and probably accounts for the absence of patching on these cells. Now that spectrin-like proteins are being found in other cells it seems possible that this is a fairly general method of restricting diffusion within the plane of the membrane. The extracellular lattice of **fibronectin** found on many cells may also limit the lateral diffusion of some membrane components.

Capping is a far more interesting process than patching in that it requires the cell to be metabolically active and the movement is non-random; the patches are brought to a specific location on the cell surface, above the lyosome-rich region near the golgi apparatus, where internalization (which requires a three-dimensional shape change, see next section) takes place. Since patch formation can occur anywhere on the cell surface the system that moves these patches to the capping region must be very versatile. Once the movement begins it is essentially linear, the patch being moved in a more-or-less straight line. The phenomenon is complex and there may be two different mechanisms available. One involves an actomyosin motor system, the other apparently depends more upon a flow of membrane components. The actomyosin-dependent system involves the development of microfilaments (or clustering of pre-existent filaments) adjacent to the patch, and also involves myosin in some way (Koch 1980). The evidence for this is largely circumstantial: the immunocytochemical demonstration of increased amounts of actin and myosin below the patch, the requirement for metabolic energy, and the sensitivity of the process to drugs which interfere with myosin light chain kinase (Sheterline & Hopkins 1981, Kerrick & Bourguignon 1984; MLCK is a likely control component in some non-muscle actomyosin systems), lead to the supposition that this is indeed an actomyosin motor system. A conflicting piece of evidence is that the process is not blocked by cytochalasin, a drug which interferes with actin polymerization, unless **colchicine**, a **microtubule**-disrupting agent, is also added. Since colchicine has no effect on actomyosin motor systems and since colchicine alone can facilitate patching and capping this appears somewhat paradoxical. The facilitation of capping by colchicine may be by releasing a microtubule-associated constraint upon lateral mobility of membrane components, allowing the actomyosin mechanism to operate more effectively, and if the actomyosin system is blocked by cytochalasin then capping will not occur. As addition of cytochalasin alone does not block capping we must invoke a second mechanism which is not actomyosin dependent but which does depend upon microtubular integrity. One suggestion for this second mechanism is that there is a continual flow of membrane phospholipids and associated integral proteins and that large 'islands' of cross-linked membrane glycoproteins, which have a low diffusion constant, would get swept back by the flow or get left behind by the more rapidly moving non-cross-linked components. In a few isolated situations (§ 4.7) there is some evidence for such a membrane flow, which was originally proposed by Bretscher to account for capping and cell locomotion (see § 6.4.3), but good experimental evidence for such a flow has proved difficult to obtain. The whole story is rather unsatisfactory, with different cellular systems showing different sensitivity to drugs, and with too great a reliance being placed on evidence from the use of pharmacological agents of doubtful specificity.

It is tempting to suppose that the rearward movement of an island of cross-linked membrane components is by exactly the same mechanism that moves particles backwards on the dorsal surface of a **fibroblast**, or that moves the cells forward if the cross-linking 'agent' is an immovable part of the surface over which the cell is crawling (see Ch. 6). Although this seems reasonable the evidence is no more than circumstantial. There is, in addition, some conflicting evidence in that the sensitivity of capping to drugs such as trifluoperazine (a calmodulin-blocking agent which may interfere with myosin light chain kinase) contrasts with the insensitivity of movement of the whole cell to this drug and others like it.

## 2.5.2 Cyclosis in algae

The giant multinucleate internodal cells of algae such as *Chara* and *Nitella* exhibit a dramatic cytoplasmic streaming phenomenon with flow rates of up to 100μm sec$^{-1}$ (Kamiya 1981). The motive force is generated at the interface between the chloroplast layer and the flowing cytoplasm, and the motor system is therefore disposed in a planar fashion (Fig. 2.17). Since these cells may be up to 50 mm long and have a diameter of approximately 1 mm, the purpose of the streaming is presumably to stir the cytoplasm and distribute the products of photosynthesis. The flow is cyclic with a distinct boundary between the flow zones. The basis for the flow is an interaction between elements of the cytoplasm and bundles of microfilaments which lie parallel to the flow and with uniform polarity (Fig. 2.17). If the microfilaments are decorated with HMM then the arrowheads point in the direction opposite to the flow, exactly as filaments are arranged in the half-sarcomere of striated muscle (Kersey 1974).

The way in which the cytoplasm interacts with these filaments is not clear although there are two possibilities (Figs 2.18a & b). Since the system is cytochalasin sensitive, one might suppose that reversible actin assembly plays some part in the process, and there do appear to be microfilaments within the streaming cytoplasm that could interact with the fixed micro-filaments of the periphery. Isolated droplets of cytoplasm contain filamentous rings which rotate (Kamiya 1981), but these may have been stripped from the stationary zone. The simplest hypothesis (Fig. 2.18b) has much to commend it, and it has been shown by Sheetz and Spudich (1983) that myosin-coated beads will move with the appropriate polarity over the exposed micro-filament plane *in vitro*, provided ATP is available. Movement of these beads is at an average rate of 2.5 μm sec$^{-1}$, comparable with the rate of sliding of filaments in striated muscle, and would be quite sufficient to drive the observed flow. Thus we would require only that some cytoplasmic elements have myosin appropriately disposed on their surfaces (in a monopolar fashion) and they would move along the tramways of microfilaments, rotating as they did so. Provided large objects are moved actively they will drag the viscous fluid cytoplasm along thus generating bulk flow and, as might be expected, the flow rate diminishes as the distance from the anchored microfilaments increases. Relatively little myosin would be required for this

[a]  [b]

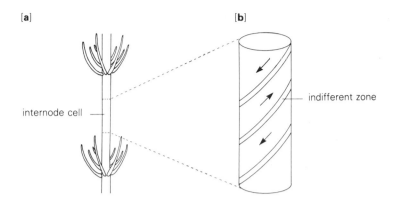

indifferent zone

internode cell

[c]

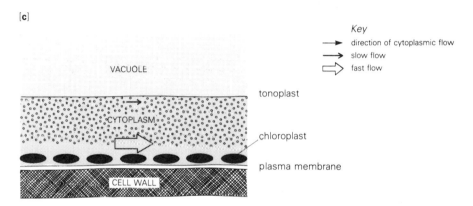

Key

→ direction of cytoplasmic flow
→ slow flow
⇒ fast flow

VACUOLE

tonoplast

CYTOPLASM

chloroplast

plasma membrane

CELL WALL

**Figure 2.17 Cytoplasmic streaming in *Nitella*** The giant internode cells of characean algae (a) exhibit a complex pattern of cytoplasmic flow (b), but for a small portion of the surface the behaviour is simple. The cytoplasmic flow is fastest immediately adjacent to the chloroplast layer (c).

mechanism to work, so the early failure to demonstrate myosin is not surprising. Myosin has now been described and characterized from *Nitella*, and the control of streaming has been shown to be calcium sensitive, although in an unusual fashion. A brief rise in free calcium induced by an action potential leads to a rather longer inhibition of streaming (Williamson & Ashley 1982).

Cytoplasmic streaming in plants is not restricted to these giant cells and may well be a very common feature. The mechanism is probably similar in most cases; actin filaments have been described in pollen tubes, and in the epidermal and **parenchymal** cells of various higher plants.

The experiment of Sheetz and Spudich described above elegantly validates the idea that the minimal motor required for movement need involve only myosin and actin, as discussed in Section 2.3. A further implication is that only the actin need be arranged in a polar fashion since the beads have myosin randomly disposed on their surfaces. Presumably myosin heads with the inappropriate polarity do not function.

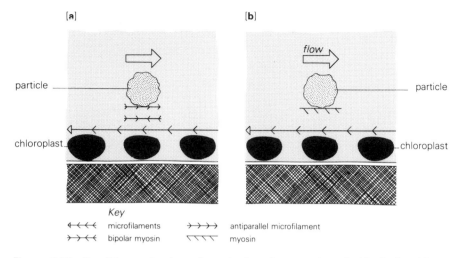

**Figure 2.18  Possible mechanisms for cytoplasmic streaming**  In *Nitella* the chloroplast layer has, on its inner face, bundles of parallel microfilaments. Particle movement could either involve the interaction of bipolar cytoplasmic myosin filaments with the fixed microfilaments of the cytoplasmic layer and a short antiparallel microfilament attached to the particle (a), or myosin could be fastened directly to the particle as in (b). The chloroplast-associated filaments have been shown to have their arrowheads pointing in the opposite direction to that of the flow, as would be required for the two models shown.

## 2.5.3  Microvilli

Although there is some dispute about the extent to which microvilli move *in vivo* their structure illustrates several important points about intracellular motile machinery, and as such they demand our attention. At first sight it might seem curious to find them being discussed in the 'planar' section of this chapter, but they are distributed over the apical surface of epithelial cells as a field or sheet and it is more likely that the shape change is over the plane of the **epithelium** rather than at right angles to this plane.

It is generally supposed that microvilli serve to increase the apical surface area of the epithelial cell, and certainly the best known microvilli are those of the brush border of epithelium from the small intestine of vertebrates. Microvilli are also found in other locations where their function is rather different, e.g. the much larger stereocilia of cochlear hair cells and in the retinal photoreceptors of the squid eye.

Within the microvilli of the intestinal brush border (Fig. 2.19) are arrays of approximately 40 parallel microfilaments, packed in a very regular fashion and stabilized with two actin-binding proteins, **fimbrin** and **villin** (Mooseker *et al.* 1980). These microfilaments are in lateral register and the core is

---

**Figure 2.19  Microvilli of the intestinal epithelial cell**  Each microvillus (only two are ▶ shown in the diagram but there are many on each cell) has a core of parallel microfilaments attached distally to an area of electron density and inserting proximally into the terminal web region. Also inserting into this region are microfilaments from the *zonulae adhaerens* of the junctional complex (see Fig. 2.20). The micrograph shows microvilli on the intestinal epithelium of a neonatal rat (photograph courtesy of Professor Scothorne, Anatomy Dept, Glasgow University). (See also Fig. 2.5 a & b.)

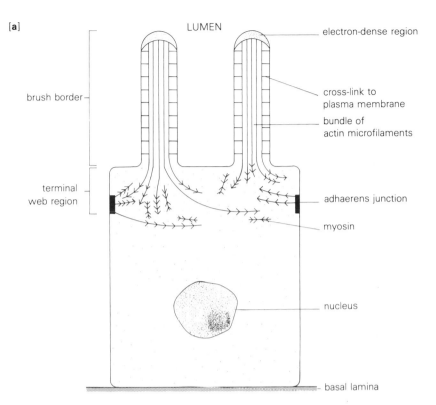

[a]

LUMEN

electron-dense region

brush border

cross-link to
plasma membrane

bundle of
actin microfilaments

terminal
web region

adhaerens junction

myosin

nucleus

basal lamina

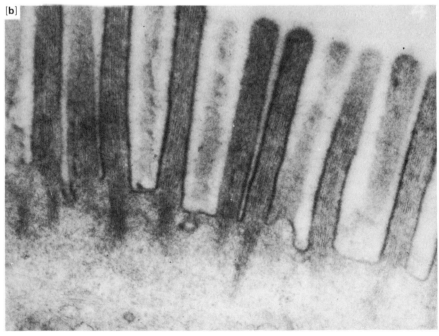

[b]

arranged with the barbed ends of all the filaments at the tip, where they are associated with the membrane at a region of cytoplasmic density (which does not seem to contain α-actinin, but has not been characterized as yet). Along the length of the core the bundle seems to be fastened to the plasma membrane by linking filaments of an 100 kd protein, which are associated with calmodulin but are insensitive to $Ca^{2+}$. The lateral links are arranged in a double-helical fashion and are inserted into the plasma membrane at sites of electron density. The cross filaments are arranged pointing away from the tip of the microvillus, just as HMM would be were it bound to the actin filaments, and can therefore serve as morphological markers of polarity (Matsudaira & Burgess 1982). The absence of myosin from the core means that we can probably discount the possibility of the microvillus being an actomyosin-based motor system, but it could potentially operate as an assembly–disassembly motor, especially since villin, which acts as an actin-bundling protein (see § 2.6.1) at low calcium levels, will fragment F-actin at high calcium concentrations (around $10^{-6}$ M). It is, however, far more probable (Mooseker et al. 1982) that the microvillus is a stiff projection supported by this complex core structure – but intuitively one would imagine that absorption from the intestine would be facilitated if the brush border were to sweep fluid to-and-fro. Potentially this could happen: the actin filaments of the core splay out proximally into the **terminal web** region where myosin *is* found (Mooseker et al. 1978) together with a spectrin-like protein (terminal web 240–260k protein (**TW 240–260k**)), α-actinin, calmodulin, and tropomyosin. The bundle microfilaments might therefore interact, through myosin, with microfilaments of the *zonula adhaerens* of the **junctional complex** that form a ring around the cell at approximately the same level in the epithelial sheet. Microfilaments of the core, which are attached to the tip of the microvillus, and microfilaments attached at the junctional complex (Fig. 2.20) will have opposite polarity and could therefore be involved in a sliding interaction.

The terminal web, the cytoplasmic layer immediately below the bases of the microvilli, although confusingly named in one respect, may appropriately be considered as a tangled web. Even if we consider the filaments derived from a single microvillus there are, potentially, antiparallel interactions in a 360° sector, as well as the interactions between the microfilaments from diametrically opposed parts of the junctional complexes. With the addition of a few more microvilli the situation becomes increasingly complex – rather like a plate of spaghetti. If the calcium level fluctuates locally in the apical cytoplasm then individual microvilli will twitch and writhe, stirring the medium; adjacent cells will contribute to the movement both through the actions of their microvilli and by serving to anchor the junctional complexes in the sheet. Isolated terminal web fragments, in which the basal parts of the epithelium have been torn away so that access of medium to the apical surface is easy, will contract if supplied with $Mg^{2+}$-ATP and calcium at $>10^{-7}$ M, which supports the view that an actomyosin motor is involved somewhere.

Although the microvilli themselves may not be active motile appendages, a consideration of the terminal web region and its properties gives some interesting insights into a planar contractile structure. Were we to shave off the microvilli then the antiparallel interaction of microfilaments derived from *zonulae adhaerens* would still remain and could, by pulling the junctional complexes together, serve to alter the relative areas of basal and apical surface of the epithelium, the sort of change that is required in morphogenesis to transform a sheet into a tube. The formation of a stabilized

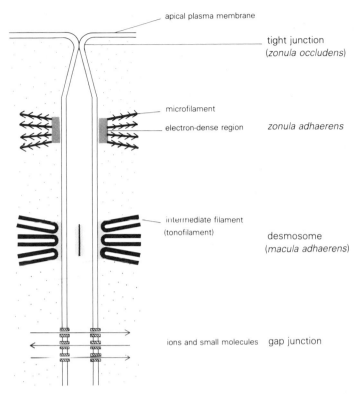

apical plasma membrane

tight junction
(*zonula occludens*)

microfilament

electron-dense region    *zonula adhaerens*

intermediate filament
(tonofilament)

desmosome
(*macula adhaerens*)

ions and small molecules    gap junction

**Figure 2.20 The junctional complex of epithelial cells** Four different junctions are involved. Moving from the apical surface towards the basal: the tight junction (**zonula occludens**) acts as a gasket preventing fluid movement across the epithelial sheet. and also preventing the lateral diffusion of integral plasma membrane proteins from apical to baso-lateral plasma membrane. The adhaerens junction is the site of insertion of actin microfilaments, and resembles the focal adhesions of fibroblasts in some respects. Microfilaments from here may interact with microvillar core filaments, which will be antiparallel. Desmosomes (*maculae adhaerentes*) provide tension-resisting linkages into which are inserted cytokeratin-type intermediate filaments (**tonofilaments**); hemi-desmosomes (not shown, but just one-sided versions of desmosomes) may link the cell to the basal lamina. **Gap junctions** serve a totally different function: they are low-resistance pathways for the passage of small molecules or ions between the cells and are therefore communication channels. (See also Fig. 7.6).

microfilament bundle as a structural element attached to a planar contractile region (the terminal web) may be considered as analogous to the formation of **axopodia** in Heliozoa, where the bundle is of microtubules (see § 3.7.3), and possibly as homologous to the formation of **microspikes** at the leading edge of some cells and to the **acrosomal processes** of certain spermatozoa (see § 4.5).

An interesting comparison can be made between microvilli and cilia. in the former the extent of movement is poorly defined, whereas the latter form the major propulsive device for swimming cells (Chs 3 & 5). Whether the microvillus could be modified to act in the same way as a **cilium** is not clear. Superficially it would not seem to be beyond the bounds of possibility, but a perfectly good solution to designing cilia already exists in the **axoneme**.

## 2.6    Non-linear motors: three-dimensional shape changes

Many of the shape changes undergone by cells are more complex than can be achieved by linear or planar elements alone, although a considerable amount might be achieved by contracting a bounded surface or a circumferential band. Among the more complex shape changes to be considered are the protrusions of pseudopodial processes in phagocytic engulfment and the extension of the leading edge of a moving cell. Because these shape changes involve both expansion and contraction we should not expect the contractile actomyosin motor system to be solely responsible and there may indeed be two systems operating: the actomyosin contractile system and an assembly system which is intimately associated with the contractile mechanism.

### 2.6.1    Actin meshwork formation

As G-actin polymerizes to F-actin the viscosity of the solution rises because the polymers tend to become tangled, and the solution changes from 'Newtonian' to 'non-Newtonian'. This comes about because long filaments when isotropically (randomly) arranged offer a greater viscous drag than they do when ordered parallel to the flow; the apparent viscosity drops as the flow rate increases and the molecules become oriented. This property alone is not particularly useful but if cross-linking agents are added the filaments become transformed to a three-dimensional matrix or gel which has viscoelastic properties. The physical properties of cross-linked polymers are complex, especially when dealing with loose meshworks of hydrated molecules.

Two things define the properties of such a gel, the length of the filaments that are interlinked and the nature of the links between them. The geometry of the linkages, the number of linkages and the permanence of the linkages all influence the properties of the gel. An analogy is to think of the properties of scaffolding and the effect of having many short rods with numerous linkages in comparison with a small number of very long rods sparsely linked with flexible couplings (although this analogy is not ideal because we are dealing with ropes and not stiff rods). Cross-linking in the case of actin gels is brought about by actin-binding proteins which can be classified under three main headings. Filaments can be linked by their ends onto other structures, to each other in end–side interactions, or to each other in side–side interactions. So many actin-binding proteins are being described that it is probably hopeless to try to produce an exhaustive list, but a recent review (Weeds 1982) may be helpful, and Table 2.1 gives an indication of the diversity.

END LINKAGE

End linkage of microfilaments is important both in gel formation and in anchoring filaments for linear contractions as in the Z-disc of striated muscle, the intermediate junction of smooth muscle and the focal adhesion of fibroblasts. Such terminal anchorages may also serve to nucleate filament assembly; if they did so then the distribution of microfilaments would be specified by the location of anchorage points. Soluble 'monovalent' forms of nucleating or anchorage sites may have such a high affinity for the ends of filaments that they generate new ends by breaking pre-existent filaments into shorter segments. Proteins which bind at the 'preferred disassembly' ends of filaments will stabilize pre-existing filaments; whereas by blocking the barbed end further assembly is inhibited and the filaments will disappear or shorten considerably. Obviously such blocking proteins must distinguish

**Table 2.1**  Proteins which interact with actin.

| | |
|---|---|
| actin | self-assembly to form F-actin |
| myosin | the motor; binding activates myosin ATPase |
| tropomyosin | stabilization of thin filament, affects myosin-binding site in striated muscle |
| troponin | troponin I interacts with F-actin to modify the effect of tropomyosin |
| leiotonin | smooth-muscle equivalent of troponin (?) |
| profilin | binds to G-actin and modifies polymerization rate; isoform in *Thyone* sperm |
| α-actinin | binds barbed end of F-actin, found in Z-disc and stress fibres |
| | |
| membrane associated: | |
| spectrin | part of nucleating site on erythrocyte plasma membrane |
| fodrin | spectrin isoform from brain |
| TW 240–260 K | spectrin isoform from terminal web of intestinal brush border |
| vinculin | membrane associated, in smooth muscle and focal adhesions |
| connectin | binds actin (intracellularly) and laminin (extracellularly) (Brown *et al.* 1983) |
| | |
| bundle forming: | |
| actinogelin | from Ehrlich ascites cells |
| fimbrin | from microvilli of brush-border |
| villin | from microvilli, possibly a gelsolin isoform (not fimbrin) |
| fascin | from sea-urchin eggs |
| | |
| gel cross linkers: | |
| filamin | forms flexible linkages — many cells |
| caldesmon | like filamin(?); from smooth muscle |
| | |
| calcium-sensitive nucleating/cleaving: | |
| gelsolin | isoforms in many cells |
| fragmin | like gelsolin; from *Physarum* |
| | |
| filament capping: | |
| acumentin | blocks pointed end of F-actin; from macrophages |
| β-actinin | isoform of acumentin from muscle |
| capping protein | binds barbed end; from platelets |
| | |
| miscellaneous: | |
| scruin | protein in coiled acrosomal process of *Limulus* |
| MAP-2 | microtubule-associated protein |
| DNAase I | binds G-actin but not F-actin; (why?) |
| | |
| others (not proteins): | |
| phalloidin | stabilizes F-actin |
| cytochalasin | blocks barbed-end assembly |
| ATP/ADP | |

See Glossary also.

between G- and F-actin or the monomer pool would compete for binding, and different proteins are required to bind the different ends. Polyvalent forms of end-binding proteins could generate meshworks and the lateral association of such proteins could produce a loosely organized bundle of filaments; the structure of the gel will depend upon the arrangement, number and orientation of the binding sites. If the properties of the linking protein(s) can he altered by changing the ionic *milieu* or by phosphorylation then the strength of the gel or the disposition of filaments can potentially be controlled.

Various proteins may act by binding to the ends of microfilaments, these are listed in Table 2.1. Cytochalasin B is thought to block filament assembly by binding to the preferred growth ends of filaments (Lin *et al.* 1980, Brown & Spudich 1981); only labile microfilament assemblies should therefore be sensitive to this drug, although it may also block end–side linkage.

### END-SIDE LINKAGE

This seems to be a relatively infrequent linkage and only **filamin** is known to act in this way (Hartwig *et al.* 1980). Linkage seems to be at the pointed end of the filament and may arise because filamin, having bound to the side of a filament serves as a nucleating site. This seems to be only one of the ways in which filamin acts as a cross-linker and may not be its physiologically most important rôle. Filamin is a large molecule of 220 kd, normally dimeric and flexible – which means that the linkage need not be of strictly defined geometry. Isoforms of filamin have been found in striated muscle, smooth muscle and various non-muscle cells.

### SIDE–SIDE LINKAGE

Such a linkage will generate filament bundles and may be important in stabilizing stress fibres and the core of the microvillus. The best known example of a stable microfilament bundle is in the microvilli of intestinal epithelial cells where fimbrin (a protein of 68 kd) seems to be the major cross-linking component. Other actin-binding proteins such as $\alpha$-actinin will also promote bundle formation. Bundle formation seems to be promoted by clustering of the membrane insertion sites, those cells which form distinct focal adhesions have well developed stress fibres. In Rous sarcoma virus-transformed fibroblasts the loss of prominent bundles is associated with an increase in the phosphorylation (at a tyrosine residue) of vinculin, a protein which is closely associated with the insertion site (Geiger *et al.* 1981). The phosphorylation of vinculin is increased considerably following viral transformation, but the proportion of the vinculin actually phosphorylated is very low: the significance of these observations is not yet clear.

## 2.6.2   Dynamics of the meshwork

Various factors influence the mechanical properties and location of actin gel meshworks and, since by assembling a mesh the cell can build a protrusion, these controls are of considerable interest. There are three levels at which control could operate:

(a)   by controlling the number of filaments available and their length,
(b)   by regulating where in the cell filaments are located, and what they are attached to,
(c)   by altering the number of cross-linking agents per filament and the type of cross-linkers available.

It is likely that a combination of all these plays a part in regulating the local properties of the cytoplasm, and to some extent there seem to be too many possibilities.

## CONTROL OF ACTIN POLYMERIZATION

This determines the availability of fibres for gel formation and their length. Anything from 30–70 per cent of the actin in non-muscle cells may be in the monomer form (Bray & Thomas 1975, Clark & Merriam 1978), a clear implication being that the meshwork is likely to be labile and in some form of dynamic equilibrium with the monomer pool. Polymerization of G-actin (shown schematically in Fig. 2.21) is sensitive to the ionic environment and to the availability of ATP. It could also be influenced by the concentration of monomer, although the critical concentration for polymerization is probably much lower than the concentration normally found in the cell as a whole (Koffer et al. 1983). It is possible that the local concentration may fall below the critical level when rapid polymerization is taking place in a restricted area, and this might be one reason for the surprisingly high proportion of unpolymerized actin in cells. Not all of the actin monomer is available for polymerization, some is 'sequestered' as a complex with a small protein, **profilin** (Carlsson et al. 1977; see also § 4.5.2). It seems likely that the interaction between actin and profilin is regulated by the cell, although the mechanism is not known. Assembly can also be affected by the availability of nucleation sites, and the number and length of filaments which are actually present will also depend on the rate at which they are being depolymerized. Stabilization of the filamentous polymer by capping of the disassembly end, preventing further addition, or by lateral association with proteins such as tropomyosin, will affect the balance of the equilibrium between monomer and polymer, and thus the number and size of the filaments present.

## LOCALIZATION OF POLYMER

By linking nucleation sites to particular structures the disposition of polymer within the cell can be regulated. The most important site is probably the cytoplasmic face of the plasma membrane, where membrane-associated nucleation sites linked directly or indirectly to integral or transmembrane proteins account for the **cortical meshwork** of microfilaments which contribute to the mechanical properties of the cell surface (Cohen & Foley 1980; Luna et al. 1981). If nucleation sites can be switched from active to inactive forms then the distribution of the meshwork can be altered (Herman et al. 1981). Since the cell surface is the interface between the cell and its environment, and since altered permeability of the membrane would alter the ionic composition of the cytoplasm, we might expect this to be an important controlling influence. Another possibility is that membrane-associated kinases activated by receptor–ligand interactions might alter the properties of an anchorage site.

## CROSS-LINKING OF POLYMER

Depending upon the number of polymers, a fixed amount of cross-linking agent will generate weak or strong gels or, if the polymers greatly outnumber the cross-links, whether a gel is formed at all. The multiplicity of cross-linking proteins that have been described suggests that various types of gel may be formed. It is of some interest that actin-binding proteins can be subdivided into those that are divalent-cation sensitive, and those that are not. By varying the proportion of the two types of binding protein, the sensitivity of the gel to changes in divalent ion concentration can be altered (Yin & Stossel 1979).

Calcium appears to be the most important cation that regulates gel properties and the sensitivity is the opposite of that which might have been

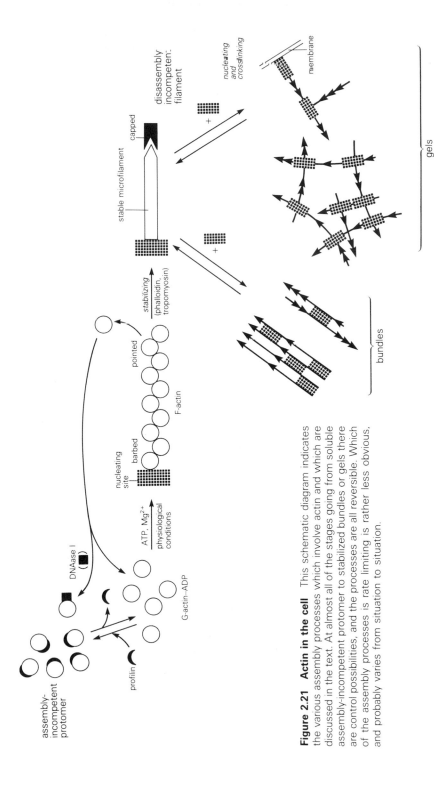

**Figure 2.21  Actin in the cell**  This schematic diagram indicates the various assembly processes which involve actin and which are discussed in the text. At almost all of the stages going from soluble assembly-incompetent protomer to stabilized bundles or gels there are control possibilities, and the processes are all reversible. Which of the assembly processes is rate limiting is rather less obvious, and probably varies from situation to situation.

expected. Actin-binding proteins of the **gelsolin** type bind actin with increasing avidity as the calcium concentration rises, and at low calcium concentration they act as nucleation sites. At high calcium concentrations, however, they cause cleavage of filaments, thereby increasing the number of filament ends available for binding, and thus weaken the gel by altering the ratio of cross-linkers and filaments. At calcium concentrations which would initiate the myosin–actin interaction in striated muscle the gelsolin-type proteins cause filament shortening by acting as additional nucleation sites. High and low concentrations, in the context of gel formation, turn out to be similar to the concentrations involved in controlling the actomyosin motor of muscle. Direct testing of the actomyosin contractile system in non-muscle cells by permeabilizing the plasma membrane and providing exogenous ATP shows that the calcium sensitivity for the contraction of the cell is effectively the same as for contraction of the permeabilized striated-muscle preparation. At $10^{-6}$ M $Ca^{2+}$ the gel will break down and the actomyosin ATPase, if regulated in the same way as in muscle, would switch on. At $< 10^{-8}$ M $Ca^{2+}$ the gelsolin will nucleate but not cleave filaments so that a gel is more likely to be formed or will be more rigid, but the contractile system is inactive. Thus the contractile and assembly systems have opposite requirements for calcium ions and could perhaps act as 'antagonists' in a mechanical sense.

## 2.6.3 Applications of the system

### SHAPE CHANGE IN GELS

Gels of actin can be formed *in vitro* by the addition of an actin-binding protein such as filamin at an appropriate stoichiometric ratio (where the filament counts as a unit; the stoichiometry relative to G-actin will be indeterminate) (Hartwig & Stossel 1975). Incorporation of myosin into such a gel will, if ATP is provided, lead to a contraction of the gel that separates a meshwork phase from a fluid phase, the overall volume remaining constant. Contraction of a spherical gel would produce a smaller sphere in which the filaments are more densely packed; the contraction is in three dimensions because the filaments are randomly disposed in three dimensions. By altering the way in which the filaments are packed the contraction can be made non-uniform; a linear organization of the filaments would give a linear shape change – at right angles to the long-axis of the filaments.

If the gel is associated with a calcium-sensitive binding protein then the outcome of raising the calcium concentration may be more complex. A calcium-switched actomyosin motor will generate a contractile force if ATP is available and the calcium concentration is raised, but this will also lead to disintegration of the gel because the filaments become fragmented, and no useful purpose will be served. The events which follow immediately from an increase in divalent-cation concentration will depend upon the kinetics of the two processes: the activation of the contractile event and the disassembly of the gel which is being contracted. There is some evidence that some reduction in the degree of cross-linking (possibly by cleaving filaments) is necessary to allow significant contraction in the case of gels from *Amoeba* – the solation–contraction coupling hypothesis (Hellewell & Taylor 1979). The system has, however, the potential for great variability simply by varying the proportion of cross-linkages which are cation sensitive.

### PHAGOCYTOSIS

The process of phagocytosis, by which a cell internalizes particles, involves

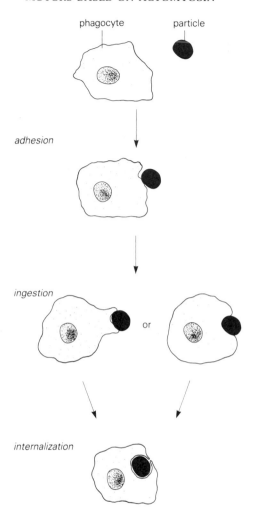

**Figure 2.22  Phagocytosis**  The process of phagocytosis, though continuous, can be subdivided into an adhesion phase, an ingestion phase and a later internal digestion phase. Internalization involves the formation of a phagocytic cup either by the protrusion of processes or by the particle 'sinking in' (or a combination of the two). The shape changes involved require the activity of the actomyosin motor system, and it is possible that a particle is just a small piece of substratum. The final stage in internalization involves membrane fusion, and the protrusions forming the cup must not be contact-inhibited.

the production of a pseudopodial cup in which the particle is engulfed (Fig. 2.22). The 'arms' which wrap around the particle seem to contain only a rather amorphous granule-free cytoplasm when viewed by transmission electron microscopy (TEM) and can be shown, by immunocytochemical methods with the light microscope, to be rich in actin and actin-binding protein (Stendahl *et al.* 1980). Thus we assume that the protrusion is generated by local formation of an actin gel linked to the plasma membrane. The particle also seems to 'sink into' the cell, and it has been suggested

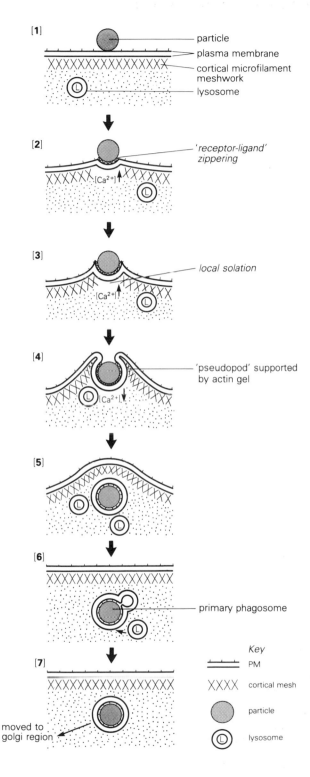

[1]
particle
plasma membrane
cortical microfilament meshwork
lysosome

[2]
'receptor-ligand' zippering
[Ca²⁺]↑

[3]
local solation
[Ca²⁺]

[4]
'pseudopod' supported by actin gel
[Ca²⁺]↓

[5]

[6]
primary phagosome

[7]
moved to golgi region

Key
PM
cortical mesh
particle
lysosome

**Figure 2.23 Intracellular processes in phagocytic engulfment** The formation of the phagocytic vacuole requires a complex series of changes in the organization of the cytoplasm, and these diagrams illustrate some of the processes that are thought to be involved. The initial adhesion event (1) may lead to the redistribution of plasma membrane components, since the particle is multivalent, and this may in turn lead to changes in the local ionic *milieu* within the cytoplasm (2). If intracellular calcium ion concentrations rise then proteins of the gelsolin type may act on the cortical microfilament gel allowing the particle to sink in (3), (4), and also permitting lysosomes (L) to approach the inner face of the plasma membrane (5). The protrusions which form the phagocytic cup may be supported by an actin gel (like the forward protrusions in moving cells), or may be pulled up by the cell–particle surface interaction (zippering). Once the particle has been enclosed within a vesicle of plasma membrane the primary phagosome is increased in volume by the subsequent fusion of lysosomes (6), which do not fuse with the normal plasma membrane because their access is hindered by the cortical meshwork. The mechanism by which the phagosome is moved to the golgi region (7) is not understood.

(Hartwig *et al.* 1980; Fig. 2.23) that the time-dependent initiation of a solation–contraction phenomenon at the base of the forming vacuole contributes to this process. Certainly myosin only seems to be found in the basal region.

There are, however, indications from the experiments of Griffin *et al.* (1975) that the interaction of the phagocyte surface with the surface of the particle contributes to the formation of the vacuole by a 'zipper' mechanism (see Fig. 2.24). It could well be that the cooperative activity of spreading adhesion to the particle and active protrusion of the phagocytic cup walls are

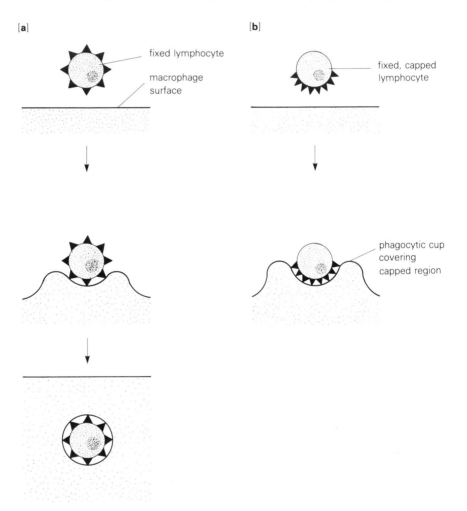

**Figure 2.24 The zipper hypothesis** Fixed B-lymphocytes with a uniform distribution of anti-IgG on their surfaces are internalized normally by macrophages which have Fc-receptors on their surfaces (a). If, however, the lymphocyte is allowed to cap the added IgG before fixation then the phagocytic cup is incomplete as in (b); only the IgG-coated portion of the particle (the lymphocyte) is invested with phagocyte membrane. This experiment provides support for the view that the phagosome is generated by an adhesive interaction between the surface of the phagocyte and the surface of the particle.

essential. Certainly the process of internalization in some systems is blocked by the cytochalasins which, by blocking microfilament extension, would inhibit gel formation.

## PSEUDOPOD EXTENSION IN CELL LOCOMOTION

The protrusion of the leading edge of a crawling cell such as a fibroblast also seems to depend upon the assembly of a microfilament meshwork, and the protrusions resemble those which surround the particles in phagocytosis. Further consideration of the topic will be postponed until Chapter 6 where the mechanism of locomotion will be discussed in detail.

## 2.7 Control of the actomyosin motor

### 2.7.1 *Introduction*

No serious engineer would design a motor system in which the output was not governed or regulated in some way; in our analogy with vehicles we must consider what sort of control system is used on the actomyosin motor.

Many engines are regulated by restricting the input of fuel, whether through carburettor jets, steam valves or current supply; crudely speaking this could also regulate the actomyosin motor since for an ATPase to work there must be ATP available. In some systems this might just be feasible, but in general the control is exercised by regulating the number of conformation changes which are occurring simultaneously, and for how long the ATPase is allowed to continue working. The minimal motor, a single myosin molecule, can only be on or off, there is no possibility of a graded response.

The control system is best understood in striated muscle; but before discussing each system in detail let us consider the possible control systems which could operate. The motor depends upon activation of a conformational change when myosin binds to actin which is in polymeric form. Therefore it *could* be stopped by:

(a)   absence of polymeric actin;
(b)   masking or alteration of the binding site on actin;
(c)   blocking or inhibition of the binding site on myosin;
(d)   inhibition of the conformational change in the myosin;
(e)   inhibition of the release of actin after the conformational change;
(f)   inhibition of the restoration of the resting configuration of myosin.

Of these we have already mentioned (e), since ATP binding is required for the release of the HMM, and also (a), under the heading of control of actin polymerization (§ 2.6.2).

Some of these possibilities will be hard to distinguish, and the remainder (b, c, d & f) can be reclassified under two broad headings: availability of the actin binding-site and regulation of the myosin binding-site (actin-associated and myosin-associated controls). Prejudging the question we might guess that if actin is in great excess then fewer control devices would be needed if the myosin were regulated and this would be a more economical solution, requiring less protein synthesis. Only in striated muscle does the primary control seem to be actin-linked, and only in this tissue is myosin present in such a high molar ratio with respect to actin.

A common feature of all the control systems seems to be that they are based upon calcium ion concentration and, indeed, upon derivatives of a single calcium-binding protein. From an evolutionary perspective this is entirely reasonable, once a calcium-binding protein has been developed it is easier to modify this protein than to redesign from scratch. Perhaps this also accounts for the ubiquitous rôle of calcium ions in control systems: calcium need not have any peculiar advantage, historical precedent and innate conservatism would suffice. With the advantages of hindsight we can see that searching for the location of the calcium-binding protein and its descendents would have been a good way to study control systems.

## 2.7.2 Calmodulin

Calmodulin, the modern version of the original calcium-binding protein, has been found in all eukaryotic cells in which it has been sought and, judging from the primary sequence data, it may well have a common ancestor with troponin C, parvalbumins (calcium-binding proteins from fish muscle) and myosin light chains (Means & Dedman 1980). It is a protein of 17 kd, with four calcium-binding sites of different affinities but with $K_d$s around $10^{-6}$ M. Binding of calcium to any of the sites causes a conformational change and it is this conformationally altered $Ca^{2+}$ –calmodulin complex that mediates the calcium sensitivity of a wide range of cellular processes. Different systems may require different fractional occupancy of the calcium-binding sites on calmodulin in order to respond, almost an essential requirement if variation in calcium levels is to bring about a range of responses. Calmodulin has been implicated in the control of cyclic nucleotide metabolism, of protein kinases, of microtubule disassembly (see Ch. 3), of glycogen metabolism, of secretion and of neurotransmitter release – among other things. We know calmodulin to be present in **stress fibres**, ciliary axonemes and the mitotic spindle. Given the evolutionary conservatism of sequence, the finding of two conservative amino acid substitutions between calmodulins from bovine brain and uterus, contrasting with only six substitutions between coelenterates and cows, suggests that there may be tissue diversity in calmodulin isoforms and multiple genes for calmodulin. This fine-tuning is, however, less interesting than the diversity of functions of calmodulin, which means that many cellular functions can be coordinately controlled by a change in the level of free $Ca^{2+}$ in the cytoplasm.

## 2.7.3 Control in striated muscle

A nervous impulse in a motor neuron triggers a complex chain of events which culminate in the contraction of the muscle. Acetylcholine released at the presynaptic membrane of the neuron diffuses across the synaptic cleft, binds to acetylcholine receptors on the plasma membrane of the muscle cell (the sarcolemma), and induces an action potential by altering membrane permeability. The electrical depolarization is transmitted over the sarcolemma, and deep into the cell as a depolarization in the invaginated **T-tubule** system. The altered ionic environment triggers the release of calcium ions from the **sarcoplasmic reticulum** (SR) by altering the permeability of the membrane of the SR. Diffusion of calcium ions from the SR to the sarcomeres is rapid because the concentration gradient across the membrane of the SR is steep and the SR is arranged in such a way as to facilitate delivery to the appropriate region of the cell. Initiation of contraction is rate-limited by this diffusion event and can therefore only be speeded up by altering the disposition of the calcium-sequestering system. The resting calcium ion concentration around the filaments of approximately $10^{-8}$ M rises by two orders of magnitude, and the actomyosin ATPase is activated. Control at the molecular level resides in the unmasking (or modification) of the binding site on the thin (actin) filament (Ebashi 1974).

Associated with actin in the thin filament are two proteins, tropomyosin and troponin. Tropomyosin spans seven actin monomers and apparently blocks or modifies the myosin-binding site so that the cross-bridge cycle cannot be initiated. Troponin has three constituent polypeptides: Tn-I, Tn-T and Tn-C. Of these only Tn-C has calcium-binding activity and, as a consequence of calcium binding, a conformational change in the troponin alters the interaction of tropomyosin and actin in such a way as to permit myosin S1 to bind and the cross-bridge cycle to start. Notice the economy of design: one cation-binding protein alters seven actin molecules; but notice too that a more obvious solution would have been to make the tropomyosin calcium sensitive. The system has been built with sub-units designed for other tasks, and the calcium-binding moiety, Tn-C, has been linked by special brackets (Tn-T and Tn-I) to the tropomyosin-stabilized actin polymer.

Some uncertainty still exists as to the precise way in which the calcium binding brings about the myosin–actin binding (see Sheterline 1983 for more detail), but one thing is clear, the actin is modified to switch the motor on; the control system is on the thin filament.

Shortening of the muscle will cease once the sarcomeres have reached their minimal length but cross-bridges will continue to be made and broken and the muscle will resist extension. Relaxation of the muscle occurs when the calcium level drops to $<10^{-8}$ M, a consequence of the pumping of calcium ions back into the SR by an ATP-driven ion pump in the membrane of the SR.

Minor modifications of this general control mechanism do exist but the details need not concern us here. Further information can be found in Squire (1981).

### 2.7.4 Control in molluscan muscle

The control of contraction in molluscan muscle is unusual but, since it illustrates a type of control which would be appropriate in many less specialized systems, it will be mentioned briefly. Unlike vertebrate striated muscle the control is myosin associated, the myosin is directly calcium sensitive and there is no troponin (Szent-Gyorgyi et al. 1973). Although tropomyosin is present on the thin filament it plays no part in regulation. Myosin from scallop (Pecten) muscle has two classes of light chain with two regulatory light chains and two EDTA-light chains associated with each heavy-chain pair. Removal of one of the two 18 kd EDTA-light chains (by EDTA chelation of divalent cations – hence the name) renders the muscle unregulated, the ATPase activity is no longer calcium sensitive; sensitivity can be restored by adding back the EDTA-light chain. Curiously, the isolated light chain does not bind calcium and removal of a single light chain affects the control of both S1 heads, implying some sort of cooperative interaction. Even more surprisingly vertebrate regulatory light chain (LC-2), which does bind calcium, will substitute for the EDTA-light chain although it does not confer calcium sensitivity in the muscle from which it is extracted.

### 2.7.5 Control in smooth muscle

As with structure, the control of contractility in smooth muscle is much less clear than in striated muscle and at least four possible control systems are known to exist Some of these coexist and control might be a cooperative interaction, or it may be that smooth muscle is in fact a more heterogeneous tissue and that, depending upon the source, the control system may vary. The latter seems quite likely, given the diverse functions of smooth muscle. The control systems, probably in order of importance, are:

(a)  myosin light chain kinase,
(b)  **leiotonin**,
(c)  troponin-like proteins,
(d)  calcium-sensitive actin-binding proteins.

All are calcium sensitive and could serve to regulate at physiologically significant levels of calcium ion concentration changes.

## MYOSIN LIGHT CHAIN KINASE

Phosphorylation of the regulatory light chain (LC-2) of smooth muscle myosin switches the myosin into an 'active' form and contraction will proceed, i.e. a control mechanism of type (c) (§ 2.7.1) (Sobieszek & Small 1976, 1977). The kinase responsible for activation is myosin light chain kinase (MLCK), a protein found in skeletal muscle, smooth muscle and non-muscle cells with specificity for the autologous tissue isoform of LC-2 in each case. Calcium sensitivity comes from the interaction of MLCK with calmodulin-$Ca^{2+}$ which is essential for kinase activity. The MLCK may in some cases also be inhibited through phosphorylation by a cAMP-dependent kinase which reduces the affinity of MLCK for calmodulin-$Ca^{2+}$ and thus decreases the extent of myosin activation (Hathaway et al. 1981).

Relaxation of a phosphorylation-switched system will require an enzymic dephosphorylation, and **phosphatases** have been discovered with specificity for LC-2 phosphate. Phosphorylation may also affect myosin self-assembly (see § 2.2.2).

Drugs which interfere with calmodulin, such as trifluoperazine, W7 and calmidazolium, will, among other things, block MLCK activation by calcium. The ubiquitous rôle of calmodulin renders such drugs rather crude probes.

## LEIOTONIN

Although phosphorylation clearly plays an important part in controlling smooth muscle, the present evidence (discussed in detail by Hartshorne 1982) does not exclude the possibility of other controls, and it may be that other control mechanisms play a part (Ebisawa 1983). The major problem seems to be the absence of a simple relationship between the extent of myosin phosphorylation and the level of tension of the muscle. A controversial and unresolved question concerns the role of leiotonin.

Leiotonin appears to be the smooth-muscle analogue, if not homologue, of troponin; like troponin it requires tropomyosin to be present, but unlike troponin it seems to activate the actin–tropomyosin complex rather than merely releasing an inhibition. Leiotonin is a dimer of leiotonin A (Ln-A), the actin-activating sub-unit, and leiotonin C (Ln-C), the calcium-binding sub-unit (Hirata et al. 1980). Leiotonin C is homologous with Tn-C and calmodulin but differs in its affinity for $Sr^{2+}$, which may prove experimentally useful (Mikawa et al. 1978).

## TROPONIN-LIKE PROTEIN

Such proteins have been identified but are not generally thought to play a major rôle. Interestingly, however, smooth muscle tropomyosin has the ability to interact with troponin from striated muscle and will do so, whereas skeletal muscle tropomyosin will not substitute for smooth muscle tropomyosin in regulating reconstituted smooth muscle actomyosin.

ACTIN-BINDING PROTEINS

The major calcium-sensitive actin-binding protein from smooth muscle is caldesmon (Sobue *et al.* 1981), although it is possible that smooth muscle $\alpha$-actinin may also lose actin-binding capacity at high ($10^{-6}$ M) calcium ion concentration. Caldesmon is a homodimer with sub-units around 150 kd, which is not itself calcium sensitive but can be dissociated from F-actin by calmodulin-$Ca^{2+}$ and is therefore sensitive at the same concentration of calcium ions as MLCK (apparent Ki for calmodulin is $10^{-6}$ M $Ca^{2+}$).

The coordinate control of actin location and the actomyosin interaction may well be important in smooth muscle where rearrangement of the components of the motor plays a part in the overall behaviour. It is also important to realize that the coexistence of MLCK and caldesmon, for example, is in the cell as a whole and does not mean that their local distribution or activity is identical. The MLCK may be semi-permanently associated with its substrate, but the distribution of caldesmon is unknown at present.

## 2.7.6 Control in non-muscle cells

Even less is known about the regulation of the motor system in non-muscle systems although there is no shortage of hypotheses; Hitchock's review (1977) is still well worth reading as a clear statement of the possibilities. It seems probable that the system is myosin regulated in that troponin has not been found, although calmodulin is present. Crude actin isolated from non-muscle systems mixed with striated muscle myosin stimulates ATPase activity which is insensitive to calcium concentration, suggesting that the regulatory system in the cell is more closely associated with the myosin. Thus control through MLCK or through direct calcium sensitivity of the myosin light chain is a serious possibility. In platelets there seems good evidence that cAMP-modulated MLCK activation is an important control (but platelets are unusual in undergoing only one contractile event in their lifespan, and once-off phosphorylation is very appropriate).

A second major element of control in non-muscle is through the availability (and distribution) of F-actin and the extent to which these filaments can be made to move, a function of the degree of cross-linking. The calcium sensitivity of proteins, such as gelsolin, although direct and not mediated by calmodulin, has a Ki similar to that of calmodulin, of around 5 × $10^{-7}$ M $Ca^{2+}$; at high $Ca^{2+}$ concentration these proteins act as nucleation sites and their effect is to cleave F-actin, destroying the mechanical properties of a meshwork based on long filaments.

In neither smooth muscle nor in non-muscle cells is there a well-defined analogue of the calcium-sequestering vesicles of the sarcoplasmic reticulum. The local level of calcium ions in the cytoplasm must be regulated by diffusion across the plasma membrane and possibly with uptake by endoplasmic reticulum and by mitochondria. Release of bound calcium from intracellular stores can occur and may be sufficient to permit operation of the motile machinery even in the absence of extracellular calcium. A free-living cell such as *Amoeba proteus* must, of course, be a closed system with respect to many ions, including calcium, because of the very low concentration of ions in the external environment.

## 2.8 Summary

This chapter has covered one of the most widely distributed of motor systems and the one about which we know most. The various protein components have been well characterized and the way in which they are assembled into contractile structures is moderately well understood. The key to understanding the actomyosin motor was striated muscle, and the knowledge gained from this system has enabled the less specialized systems to be tackled from a position of strength. Reduced to its bare essentials the motor requires a filament of actin and a pair of myosin molecules which drag themselves hand-over-hand along the rope-like filament. The various ways in which this motor can be organized and used have been discussed: the highly organized sarcomeres of striated muscle, the more flexible arrangement in smooth muscle, and the very labile organization in cells such as fibroblasts which are not wholly devoted to contraction. The applications of the motor can be considered on the basis of the activity for which the motor is adapted: linear, planar and three-dimensional shape changes are possible. In systems where the motor is rearranged frequently, the components are more labile and the assembly of actin into a cross-linked gel may almost be considered a motor in its own right. The control of the motor varies, sometimes being through control of the extent to which it is assembled, in other cases relying on control of the actin–myosin interaction.

From the actomyosin motor we move, in the next chapter, to the dynein–tubulin motor, another well characterized and ubiquitous system in which we will find several 'design-problems' solved in a similar way.

## References

The biochemistry of actomyosin systems is moving fast; the following will remain useful for a limited period.

Squire, J. M. 1981. *The structural basis of muscle contraction.* New York: Plenum. All you ever wanted to know about striated muscle.

Sheterline, P. 1983. *Mechanisms of cell motility. Molecular aspects of contractility.* London: Academic Press. A detailed coverage of the actomyosin motor of muscle and non-muscle systems, with a biochemical flavour.

Weeds, A. 1982. Actin-binding proteins – regulators of cell architecture and motility. *Nature* (Lond.) **296**, 811–16. Although many more will have been named by the time you read this, the basic principles are here.

Hitchcock, S. E. 1977. Regulation of motility in non-muscle cells. *J. Cell Biol.* **74**, 1–15. Like many good reviews this remains worth reading. Many of the ideas have been fleshed out by hard data now, but the story is much the same.

Kamiya, N. 1981. Physical and chemical basis of cytoplasmic streaming. *Ann. Rev. Pl. Physiol.* **32**, 205–36. Covers many areas that I have neglected.

# 3
# MOTORS BASED ON MICROTUBULES

## 3.1 Introduction

The association of microtubules with movement is, with one important exception, rather more an article of faith than a well documented reality but the exception is a major motor system, the microtubules of eukaryotic cilia and flagella. To minimize confusion between bacterial flagella, which are totally different both in structure and mode of action, I will use cilia as the general term for the eukaryotic appendage. The ciliary motor system, as the best understood microtubule-associated movement, depends upon two classes of proteins, **tubulins** and **dyneins**, whose rôles are analogous to those of actin and myosin respectively. Tubulin, when assembled to form a microtubule, provides a linear element with which a discontinuous inter-action occurs, involving conformational change in the dynein-ATPase. We will discuss this motor system in detail in Section 3.5. The reason for supposing that there are other microtubule-associated motor systems derives from their location within cells and the sensitivity of certain cellular movements to microtubule blocking agents such as colchicine. The classic example is, of course, the microtubules of the mitotic spindle which lie parallel to the movements of chromatids at anaphase. If the assembly of these microtubules is blocked then chromatid separation occurs, but not poleward movement, and mitosis halts at metaphase – a convenient phenomenon for the karyologist. Other movements within cells may also be microtubule

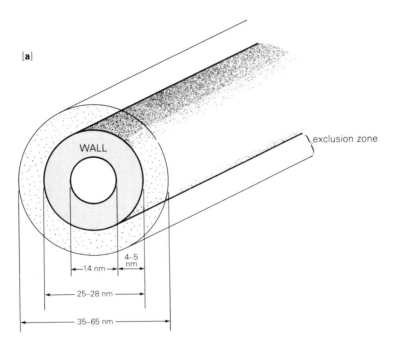

**Figure 3.1 Microtubule structure** The dimensions of microtubules are shown diagrammatically in (a). The ultrastructural appearance of microtubules assembled *in vitro* from pure tubulin (b) and from tubulin and purified MAP-2 (c) are shown in longitudinal- and cross-section (b & c from Kim *et al.* 1979). The distribution of microtubules in a glial cell grown *in vitro* is shown in (d).

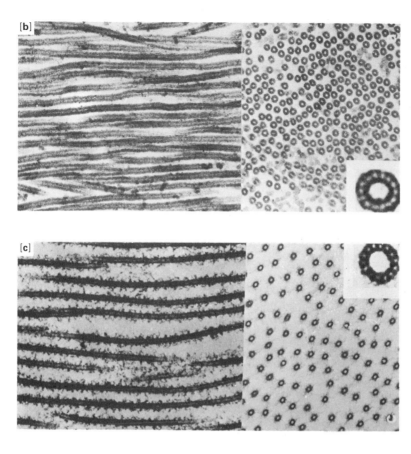

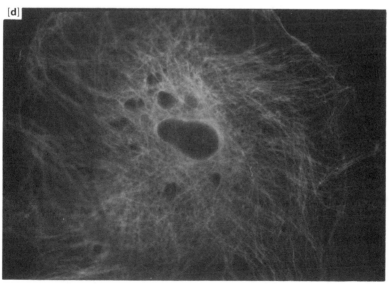

associated but the evidence becomes more circumstantial as we leave the more highly ordered systems of cilia and the mitotic spindle. Generally speaking, the evidence is mainly based upon the occurrence of microtubules oriented with their long axes parallel to the direction of movement, and the inhibition of movement when the microtubules are absent. At first sight this evidence might seem quite convincing, but consider the implications for movement if the limb bones of a vertebrate were removed; movement is not driven by the skeleton, but depends upon it.

Another aspect of microtubular systems that we must consider is the possibility that the linear element, the microtubule, may resist compression; the opportunity exists, in principle, for an assembly process to serve as a motor force. Certainly the cylindrical structure of the tubule is much more suitable for resisting compression than the solid microfilament, although one should be clear that this is not because tubulin is particularly incompressible but because a cylinder resists lateral stress or bending movements better than a solid rod of comparable cross-sectional area.

Before discussing the motor system(s) we need to describe the components of the motor.

## 3.2  Structure of microtubules

Microtubules are found in most eukaryote cells (one notable exception being the mammalian erythrocyte) and have a characteristic ultrastructural morphology, a tube of indeterminate length, having an external diameter of 25 nm and a wall thickness of 4–5 nm (Fig. 3.1). Their visualization in transverse section is facilitated by the electron-lucent zone, some 5–20 nm wide, which surrounds them (Stebbings & Bennett 1976). This zone is probably occupied by a fine, filamentous protein coating which prevents the more densely-staining cytoplasm from impinging directly upon the outer wall. The constancy of the ultrastructural morphology is the first indication that we are dealing with a highly conserved structure, an impression reinforced by biochemical studies (below). There are, however, differences in the stability of microtubules, which gave rise to some problems in early ultrastructural studies particularly of cytoplasmic microtubules. It is customary to separate microtubules into three classes and, since these classes also divide the topic of microtubule-associated movement into appropriate sections, the subdivision is useful here. These classes are

(a)  microtubules of the axoneme of cilia,
(b)  microtubules of the spindle,
(c)  cytoplasmic microtubules.

This classification is more than merely taxonomic and, as with actomyosin systems, the degree of order is associated with the extent of specialization for specific function, and the stability of the structure. We might view the ciliary microtubules as the analogue of striated muscle, the spindle more like smooth muscle and the cytoplasmic microtubules the equivalent of the isotropic actin meshwork of non-muscle cells. But this is *only* an analogy and should not be taken too seriously.

The basic sub-unit of the microtubule is the protein tubulin, which self-assembles to form microtubules, although other proteins probably contribute to the structure.

## 3.2.1 Tubulin

Tubulin forms a dimeric 'protomer' composed of α- and β-tubulin sub-units both of approximately 54 kd as judged by SDS-gel electrophoresis, gel filtration and other methods. The two tubulins have related primary sequences but differ conservatively in approximately half the amino acid residues: they can be separated on SDS-gels at low ionic strength, with β-tubulin migrating slightly faster. Sulphydryl reduction is unnecessary for their separation, suggesting that the dimer is not held together by disulphide linkages. The isoelectric points of the two monomers differ slightly but both are very highly conserved (Luduena & Woodward 1973) in their primary sequence and individual properties such as colchicine binding. The similarity once postulated to exist between tubulin and actin has been clearly shown to be incorrect, although there may be some analogies between the rôles of the two proteins. The sub-unit for microtubule assembly, the **protomer**, is a **heterodimer**, that is, a dimer composed of one α- and one β-tubulin. The evidence for this comes from experiments in which micro-tubules were solubilized (disassembled into protomers) and then covalently cross-linked by a bifunctional reagent; only a single species of dimer was subsequently found using gel electrophoresis. There is also a strict 1 : 1 ratio of α- and β-tubulin from various microtubules when disassembled (Meza *et al.* 1972): the implication must be that the two genes are coordinately controlled.

Tubulin can be phosphorylated at a single serine residue per dimer, although more highly phosphorylated forms have been isolated from ciliary microtubules. The significance of the phosphorylation *in vivo* is unclear but it is not thought to be important for assembly (Piras & Piras 1975; but see Bershadsky & Gelfand 1981). An unusual post-translational modification is tyrosinylation at the C-terminal end by an enzyme, tubulin–tyrosine ligase. Recent work suggests that tyrosinylation may be abnormal in certain pathological states and may be involved in the control of assembly (Nath & Gallin 1984). Tubulin does not seem to be glycosylated to any significant extent.

The heterodimer has a variety of binding sites (Fig. 3.2) in addition to those sites which permit self-assembly by interaction with other heterodimers. Historically, the most important site was that for colchicine binding which was a crucial marker in purifying tubulin (Weisenberg *et al.* 1968). The binding of colchicine (and its derivatives such as **colcemid**) is fairly specific for tubulin but, as with all drugs, there are other colchicine-binding sites within cells, notably those associated with membranes. Fortunately the latter can be distinguished by virtue of their binding of **lumicolchicine**, a photo-rearranged form of colchicine, which does not bind to tubulin but which will bind to the membrane sites (Stadler & Franke 1974). Various microtubule-blocking agents (see Table 3.1) also bind to the heterodimer, amongst which the most important are probably the **Vinca alkaloids**, vinblastine and vincristine (Wilson 1970), which are used as antimitotic agents in tumour chemotherapy. A single phosphorylation site has already been mentioned, as has the site for tyrosinylation on the α-tubulin sub-unit. Two nucleotide-binding sites are present, one non-exchangeable (N-site) which is normally occupied by GDP, the other being exchangeable (E-site) and having a greater affinity for GTP than GDP (Jacobs *et al.* 1974). Thus, as with actin, there is a nucleotide-binding capacity, but for guanosine rather than adenosine nucleotide.

In addition to these binding capacities the heterodimer interacts with calmodulin-$Ca^{2+}$ (Job *et al.* 1981), and with various microtubule-associated

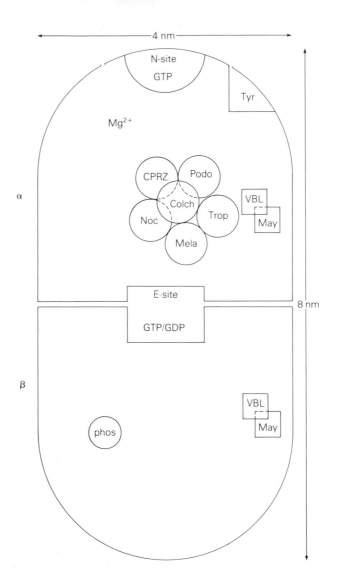

**Figure 3.2 The tubulin heterodimer** The heterodimer formed by α- and β-tubulin has a number of binding sites for agents which affect its assembly capabilities. In many cases it is not clear whether the binding site is on α or β and I have assigned some sites on a random basis. The location of the exchangeable (E) and non-exchangeable (N) sites for GDP/GTP is probably correct, the E-site being a joint venture between both monomers. The α-tubulin certainly has the site for tyrosinylation (Tyr) and a site for $Mg^{2+}$. Both α and β have sites for vinblastine and the other vinca alkaloids (VBL), and the binding of vinblastine is partially blocked by maytansine (May), although maytansine does not produce the same effects as VBL and so the sites are shown as overlapping but not coincident. The location of the colchicine site (Colch) is uncertain, but no matter where it is, the sites for chlorpromazine (CPRZ), podophyllotoxin (Podo), nocodazole (Noc) are there too, and overlap partially. The nocodazole site is also probably the site for binding of reserpine, yohimbine and rotenone. The sites for melatonin (Mela) and tropolone (Trop) are thought to be adjacent to the colchicine site. There is a serine residue which becomes phosphorylated (phos) on the β-tubulin, and there are in addition sites for the interaction of the heterodimer with other heterodimers, with MAPs, and presumably with MTOCs.

**Table 3.1**  Proteins which interact with tubulin.

| | |
|---|---|
| tubulin | heterodimer and tubule formation |
| dyneins | at least two kinds |
| MAP-1 | stabilizing of tubule (?) |
| MAP-2 | linkage to microfilaments (?) |
| Tau | promotes assembly (?) |
| $Ca^{2+}$-calmodulin | control of disassembly |
| tubulin-Tyr-ligase | post-translational modification |
| nexin | linkage of outer ciliary doublets |
| spokein | linkage of outer and central axonemal tubules |
| core protein | linkage of central axonemal pair |
| MTOCs | nucleation or site-specification |
| kinetochore | not a normal MTOC |
| membrane proteins | disassembly of microtubules facilitates capping |
| intermediate filaments | disassembly of microtubules causes rearrangement of IFs |
| kinase | phosphorylation (cAMP-dependent?) |
| | |
| others (not proteins): | |
| taxol | stabilizes assembled tubules |
| colchicine | blocks heterodimer polymerization |
| vinblastine | binds to heterodimer, will promote paracrystal formation |
| vincristine | another vinca alkaloid |
| melatonin | blocks assembly |
| podophyllotoxin | blocks assembly |
| griseofulvin | blocks assembly |
| GTP/GDP | exchangeable and non-exchangeable sites |

In some cases the binding site is on the heterodimer but not on the isolated monomer. See also Glossary.

proteins (**MAPs**) (Murphy & Borisy 1975, Kim *et al.* 1979). Some of these interactions are likely to be important in controlling the assembly of the heterodimer into the microtubule, although at present there is still considerable uncertainty about the relative importance of the various interactions.

## 3.2.2  *The multitubulin hypothesis*

Microtubules of the axoneme, the spindle and the cytoplasm have different properties; does this arise because the sub-units are different isoforms of tubulin or are the differences a consequence of variation in, for example, associated proteins? As will be described in the next section, there is an equilibrium between protomer assembly and microtubule disassembly, the two being balanced in some way. It is then reasonable to ask if there is a single sub-unit pool. Multiple sub-unit pools might arise either by post-translational modification of products of a single gene-set, or by multiple gene-set transcription. (Since α- and β-tubulin are always found in strict 1 : 1 ratio we must assume a strict equivalence in transcription of the α- and β-tubulin genes and hence the choice of the term 'gene-set'.) No doubt this will eventually be resolved by genetic analysis, but there are already strong indications that multiple pools exist, with antigenic variation even within the cytoplasmic microtubules, and it seems likely that different forms of α- and β-tubulin may exist within a single cell (Fulton & Simpson 1976). Probably the strongest evidence comes from studies of the amoeboflagellate *Naegleria gruberi* which, under certain conditions, changes from an amoeboid form to a flagellate by the *de novo* production of flagella. The amoeboid form has a considerable pool of tubulin (approximately 12 per cent of the total protein) yet the flagella are assembled from newly synthesized tubulin which has different antigenic determinants. *In vitro* translation in a rabbit reticulocyte system of messenger-RNA from differentiating *Naegleria* produces flagellar tubulin – this makes the idea of post-translational modification as the source of the difference less probable. Additional support for the multitubulin hypothesis comes from the work of Stephens (1975) who showed that the A and B sub-units of sea-urchin sperm flagellum outer doublets contained different α- and β-tubulins judged on the basis of tryptic peptide mapping.

## 3.2.3   *Tubulin assembly*

An important aspect of the properties of microtubules is their formation by a self-assembly process. The assembly process is an equilibrium reaction which can be shifted to favour either assembly or disassembly, so, like F-actin in non-muscle cells, the microtubule can be a labile structure. Much of our understanding of microtubule assembly comes from studies on assembly *in vitro*. Once this was achieved it became possible to purify tubulin much more easily, by repeated cycles of assembly and disassembly; and the persistent association of other proteins with tubulin prepared in this way indicates that these microtubule-associated proteins (MAPs and **tau**) probably play a part in the assembly process, or in stabilizing the assembled microtubule.

STRUCTURE OF THE ASSEMBLED MICROTUBULE

A variety of approaches to determining the sub-unit organization within the tubule have produced a well accepted model, shown diagrammatically in Figure 3.3 (Roberts 1974). Optical diffraction of electron micrographs of negatively-stained isolated microtubules, X-ray diffraction on flagellar axonemes, and tannic acid fixation of microtubules *in situ* all lead to the view that most microtubules are composed of 13 protofilaments axially disposed with the long axis of the heterodimer parallel to the microtubule axis. As with any cylindrical structure built of regular sub-units it is possible to describe this as a helical structure; in this case a 5- or 8-start left-hand helix, or a 3-start right-hand helix. However, the lateral association between sub-units is weaker than the axial association, hence the tendency for the protofilaments to fray out; and, since the assembly process may involve longitudinal additions of axial arrays, it is probably better to think of the tubule as composed of 13 strands arranged as a (slightly) twisted rope. Exceptions to this do occur, with examples of 12, 15, and 16 protofilaments per tubule being known. The B tubule in the outer doublets of the axoneme has, of course, a different number of protofilaments since some are 'shared' between A and B tubules.

THE ASSEMBLY PROCESS

In simple terms one could consider the assembly as a reversible polymerization reaction as shown in Figure 3.4. The progress of this reaction can be followed by changes in light scattering. The following conditions must be fulfilled if the polymerization is to proceed *in vitro* (using calf-brain tubulin):

(a)   a sufficiently high concentration of tubulin;
(b)   GTP at more than 1 mmole mole$^{-1}$ dimer;
(c)   pH 6.0–7.5, optimum 6.7;
(d)   ionic strength, optimum at physiological level (150 m$M$);

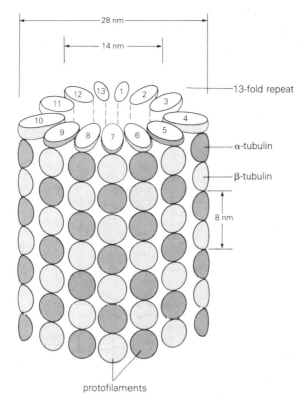

28 nm

14 nm

13-fold repeat

12 13 1 2
11 3
10 4
9 8 7 6 5

α-tubulin

β-tubulin

8 nm

protofilaments

**Figure 3.3  Structure of the microtubule**  The microtubule is built of heterodimers arranged in a very regular fashion with the long axis of the heterodimer parallel to the long axis of the cylinder. Structures such as this can be described in various ways: in terms of the helical packing, or in terms of the rotational symmetry. I find it easiest to think of the microtubule being built of thirteen strands (protofilaments) which lie along the axis of the tubule, so that there are thirteen protomers visible in a cross-section.

(e)  $Mg^{2+}$  0.1 μM–1 mM
(f)  $Ca^{2+}$, low concentration (around $10^{-7}$ M, usually accomplished by adding EGTA, which does not bind $Mg^{2+}$ appreciably);
(g)  temperature 15–45°C, optimum at 37°C.

There are, however, some problems associated with this apparently defined set of conditions. Depending upon the isolation procedure the critical protein concentration may vary: the 'contamination' of the tubulin by accessory proteins may differ depending upon the procedure and source material, and may cause some of the variation. The requirement for a low calcium ion concentration was an early stumbling block – the serendipitous use of pH buffers with calcium chelating properties provided the key – but *in vivo* the calmodulin-$Ca^{2+}$ complex will normally mediate calcium sensitivity. The GTP requirement is also disputed, but it seems likely that hydrolysis of GTP at the E-site is essential under physiologically normal assembly conditions.

With the exception of the nucleotide species this process is not dissimilar from actin polymerization, and in its calcium sensitivity resembles actin

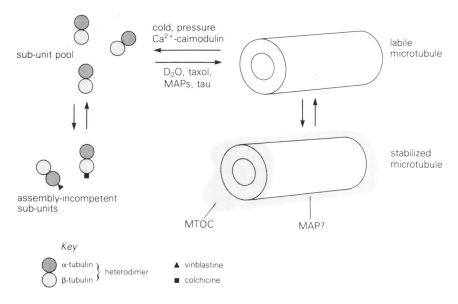

**Figure 3.4 Dynamics of microtubule assembly** Microtubules in the cytoplasm are in a constant equilibrium with a sub-unit pool, and the balance may be shifted by the experimenter or, more importantly, by the cell (some biochemists forget that the cell matters more than the test tube). The equilibrium can be altered by removing sub-units from the pool (the formation of tubulin–colchicine complexes, for example) or by stabilizing the microtubule by blocking assembly and disassembly ends (as in the cilium, for example) or by using the drug taxol.

meshworks which are formed with gelsolin as a binding component (see § 2.6.2).

Considerable controversy surrounds the rôle of various microtubule-associated proteins (MAPs) in the assembly process. Three species of protein are candidates: the high molecular weight proteins MAP-1 and MAP-2 (350 kd and 300 kd respectively) and tau, a heat-stable protein of around 70 kd which has a length–width ratio of 20 : 1. Tau has been further fractionated into tau-I and tau-II. All of these proteins remain associated with tubulin through multiple assembly–disassembly cycles and are in stoichiometric ratio to tubulin. In some conditions they may be indispensible for assembly *in vitro* and do seem to influence the formation of the intermediate multimers which add on to the growing microtubule (Jameson & Caplow 1981). Assembly of tubulin in the absence of MAPs is promoted by glycerol or dimethylsulphoxide (DMSO) and microtubules assembled under these conditions lack the 'halo' of fuzzy material. Various polycations help assembly whereas polyanions are inhibitory, possibly because they compete for basic proteins which would otherwise assist assembly.

The MAPs may confer lateral stability on the outer wall of the microtubule through an electrostatic interaction with the tubulin (notice the analogy between this and the rôle of tropomyosin in microfilaments). Colchicine binding to the dimer, which blocks incorporation of the protomer into the microtubule, involves a disulphide bond cleavage: the balance between -SH and -S-S- may well be important, and *in vivo* this is regulated by glutathione.

Most of the other microtubule-blocking agents, such as the vinca alkaloids, which also bind to the dimer, inhibit assembly of the protomer to form a microtubule; although the vinca alkaloids promote another assembly process, the formation of tubulin paracrystals (Wilson 1970). A natural inhibitor of assembly, a protein found in *Dictyostelium*, may be the first example of a class of proteins within the cytoplasm which give the cell further control over assembly (Weinert & Cappuccinelli 1982).

The preceding discussion has been on the premise that sub-units add directly to the growing microtubule. Careful analysis of the kinetics of the increase in turbidity as microtubules are formed *in vitro* from soluble sub-units shows, however, an early lag phase which is interpreted as a period in which intermediate assemblies are formed to act as nucleation sites. The dimer itself has a sedimentation coefficient of 6S, and tubulin preparations which contain solely 6S tubulin are incompetent at assembly (Borisy & Olmsted 1972). The addition of homologous or heterologous fragments of pre-formed microtubules will permit assembly to proceed. Rather smaller initiation assemblies will suffice and there seem to be disc- or ring-shaped oligomers with a sedimentation coefficient of around 36S (although other sizes are found under other conditions). These oligomers may uncoil to form protofilaments which add on to the pre-existing microtubule in an axial manner (Fig. 3.5) and a variety of schemes have been proposed. It seems probable that these oligomers are stabilized by MAPs and that whilst they are competent to assemble they are not obligatory intermediate stages. The details of assembly are in many ways less interesting than the polarized nature of the assembly, which defines a polarity in the microtubule itself (Margolis & Wilson 1978). Within the cell nucleation probably occurs at specific sites, which may be arranged in a very specific fashion relative to one another, and assembly proceeds by distal addition of sub-units. Disassembly will occur proximally, and if there is indeed an equilibrium between assembly and disassembly then the phenomenon of **treadmilling** (Kirschner 1980) will occur (see Fig. 3.6, also § 2.2.1).

## REGULATION OF ASSEMBLY AND LOCATION

That the components of the microtubule, tubulin and its associated proteins, can self-assemble and that microtubules are localized spatially and temporally implies that there is some control over the extent and location of assembly within the cell. Control of the assembly process itself could be mediated either through regulation of local conditions for the equilibrium reaction, most probably involving calmodulin-$Ca^{2+}$ (Marcum *et al.* 1978, Job *et al.* 1981), although other controls could be involved, such as modification of MAPs, or by the availability and status of nucleation sites (Schmitt *et al.* 1977). The exact rôle of cAMP in regulating assembly remains unclear, although phosphorylation of MAPs may be cAMP dependent (Sloboda *et al.* 1975), and a cAMP-sensitive protein kinase is associated with MAP-2 (Vallee *et al.* 1981). The availability of competent nucleation sites would also serve to localize the assembly process.

A variety of cellular structures appear to serve as nucleation sites and have become known generically as microtubule organizing centres (**MTOCs**). Some are distinct structures, such as **basal bodies**, but most are rather disappointingly amorphous, being merely densely granular regions from which microtubules emerge. In animal cells the major MTOC, other than the basal body in cells with cilia or flagella, is the area surrounding the **centriole**, itself a microtubule-based structure (Brinkley *et al.* 1981, Brooks &

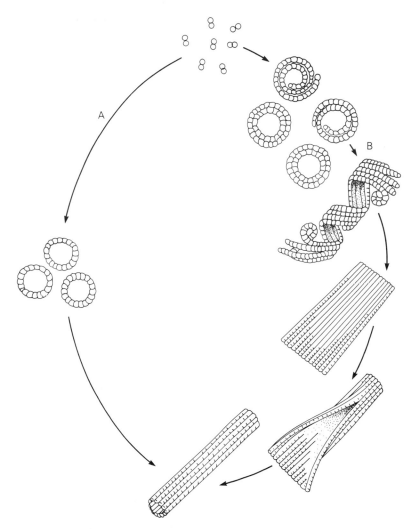

**Figure 3.5  Assembly of microtubules *in vitro***  In the absence of glycerol, protomers are assembled into discs (A) onto which further addition occurs; with glycerol present in the medium more complex intermediate structures are found (B). It is not clear which route of assembly is used *in vivo* (from Stebbings & Hyams 1979, with permission).

Richmond 1983). Cytoplasmic microtubules do not insert directly into the centriole, only into the pericentriolar region; in plants the pericentriolar region is a misnomer since centrioles are generally absent.

Microtubule organizing centres can be isolated from cells and used *in vitro* as organizing centres upon which homologous or heterologous tubulin will assemble (Binder *et al.* 1975). There is some evidence of restriction in the range of heterologous interactions but, for example, tubulin from chick brain will assemble onto MTOCs from yeast (*Saccharomyces*), algal and HeLa cells. **Kinetochores** of chromatids do not seem to act as MTOCs; although this has

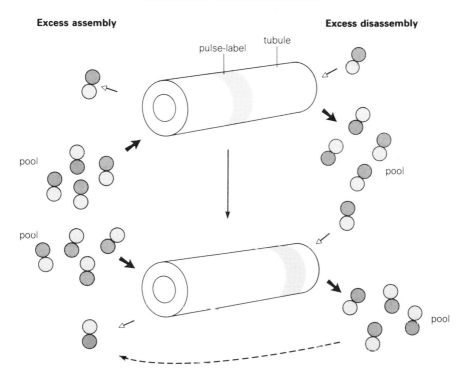

**Excess assembly**

**Excess disassembly**

pulse-label

tubule

pool

pool

pool

pool

**Figure 3.6 Treadmilling** If assembly and disassembly occur at different ends of a microtubule then at the equilibrium length, with assembly exactly counteracting disassembly, there will be a constant flux of material along the length of the tubule. Thus, if a pulse-label could be applied the label would start at the assembly end and move to the disassembly end. Although illustrated here for a microtubule, the same phenomenon will occur in any linear structure where there is constant assembly–disassembly, such as a microfilament (see Fig. 2.2).

been somewhat controversial, it now seems generally agreed. Perhaps the clearest example of control of MTOC status or competence comes from a consideration of the sequence of events in mitosis, where microtubule organization is changed spatially and temporally in a well rehearsed pattern. Some MTOCs may specify rather specific patterns of microtubules by acting as templates, whereas others may generate such patterns by specifying the number of microtubules to be initiated (Fig. 3.7), there being sufficient lateral mobility of the initiation sites that lateral interactions between microtubules themselves generate the pattern (Tucker 1977, Jones & Tucker 1981).

### 3.2.4 Polarity of microtubules

The assembly process defines a polarity in microtubules and this can be demonstrated in various ways. Generally speaking the microtubule is associated with an MTOC at its disassembly end, but it is not always possible to determine the site of the MTOC. A labelling method analogous to that used for demonstrating the polarity of microfilaments would be a great help, and two such methods have recently become available. One depends upon the

**[a] Random**

MTOC

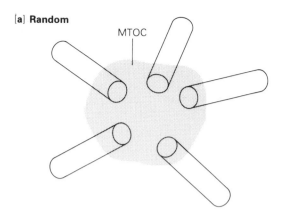

**[b] Lateral interaction between tubules**

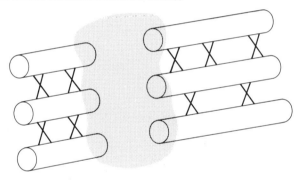

**[c] Interaction in MTOC**

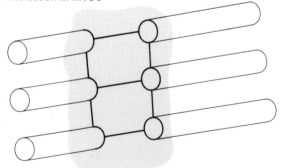

**Figure 3.7  Microtubule organizing centres**  Three kinds of organizing centre might exist (in theory at least): in (a) the number of microtubules is specified, but apart from the fact that they all originate from the MTOC, their spatial arrangement is random. Alternatively, the assembled tubules could be organized into complex arrays by lateral interactions between the tubules (b), or the array could be specified by the distribution of the initiation sites within the MTOC (c).

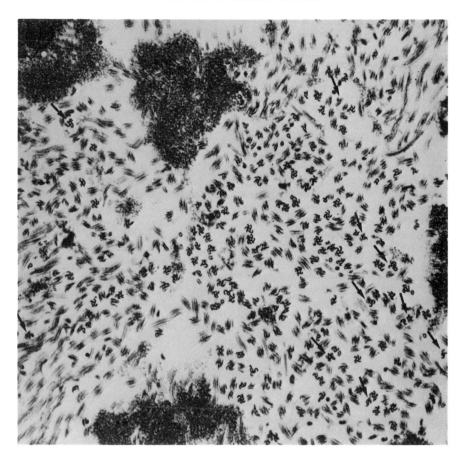

**Figure 3.8  Demonstrating the polarity of a microtubule**  The outer doublets of the ciliary axoneme have the dynein attached in such a way that it is possible to deduce the polarity of the tubule because even in the resting position, the dynein 'points' towards the basal body (see Fig. 3.9). The micrograph shows an alternative method to demonstrate polarity. If assembly competent tubulin is added in excess to a permeabilized cell, it will decorate cytoplasmic microtubules, and the handedness of the partial tubule sections (in cross-section) indicates the polarity. The picture is of a cross-section through an early anaphase cell, close to the kinetochore (on the chromatid), but between the kinetochore and the pole to which the chromatid is moving. Most microtubules have hooks, 92 per cent of which are anticlockwise, indicating that most microtubules are parallel (from Euteneuer & McIntosh 1981).

association of dynein with microtubules (Telzer & Haimo 1981). In cilia the dynein arms are directed towards the proximal (MTOC-associated end), even in the relaxed state, and dynein has been used to demonstrate polarity in the cytoplasmic microtubules of permeabilized cells: that it will bind to cytoplasmic microtubules is of interest in itself. The other method depends upon the assembly of incomplete tubules on the pre-existing tubule when excess dimer is added (Euteneuer & McIntosh 1980). The handedness of the incomplete profiles (Fig. 3.8) defines the polarity.

## 3.3 Dynein

Dynein, an ATPase isolated from the axoneme of various types of cilia and flagella (see §§ 3.4 & 3.5) has also been found in bovine brain and probably occurs in other non-ciliated cells in small amounts (Pallini et al. 1982, Hisanaga & Sakai 1983). The best-studied forms are from cilia where two types are found representing the inner and outer arms respectively (Gibbons et al. 1976). The diversity of dynein is far greater than that of tubulin, and even within a single species several isoforms may be found in different cell types. Dynein, like myosin, is a complex molecule and its structure is not well understood. Dynein 1, a 14S form which corresponds to the outer arm, probably consists of two heavy chains ($A_\alpha$ & $A_\beta$), three intermediate chains (IC-1, IC-2 & IC-3) and at least four light chains, although not everybody would agree. It is the more easily extracted form and presumably differs in some respects at least from the inner arm dynein. The ATPase activity of dynein 1 is sensitive to extraction procedures but can be restored to a much higher level by various treatments in vitro and has been referred to as latent activity dynein. Of the various methods of activating this dynein the most interesting is by allowing it to rebind to flagellar tubules, its normal site. The overall molecular weight of the dynein 1 complex is approximately 800 kd with two large sub-units of 350–400 kd. The properties of dynein 2 (MW 650–720 kd) are much less well characterized and it is not clear whether its function is the same as that of dynein 1.

The protein derived from bovine brain has been classed as a dynein isoform on the basis of its size (ca 330 kd), enzymatic characteristics and co-purification with brain tubulin. It differs from ciliary dynein in having a substrate preference for CTP and in being divalent-cation dependent. Whether this is indeed a dynein-ATPase remains uncertain, however, partly because there is no unambiguous and generally accepted way in which dynein can be recognized.

Probably the strongest indication, although rather circumstantial, that dynein-like molecules are cytoplasmically located comes from studies of patients with immotile cilia syndrome, who have a deletion in flagellar and ciliary dynein. Their leucocytes show anomalies of behaviour, despite being non-ciliated cells (Englander & Malech 1981) (see also § 7.2.2).

A new approach to studying possible dynein-mediated events comes from the finding that vanadate ions inhibit the dynein-ATPase with some degree of specificity. Other ATPases are inhibited, although less markedly, and it is not therefore a very precise probe. It may, however, indicate likely systems for further investigation.

## 3.4 The basic motor

It has been clear for some time that the mechanism of ciliary bending relies upon a sliding interaction between adjacent axonemal doublets restrained by proteolytically sensitive linkages. Thus in the classic experiment of Summers and Gibbons (1971), demembranated sperm tails of sea urchins would curve and wriggle if $Mg^{2+}$-ATP was provided, but prior trypsinization of the axoneme allowed the structure to slide apart when the motor was supplied with fuel. Before describing the mechanism of ciliary beating in detail, the basic mechanochemical cycle proposed by Satir (1982), Witman and Minervini (1982), and others, should be described. In abstracting the molecular motor in this way it can more easily be compared with the actomyosin mechanochemical cycle to which it is analogous.

The cycle, which is of course really a continuous progression, can be subdivided into four phases. Dynein, attached irreversibly (in physiological terms) to one microtubule, the A tubule of the outer doublet (or, in principle, to any suitable object), binds to the B tubule of the adjacent doublet (or any

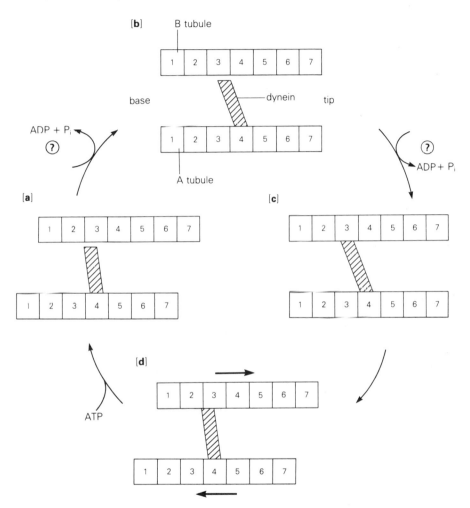

**Figure 3.9  The dynein mechanochemical cycle**  The cycle of attachment, conformational change, and detachment of dynein which brings about the sliding of adjacent ciliary doublets, can be subdivided into four stages. The dynein molecule is attached permanently (to an A-tubule) at one end and is slanted towards the base at an angle of 82° (a). A conformational change in the molecule, which increases the tilt to around 70° (b), allows the attachment of the dynein to the B-tubule of the adjacent doublet (c). Following attachment, there is another conformational change restoring the dynein to the 82° position (d). If ATP is available to bind to the dynein then the B-tubule is relinquished and position (a) regained. Whether ATP hydrolysis accompanies baseward tilting of the dynein (a–b) or follows attachment to the B-tubule (b–c) is unresolved. Each attachment/detachment cycle causes a lateral displacement of one doublet relative to the other.

other microtubule). When dynein binds ATP it releases the reversibly-bound tubulin (Fig. 3.9b) and shortens. Hydrolysis of ATP leads to a conformational change which involves both extension and tilting proximally towards the disassembly ends of the microtubules (Fig. 3.9c) and that leads to the re-binding of the next tubulin molecule (Fig. 3.9d). Binding is followed by another conformational change which restores the position in Figure 3.9a, except that the two linear elements have moved relative to one another. Release of the ADP occurs at an undefined stage. The net movement is approximately 16 nm cycle$^{-1}$ although some authors have suggested smaller movements. The minimum movement would be 4.5 nm, the monomer–monomer axial spacing, assuming both $\alpha$- and $\beta$-tubulin will serve for dynein attachment. Vanadate blocks the hydrolysis of ATP and the arms remain in the detached, 'perpendicular' position (Fig. 3.9b). Detachment does not require the hydrolysis of the ATP but the absence of ATP will lock the cycle in the starting/rigor position (Fig. 3.9a).

The major differences between this system and the myosin cycle lie in the use of identical parallel sliding elements and the difference in the control – the dynein-ATPase does not seem to be calcium sensitive.

## 3.5 Cilia and flagella

### 3.5.1 The ciliary motor

The basis of movement in cilia has already been described, the effect of conformational change in the dynein-ATPase is to slide linear elements to which the dynein is attached relative to one another, but the way in which this motor is used to generate a bend in a cilium involves a rather complex organization. The pattern of bending in cilia and flagella differs, as will be discussed in Chapter 5, but the mechanism is the same. Both cilia and flagella are characterized by the so-called '9 + 2' arrangement of 9 outer doublets with a central pair of microtubules which is found almost universally in eukaryotes. Figure 3.10 shows a schematic transverse section through a cilium and Figure 3.11 a detail of the longitudinal arrangement of dynein and other elements. The whole axoneme is held together with protein cross-links of various sorts, the number of interactions involved probably accounts in part for the highly conserved nature of the tubulin primary sequence. Bending could be achieved by shortening of some doublets or by sliding, the latter being somewhat harder to visualize. That sliding is the mechanism is shown by the experiment of Summers and Gibbons (1971, see also Sect. 3.4) and by the work of Satir (1974), who showed that the outer doublets remained of

**Figure 3.10 Transverse section of a cilium** (a) A diagrammatic cross-section. An ▶ important point to notice is that the classical '9+2' arrangement of microtubular doublets is enclosed in a membrane, the complexity of the interior often leads students to neglect this point. In this section the observer is within the cell looking towards the tip of the cilium, the dynein arms are arranged in a clockwise fashion. (b) Transmission electron micrograph to illustate the appearance of a ciliary axoneme. The photograph is of rather an unusual organism, the blood-stream form of the parasitic protozoan, *Trypanosoma brucei*: the 9+2 microtubule array of the axoneme lies next to the paraxial rod (p), a structure which probably contributes to the mechanical properties of the flagellum. The flagellum is adjacent to a portion of the cell body, which has a very conspicuous glycocalyx (g); the sub-pellicular microtubules, which play a cytoskeletal rôle, can also be seen in section (s) (photograph courtesy of Keith Vickerman and Lawrence Tetley, Zoology Dept, Glasgow University).

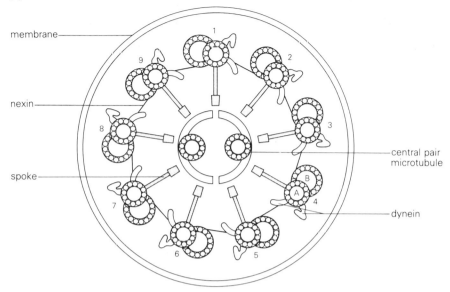

membrane

nexin

spoke

1
2
3
4
5
6
7
8
9

A
B

central pair
microtubule

dynein

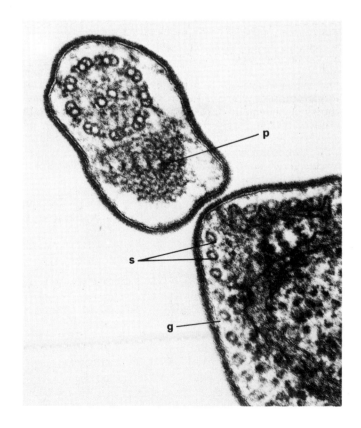

p

s

g

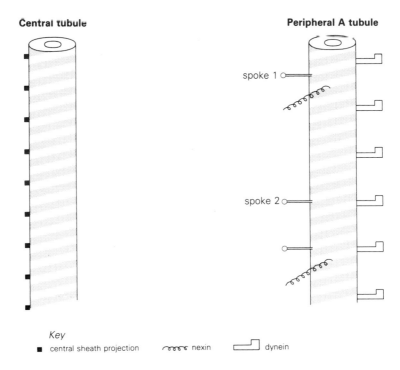

**Central tubule**

**Peripheral A tubule**

spoke 1

spoke 2

Key

■ central sheath projection　　∿∿∿ nexin　　⌐¬ dynein

**Figure 3.11 Periodicity along the axoneme** In considering the periodicity of structures along the ciliary axoneme, it is convenient to remember that the heterodimers are arranged longitudinally and are approximately 8.3 nm long; the repeating unit is therefore likely to be integer multiples of this value, although it may not be possible to distinguish whether the attachment is to one heterodimer or to two adjacent dimers. The central pair of microtubules in the ciliary axoneme each have projections every 2 units, the A-tubule has dyneins attached every 3 units, and nexin every 12. The spokes are arranged in pairs, separated by 3 units, the pairs being 9 units apart. In the diagram the relationship between attachment points should not be taken too literally, the various projections are disposed in three dimensions on a tubule which has helical symmetry.

constant length (see Fig. 3.12). Not only does the proteolytic digestion of the radial spokes show that sliding *can* occur, it also shows that integrity of the axonemal structure is essential to generating a curvature. The importance of the radial spoke structures is also shown by the occurrence of mutants in *Chlamydomonas* which have 'paralyzed flagella' (Luck *et al.* 1982). Some of these (such as pf17) lack radial spoke protein and, from complementation studies, it seems that there are a number of structural proteins which are essential to normal flagellar function (Witman *et al.* 1978). The effect on the pattern of bending of restraints on sliding which might arise from radial linkage is shown diagramatically in Figure 3.13; note that the restraint must be reversible, since in the region of sliding the doublets are less restricted. Bend formation and propagation do not, however, depend simply upon radial spoke integrity because pf17 mutants of *Chlamydomonas* form recombinants with suppressors of the paralysed flagella phenotype ($sup_{pf} -1$ & $-3$) which show symmetrical flagellar beating, even though the spokes are still absent. Both the forward (principal) and the reverse bending patterns in these

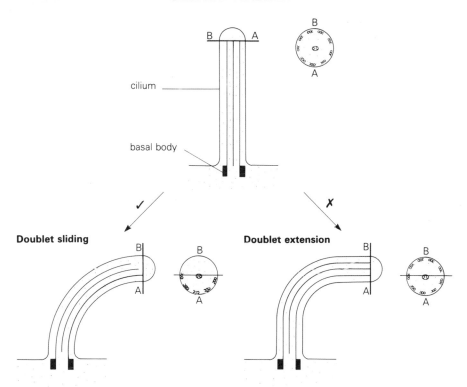

**Figure 3.12  Ciliary bending**  If a straight cilium with a cylindrical core of microtubule doublets is bent then there are two possibilities (since the doublets are anchored at the base): either doublets slide, and at the tip only those on the inside of the bend will be seen; or the doublets might shorten and the tip would retain its symmetrical doublet pattern. The important observation made by Satir was that the former prediction was correct, and that only some doublets were visible in sections taken near the tip. That the doublets do not shorten is also confirmed by the retention of regular spacing of projections on both inside and outside curves.

recombinants resemble the wild-type principal or effective stroke: it seems that the rôle of the radial spokes may be in constraining the reverse bend to give an asymmetric beat (Brokaw *et al.* 1982) (see below). The restraint may therefore be through the peripheral interdoublet linkages rather than through radial linkage to the central core. The constraints placed on sliding are, nevertheless, essential for the system to work.

For convenience the description below is for a cilium which has a simple beat restricted to a single plane, rather like an oar on a rowing boat. Far more complex beat patterns can be observed but can be accounted for quite easily (in outline though not perhaps in detail) (Sleigh & Barlow 1982).

Perhaps rather surprisingly only half the doublet–doublet interactions contribute to a single stroke of the cilium and the axoneme may be considered as a pair of motors coupled in opposition; the return stroke of the cilium (generating the opposite curvature) is as important as the forward stroke. The subdivision occurs in the same plane as the plane of beating, as though the upper surface (half-cylinder) of an oar moved it forward and the

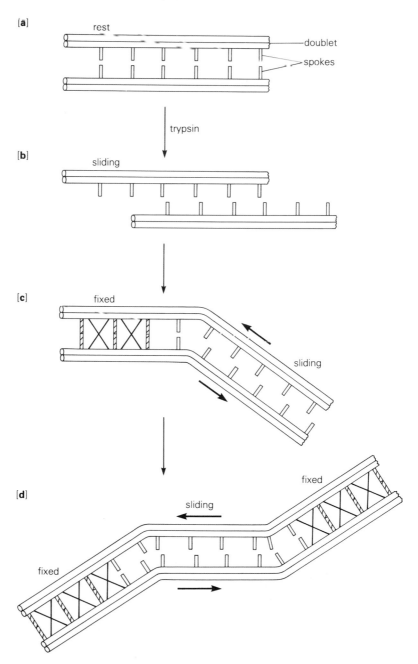

**Figure 3.13 Restricted sliding and its consequences** If the isolated axoneme is treated with trypsin then the doublets will slide apart if ATP is provided (a), (b). If the doublets are linked at one end, as in (c) then the outcome of sliding will be to generate a bend, and if both ends are linked, as in (d), then the cylinder will be thrown into an S-bend. The diagrams illustrate only two diametrically opposed doublets and the axoneme (9 doublets) is more complicated (Figs 3.10 & 3.14), but the same principle applies.

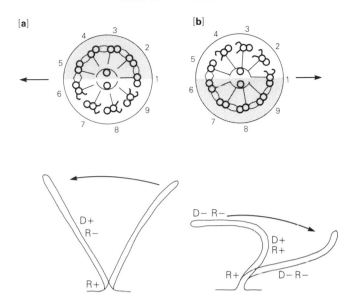

**Figure 3.14  The doublet interactions in a ciliary beat cycle**  In the effective stroke of a ciliary beat only the doublets in the upper half of the section shown in (a) are active (doublets 1–5) and the cilium is straight because all doublets are synchronously activated. In the recovery stroke the other doublets are active (b), and a bend is propagated along the length of the cilium because the doublets are activated metachronously in a wave which passes from the cell body towards the tip. Regions where the doublets are active are indicated 'D+'; linkage by radial spokes 'R+'. Inactive doublets and unlinked spokes are shown as 'D–' and 'R–' respectively (from Sleigh & Barlow 1982).

lower surface moved it back. Reference to Figure 3.14 will perhaps clarify matters, although bundles of pencils or of flexible tubes may be necessary aids to visualization.

Two curious aspects of this model require further comment. First, not all of the doublet pairs seem to be working efficiently, some slide much more than others, yet individual doublets do not differ in sliding ability. The behavioural differences of doublets in the intact axoneme may perhaps arise by 'wasted cycles' in the slower doublet interactions. That doublets are identical is shown by the experiments of Takahashi *et al.* (1982) on the disintegration of trypsinized axonemes. Sliding rate can be measured as a change in length as the axoneme extends or 'telescopes', and the length increases linearly as a function of time (Fig. 3.15). Sliding speed can also be shown to be related to $Mg^{2+}$-ATP concentration. Under the conditions of this experiment the speed is unrelated to the number of pairs of dynein cross-bridges which are interacting, and very few are involved once the two doublets have slid almost wholly apart; it is, however, a system in which there is virtually no resistance to overcome. Clearly, the power output from even a few dynein molecules is quite sufficient to move 10 μm lengths of doublet, something to be borne in mind when considering movements potentially associated with a dynein–tubulin motor in other systems. If half the dynein arms are removed (the dynein 1 of the outer arm is more easily extracted than dynein 2 of the inner) the speed of sliding approximately halves. This may be because each of the arms of a pair act 'out-of-phase', rather like a pair of legs.

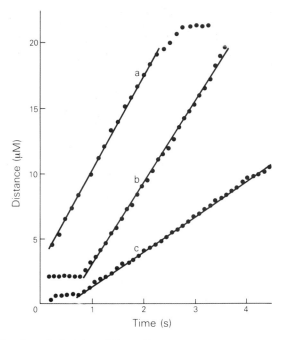

**Figure 3.15  Kinetics of doublet sliding**  Trypsin-treated axonemes will slide apart if supplied with $Mg^{2+}$ATP, but the speed of sliding remains constant even though the extent of overlap progressively diminishes and fewer doublets are acting effectively. The constancy of sliding speed is shown by the linearity of the plots of distance *vs* time (the distance measurement is simply given by the length of the trypsin-treated axoneme). The three lines are the rates at different $Mg^{2+}$ATP concentrations: a = 220 μM, b = 150 μM, c = 15 μM) (from Takahashi *et al.* 1982).

Two speeds of beating, which may depend upon the two gaits (hopping and running as it were), have been observed in intact cilia of the bronchial tract of the newt (R. Hard, personal communication).

The second curious feature is the problem of controlling the activities of the two halves of the axoneme. If both halves were to act simultaneously there would be no net movement; they must act alternately. What, therefore, causes switching between the opposed motors and how can this be arranged in an apparently symmetrical structure without obvious compartmentalization? Reversal does not depend upon rotation of the axoneme, which would have been one solution to the problem (Tamm & Tamm 1981) (although it would make one half of the doublets redundant). The two motors (the two halves of the axoneme) are *not* symmetrical of course, otherwise the rearward beat would have the same pattern as the forward beat and no net work could be done. This does indeed seem to be a problem in recombinants of pf17 mutants of *Chlamydomonas* where defects in the radial spokes seem to lead to a simple symmetrical beat rather than the asymmetric beat which is required for efficient propulsion (Brokaw *et al.* 1982). Pharmacological studies with the lateral cilia of the gill of *Mytilus* have shown that vanadate blocks the cilia at the end of their downward stroke, whereas treatment with 12.5 mM calcium chloride and the calcium ionophore A23187 blocks them at the top of the upstroke (Fig. 3.16). The latter blockage by high calcium ion

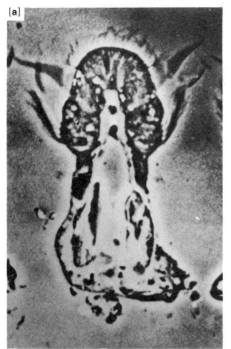
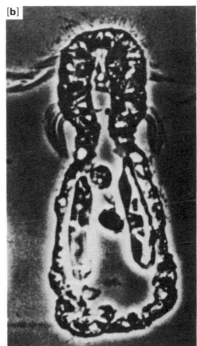

**Figure 3.16  Ciliary reversal**  Mussel gill cilia locked in 'up' and 'down' positions by the inhibitory action of 12.5 mM $CaCl_2$ and ionophore A23187 (a) or 20 mM vanadate (b) (from Wais-Steider & Satir 1979, but reproduced in Satir 1982).

concentration does not seem to be a direct effect on the dynein cycle, but does seem to be calmodulin-$Ca^{2+}$ mediated, since pharmacological blockade of the calmodulin with trifluoperazine (Stelazine) prevents the high-calcium blocking of ciliary beat (Satir 1982, Blum et al. 1980).

Whatever the switching mechanism, the net effect is to cause the cilia to beat, provided ATP is available. If the subdivision of activity were helically arranged or if the stiffness of the cilium were altered at different phases of the beat-cycle, then more complex patterns of beating would be generated; and if the switching moves along the axoneme metachronously rather than being synchronous along its length, then the axoneme will be thrown into 'waves' rather than remaining stiff – as happens in the flagellum.

## 3.5.2  Applications

By waggling an appendage such as a cilium, the cell can interact mechanically with the environment. In Chapter 5 we will discuss how cells utilize these appendages to move around from place to-place, but this is not the only use for cilia. Cells which are anchored by junctional complexes within epithelial sheets do not change position, but cilia protruding from these cells cause the adjacent fluid to move over the surface. In mammals the tracheal epithelium is responsible for sweeping mucus-trapped particles from the lungs (unless smoke inhalation poisons them), and ciliated epithelia in the female genital tract (Fallopian tube) move the oocyte from the ovary towards the uterus. The ciliary tracts of the gill in bivalve molluscs are another particle-moving device, and a sophisticated organization of ciliary tracts on the gill permits the sorting of food particles, the rejection of detritus, and the generation of a fluid flow over the gill surface for gaseous exchange.

In these examples it is clear that the ciliary beat in adjacent cells must be controlled in a coordinated fashion so that particles or fluid are swept in the appropriate direction. The electrical coupling of epithelial cells means that such control could be through a depolarization wave, and membrane permeability changes play an important part in controlling ciliary beat (Bessen *et al.* 1980).

## 3.6    Movement in the mitotic spindle

The mechanical basis of the movement of chromosomes and chromatids in mitosis remains a confused and controversial area. There is no shortage of models but the evidence for most of these is, at best, inconclusive. The spindle itself is composed of microtubules that originate from the divided pericentriolar MTOCs, which lie at opposite poles of the spindle (Fig. 3.17). Three different movements must be explained by any model:

(a)    the equatorial distribution of chromatid-pairs which leads to the metaphase distribution,
(b)    the separation of chromatids in anaphase,
(c)    the separation of poles at anaphase.

In addition to these major movements, there are also aberrant movements of chromosome fragments in abnormal nuclei (unpaired chromatids may move poleward before anaphase), and the colchicine-insensitive movements which give some order to the metaphase of the colchicine-blocked mitotic cell.

Two main classes of hypothesis for chromosomal movement have been proposed, those that depend upon controlled assembly–disassembly of microtubules, and those that rely upon sliding interactions between microtubules. Combinations of these methods are obviously possible and there are sceptics who doubt that microtubules provide anything more than the framework, and who point to the occurrence of actin within the spindle in sufficient quantity to achieve the movements observed with a realistic ATP consumption.

Formation of the spindle does, of course, require the assembly of a complex microtubule framework and presumably MTOCs are activated at the appropriate time, in the appropriate place, and with the capacity to compete successfully with cytoplasmic microtubules for the sub-unit pool. It is known that the formation of the spindle is blocked by agents such as colchicine, which interfere with microtubule assembly, and that low temperature or high pressures will cause disruption of the spindle once it is formed (Salmon & Segall 1980). If assembly–disassembly processes are involved directly in movement, rather than just in the formation and later disappearance of the spindle framework, then control of these processes must be localized within the spindle. The polar distribution of calmodulin lends some credence to this view (Welsh *et al.* 1978), as does the inhibitory effect of **taxol**, which stabilizes assembled microtubules (Baum *et al.* 1981). The discussion of mitosis which follows will be restricted in general to the mitotic division of cells of higher plants and animals, although considerable variability in the pattern of division exists, particularly in protozoa and fungi.

### 3.6.1   Microtubule organization within the spindle

Two sets of microtubules are involved in the structure of the spindle, those derived from the organizing centres located at the poles of the spindle, and those inserted into the kinetochores of the chromatids. The pole-associated microtubules are more sensitive to colchicine (and to other microtubule-disrupting agents) and arise from an area which is known to be an organizing centre; the kinetochore microtubules are more stable and the kinetochore is not a typical MTOC (indeed, many would not consider it to be an MTOC at all) in that the association with microtubules is at the assembly end, which is normally distal to the MTOC (Euteneuer & McIntosh 1981).

Large numbers of microtubules are involved in the formation of a spindle, for example, 5000–10 000 in the spindle of *Haemanthus catherinae* and 900–1500 in **PtK1** cells, and they are distributed in a regular pattern. In some spindles the microtubules are curved to generate a cylindrical or barrel-shaped structure and there must, presumably, be lateral interactions between the microtubules. This view is supported by the experimental observations of Snyder and McIntosh (1975) who found that isolated mitotic centres placed artificially in proximity did not interact normally, even though microtubule assembly was facilitated by the presence of excess tubulin. Overlap of microtubules from the two centres was not sufficient to generate a pseudo-spindle, nor was there any separation of the centres. The lateral interaction could be through dynein or MAPs but there is no evidence to distinguish (or support) these possibilities at present. Some ultrastructural evidence for cross-bridges 45–54 nm long does, however, exist.

Recent application of the methods for demonstrating the polarity of microtubules (Euteneuer & McIntosh 1980) have shown that in PtK1 and **HeLa** cells the microtubules in each half-spindle have identical polarity, that is, they are arranged in a parallel fashion and not antiparallel as was required by some early models for movement. If this is generally true then the equatorial region, besides being the site where chromosomes are ordered at metaphase, is the only region where microtubules could interact in an antiparallel fashion. In the late stages of mitosis the remnants of this area, the telophasic body, retain a distinct electron density, and the protein density is sufficient to prevent the access of antibodies directed against tubulin (Fig. 3.18).

In polarized light the spindle shows a marked **birefringence**, and this property has been utilized extensively in studies on assembly and disassembly under various conditions. The birefringence disappears when, for example, the pressure is raised, and reappears when normal atmospheric pressure is restored. Even if the microtubules themselves are not the sole contributors to the form birefringence, their presence is essential in producing the anisotropy of optical properties. Not only does an increase in pressure lead to a loss of birefringence but the two poles move closer together. Subsequent reassembly of the microtubules proceeds at around 1–2 $\mu$m min$^{-1}$ (in *Chaos carolinensis*), a rate that is, interestingly, the same as the rate of movement of chromatids at anaphase.

All the structural and assembly–disassembly evidence points to microtubules as the major contributors to the shape of the spindle, and suggests that the poles are held apart by the compression-resistance of the microtubular array.

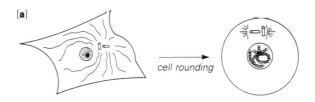

[a]

*cell rounding*

[b]

(i)

(ii)

(iii)

(iv)

(v)

(vi)

(vii)

(viii)

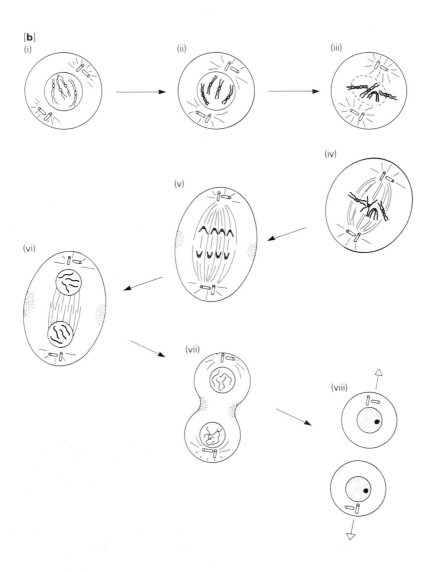

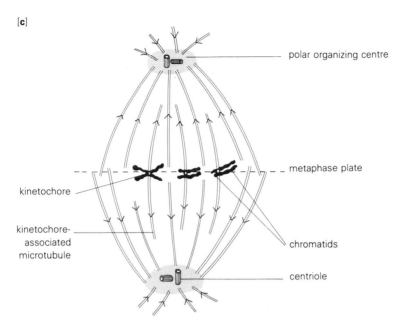

**Figure 3.17 Microtubules in mitosis** The interphase cell has microtubules dispersed through the cytoplasm, all derived from the pericentriolar MTOC (a) (see also Fig. 3.1). The rounding-up prior to mitosis is accompanied by changes in the microtubular pattern, and in prophase (b(i)) the daughter centrioles (and associated MTOCs) are diametrically opposed (at opposite poles). From the poles spindle microtubules are assembled (b(iii)), and by metaphase the spindle consists of pole-derived and kinetochore-associated microtubules, and the chromosomes are located on the equatorial metaphase plate (b(iv)). At anaphase, chromatid separation is occurring, and there is an increase in the spacing of the poles (b(v), b(vi)). The spindle microtubules linger on in telophase (the telophasic bundle) (b(vii)), and at this stage the equatorial band of microfilaments is beginning to contract, constricting the cell and eventually separating two daughter cells as cytokinesis proceeds (b(viii)). The details of microtubule organization in the spindle are shown in (c), where the arrows indicate the polarity of the microtubules (see also Fig. 3.8). Assembly occurs at the barbed ends; notice that the kinetochore microtubules are inserted by their points and that all the microtubules in a hemisphere are parallel.

## 3.6.2   Mechanism of movement

Even if microtubules are essential to spindle structure, as described in the previous section, they need not necessarily contribute to the movement of chromatids. Disassembly of the microtubules permits the poles to come closer together, but it is difficult to visualize a mechanism by which disassembly could pull upon the poles. It is far easier to suppose that the separation of the poles, which is blocked by colchicine, pushes the poles apart against some sort of elastic resistance which is not microtubule-generated.

Another factor to be borne in mind when considering possible force-generating mechanisms is the rate at which movement actually occurs. The fastest movements in mitosis are of 1–2 $\mu$m min$^{-1}$, comparable to the rate of slow neuroplasmic flow (see § 3.7.2) and some two orders of magnitude slower than the movements of pigment granules in teleost chromatophores

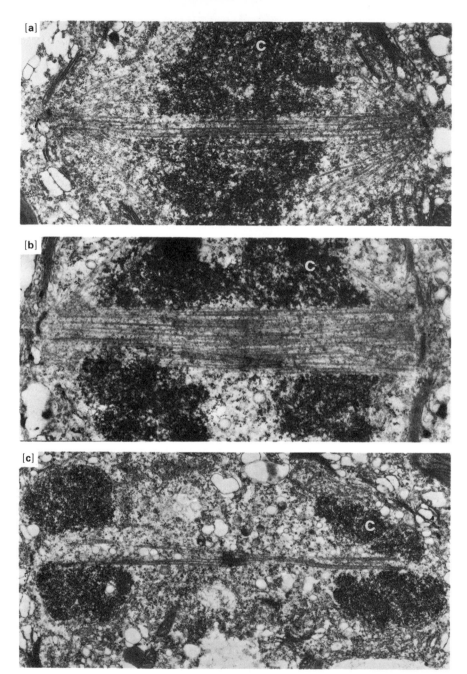

**Figure 3.18 Microtubules in the elongating spindle** The ultrastructure of the spindle at three stages of the mitotic process is shown in these electron micrographs of *Diatoma vulgare*. (a) A longitudinal-section through a metaphase spindle with the chromatin (C) centrally located. The mid-region of the spindle, where there may be overlap of antiparallel microtubules, stains rather more densely. At this stage the spindle is 5.3 μm long. (b) The spindle in early anaphase: chromatids have begun to separate but the spindle is still about the same length. (c) At telophase the chromatids are at the poles and have separated from the microtubules somewhat. The whole spindle has elongated, and is now 7.3 μm long, but the equatorial overlap region is still densely stained (from McDonald *et al.*, 1977).

(§ 3.7.4). If the spindle microtubules had the number of dynein molecules that are associated with microtubules in the axoneme, then chromosomes would move at around $1 \ cm \ min^{-1}$. Of course this argument involves a lot of assumptions, but the force required to move a chromosome has been calculated (with assumptions regarding the viscosity of the spindle matrix which have been partially tested) at around $10^{-13} \ N$, which could come from the hydrolysis of as few as 20 ATP molecules. Even if these calculations are wrong by a factor of ten then a very clear prediction is that the motile machinery would be a small proportion of the total spindle structure. If, in our overall analogy, we were to grub up a 20 mile length of dual-carriageway road, complete with the cars which are moving in opposite directions, and were then to grind up this material, what would our chances be of discovering the internal combustion engine, never mind its mechanism?

Under these circumstances, model-building need not be too tightly constrained by factual considerations and may be useful if it can indicate experimental approaches.

One of the intellectually more attractive of these models, and one which has survived some testing of its predictions, is that proposed by Margolis et al. (1978). This model is illustrated in Figure 3.19, where it will be seen that the model predicted (successfully) that the microtubules of each half-spindle should be parallel and that disassembly should be restricted to the poles of the spindle. The latter prediction is certainly consistent with the demonstration, using fluorescently-labelled antibody, that calmodulin is localized in the region of the pole (Marcum et al. 1978). The motive force for chromatid movement is not explicit in this model although the treadmilling (see § 3.2.3) of microtubules is supposed to carry unattached material poleward at metaphase. Metaphase on this model, as on many others, is a balance of pole-directed forces. Chromatids are brought to the equatorial plate by the inequality of the forces exerted on the two kinetochores if they are located elsewhere in the spindle; the chromatids move apart once their linkage breaks down and the two equal and opposite forces can act separately. The separation of poles could be accounted for by increased assembly at the distal ends of pole-associated microtubules facilitated by the increased size of the sub-unit pool as kinetochore-associated microtubules are disassembled.

An essential part of this model is that the assembly and disassembly of different microtubule-sets should be controlled independently. In that some microtubules of the spindle are more resistant to disruption by assembly inhibitors than others, this is not unreasonable, although the molecular basis of their differential stability is uncertain.


### 3.6.3   Other aspects of mitosis

No matter which model for mitosis eventually proves to be the most successful there are some further aspects that must be taken into consideration. Separation of two chromosome-sets is only part of cell division and must, of course, follow replication of the DNA. The separation of the nuclei must be linked to the separation of two cells. The process of cytokinesis in animal cells (see § 2.4.3), which depends on an actomyosin motor system, must be linked spatially to the process of mitosis: the circumferential band of microfilaments must constrict the cell equatorially and not meridionally. In plants the cell plate, which gives rise to the dividing wall between daughter cells, must similarly be in the equatorial plane or, in those cases where

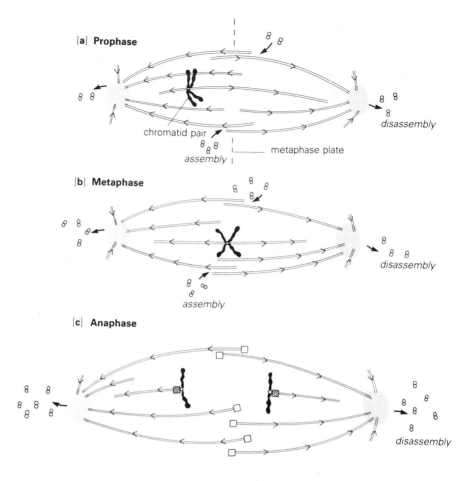

**[a] Prophase**

chromatid pair

metaphase plate

assembly

disassembly

**[b] Metaphase**

disassembly

assembly

**[c] Anaphase**

disassembly

**Figure 3.19 Model for mitosis** This model is based on that proposed by Margolis *et al.* (1978), but has been modified to take account of the information now available about microtubule polarity. In (a) a single chromatid pair is shown off-centre: the assumption is that the asymmetry of forces generated by the assembly of the microtubules will bring the pair to the metaphase plate. At this stage there is net assembly. Treadmilling would account for the poleward movement of small particles which is sometimes observed. The hypothesis depends on the assumption that at the end of metaphase the assembly, but not disassembly, of kinetochore microtubules is blocked (■), and (perhaps slightly later) the assembly of pole-derived microtubules is also blocked (□) (c). Movement of the chromatids towards the poles could depend either on the disassembly of kinetochore microtubules, or on a sliding interaction between parallel microtubules. At metaphase the balance of forces will keep the chromatid pairs at the equator (b), but once the chromatids have separated they are free to move poleward.

division is unequal (as in the production of stomatal guard cells), the cell plate must lie between the daughter nuclei. The cell plate of plant cells is known to be microtubule associated, although the detailed specification of position remains somewhat unclear. Localization of the microfilament ring of cytokinesis may depend upon the divided microtubular cytoskeleton in some way and is another example of interaction between microtubules and other cytoplasmic elements. Perturbation of the behaviour of the MTOCs of the centriolar regions has been implicated in the formation of syncytia by **BHK** cells infected with simian virus 5 (Wang *et al.* 1979), and the alignment of nuclei seen in these syncytia seems to depend on microtubules and intermediate filaments.

## 3.7 Movement associated with cytoplasmic microtubules

### 3.7.1 General

Under this heading we will consider a variety of intracellular movements which are in some way associated with microtubules. The choice of the word 'associated' is a deliberate one, the association may be no more than that of the locomotive with its rails or may be more intimate. Before embarking on a detailed discussion let us consider the possible mechanisms.

A dynein–tubulin motor is clearly possible: the dynein-like ATPases might well be linked to structures other than microtubules and, in much the same way as myosin linked to latex spheres works to generate movement on actin cables (§ 2.5.2), the ordered array of microtubules might provide the rack of a rack-and-pinion railway. Dynein will certainly interact with cytoplasmic microtubules (Haimo *et al.* 1979) and this possibility must therefore be taken seriously.

Alternatively, the object to be moved might latch onto the treadmilling of microtubules, either linked by side arms, or intercalated axially. The latter would imply an MTOC diametrically opposed to a socket for the assembly end of the 'pushing' microtubule; the former seems perhaps more plausible. Lateral linkage could be directly with MAPs, or might rely upon the interaction of F-actin with MAPs that is suggested by the experiments of Griffith and Pollard (1978). In principle there might also be microtubule–microtubule lateral interaction (dynein-mediated perhaps) with a short piece of microtubule on the moveable object.

Less well defined associations are also possible. A linear arrangement of microtubules might constrain the movement of large particles or vesicles (§ 3.7.4) or might act as a fence, preventing access of vesicles to the plasma membrane for example. An axial supporting-rod composed of microtubules might permit other movement systems to extend their activities along a thin cytoplasmic protrusion (§ 3.7.3). There also seems to be evidence for microtubules acting to restrict the lateral mobility of integral plasma membrane proteins (see § 2.5.1).

In attempting to link movement with microtubules we might bear in mind:

(a) that sliding interactions in cilia occur between parallel microtubules;
(b) that the polarity of microtubules is likely to impose polarity on movement, and if movement is bidirectional then there should be antiparallel microtubules;
(c) that lateral interaction will require a constant separation distance with some form of cross-linking structure.

## 3.7.2 Axoplasmic flow

The cell body of a neuron may be separated from the efferent synapse by a considerable length of thin axon, up to several metres in some motor neurons. Supplying material to this distant region cannot depend solely upon diffusion and some form of transport mechanism is required. Physical or

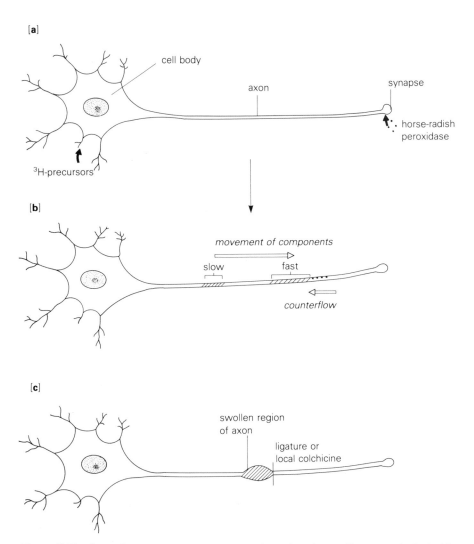

**Figure 3.20 Axonal transport** In an axon, such as that shown diagrammatically in (a), synthesis in the cell body is remote from the site of release, the synapse. If the cell is provided with a pulse of radioactively labelled amino acids, then at later times the label can be found in the axon, the distance moved depending on the time which has elapsed. The label moves in two parts, a fast component, and a slow component. Horse-radish peroxidase, taken up during the process of membrane recycling in the synaptic region, travels in the opposite direction, towards the cell body. Transport can be blocked by literal or by pharmacological ligatures, and an upstream swelling may occur (c).

pharmacological ligation of an axon leads to an upstream accumulation of material. Colchicine or vinblastine locally applied seems almost as effective as string, implying that labile microtubules are involved. Neurons contain microtubules (brain is a standard source material for cytoplasmic microtubules) and there are reports of a dynein-like protein in nervous tissue (Pallini et al. 1982).

The movement along the axon can be divided into two categories: a slow movement (around 1 mm day$^{-1}$) of the whole axoplasm, apparently as a unit, and a much more rapid (250–400 mm day$^{-1}$) transport, which may be bidirectional (Dahlstrom et al. 1974, Heslop 1974). After pulse-labelling the cell body the slow movement can be followed as a bolus of radiolabelled material moving down the axon: the bolus remains discrete, implying a bulk movement of the axoplasm (Black & Lasek 1980). The rapid movement is superimposed upon this slow movement and delivers material en route as well as to the terminal. A range of different proteins and membrane components is transported in this fast flow, and transport is ATP dependent (Adams 1982). Bidirectional movement does occur, thus horse-radish peroxidase internalized at the terminal appears eventually in the cell body and vice versa (Schwab et al. 1982).

The evidence for rapid movement being microtubule associated comes from the colchicine sensitivity of movements and the non-random association of vesicles and microtubules in cross sections of the giant axon of Petromyzon (Smith et al. 1970). The latter, even with slight indications of cross-bridges, is no more than circumstantial, and the colchicine sensitivity has been disputed. Foreign particles injected into a crab axon are transported with some selectivity in rate depending upon the size and surface charge of the particle (Adams & Bray 1983). Alternative models for axoplasmic flow, which involve myosin linked to the extensive neurofilament network and actin associated with the plasma membrane, have also been proposed (Lasek & Hoffman 1976: these authors also review this area). Fast axonal transport in Aplysia neurons is inhibited by the microinjection of macromolecules (specific IgG, DNAase I) which affect actin polymerization, but is not affected by stabilization of microfilaments by phalloidin. Whether this can be taken as convincing evidence for an actin-dependent motor system is rather doubtful (Goldberg et al. 1980). An interesting problem arises in view of the bulk movement of axoplasm: in a growing axon the delivery rate is comparable to the rate of extension (approximately 1 mm day$^{-1}$), but in an adult neuron of fixed length much of this material must be disassembled and either returned or degraded. Only a very small proportion will be released as neurotransmitter.

## 3.7.3   The axopodia of heliozoa

The Heliozoa, the 'sun-burst animalcules', have many long projections from their cell bodies which serve to trap the bacteria on which they live. These axopodia are supported by a microtubular bundle of exquisite symmetry (shown in Fig. 3.21) and conditions which cause disassembly of microtubules lead to the collapse of these projections. Food particles trapped distally by an adhesion mechanism are transported toward the cell body at a rate of 0.3–1.0 μm sec$^{-1}$. In this case the microtubules are almost certainly serving a purely supporting function since artificial axopodia supported with a glass-fibre rod transport material equally well (Edds 1975).

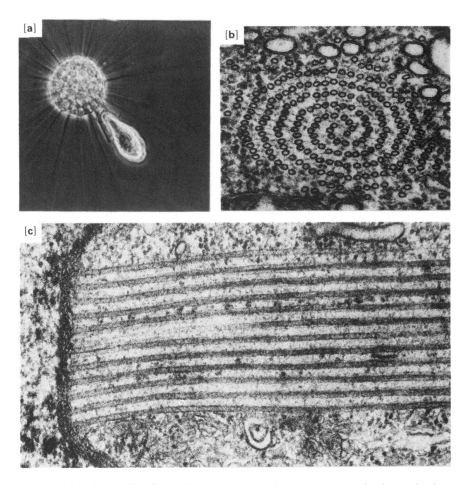

**Figure 3.21 Axopodia** The heliozoan protozoa have numerous slender projections which protrude radially from the cell. These processes are supported by a microtubular bundle that resists bending. The illustrations are of axopodia from *Actinophrys sol*, but the morphology is very similar in other species. In (a) the heliozoan has captured a ciliate (*Tetrahymena*) almost as large as itself; the attached axopodia will shorten and the prey will be phagocytosed by the cell body. Phase contrast: the cell body has a diameter of approximately 50 μm (photograph courtesy of Colin Ockleford, Anatomy Dept, Leicester University). (b) A transverse section through a large axoneme showing the exquisitely symmetrical arrangement of microtubular profiles. Some links between tubules in adjacent rows can be seen (from Ockleford & Tucker 1973). (c) A longitudinal section of a large axoneme contacting the nuclear envelope on the left, and showing the regular longitudinal spacing of the microtubules (from Ockleford & Tucker 1973).

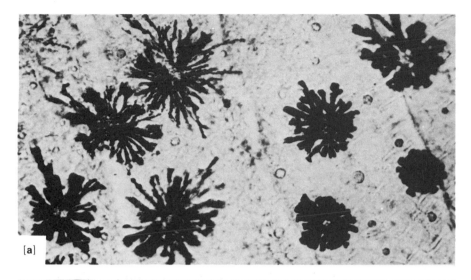

[a]

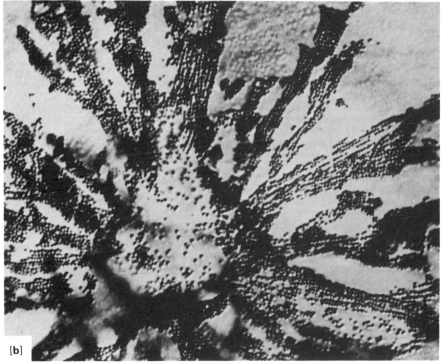

[b]

**Figure 3.22  Teleost melanophores**  Melanophores (chromophores with melanin as the pigment) of *Fundulus heteroclitus* (the killifish)  In (a) the pigment granules in the cells on the right are aggregated in the centre of the cell, whereas those in the cells on the left are dispersed. In (b) a melanophore with dispersed pigment is shown in rather more detail, and the central granule-free zone can be seen. Within the radial processes are hundreds of granules arranged in longitudinal rows which lie parallel to cytoplasmic microtubules (both pictures from Murphy & Tilney 1974).

### 3.7.4 Teleost chromophores

Many fish, particularly the bottom-dwelling flatfish, have the ability to change their spots so as to blend with their surroundings. This skill in camouflage comes from their ability to alter the distribution of pigment granules in large stellate cells which lie in the epidermis. With the pigment centripetally located as a small discrete spot of colour it contributes little to the appearance of the fish; when dispersed it becomes conspicuous. Neuronal control of pigment dispersal permits the fish to change colour within a repertoire delimited by the range and colour of its chromophores.

The accessibility of these cells and the ability to trigger pigment granule movement by artificially applying neurotransmitters has provided much enjoyment for film makers and audiences. Granule movement is rapid (up to 5 $\mu$m sec$^{-1}$ centripetally, although only around 2 $\mu$m min$^{-1}$ centrifugally) and can be blocked by treatments which disrupt microtubules. Sadly, centripetal movement does not, however, rely upon microtubule disassembly, nor does the converse hold true (Murphy & Tilney 1974). As Schliwa (1978) has demonstrated, by painstaking electron microscopy, the number of micro-tubules in each 'arm' of the cell remains constant whether the pigment is dispersed or concentrated. Removal of the microtubular cytoskeleton does prevent dispersion, but the rôle of microtubules seems to be supportive rather than as active participants. This said, the mechanism remains obscure, although it has been proposed that assembly and disassembly of the **microtrabecular network** of the cytoplasm may drive the granules (Schliwa 1979).

### 3.7.5 Other microtubule-associated movements

The three systems discussed above have received rather more attention than most and yet cannot be said to be clear examples of microtubule-associated movement in which microtubules play any more than a cytoskeletal rôle. Other systems have been studied, notably the suctorian tentacle (Bardele 1974), the complex movements of the cytopharyngeal basket of certain ciliates (Tucker 1972), also the movement of particles proximally on the surface of flagella (Bloodgood 1982), and along the nutritive tube of oocytes of insects (Hyams & Stebbings 1977). There is also evidence of an association between mitochondria and microtubules (Heggeness *et al.* 1978); and the intimate association between microtubules and the plasma membrane (Albertini & Clark 1975) lends some credence to the idea that there may be a linkage between some membranous organelles and microtubules. Whether the association with microtubules implies that there is a functional relationship involving movement is uncertain. All these systems provide further circumstantial evidence of an association between movement and microtubules aligned along the path that is taken. The general topic is dealt with very nicely in Stebbings and Hyams (1979), and since progress in this field seems to be relatively slow I do not propose to go into any further detail.

### 3.8  Summary

The microtubule–dynein motor system is much less well understood than the actomyosin motor discussed in Chapter 2. In cilia the principles of the mechanochemistry have only recently been clarified and our ignorance of the biochemistry of dynein(s) contrasts very unfavourably with the detailed information available for myosin. The control systems, even for the ciliary motor, have not yet been worked out, but the field seems to be awakening. Perhaps the next decade will see the dynein–tubulin system receive the sort of attention that was lavished on the actomyosin motor in the seventies.

When we move away from the specialized ciliary motor system the picture becomes even less satisfactory, and in none of the examples has the movement been unequivocally linked to microtubules. It seems very probable that mitosis depends on microtubules, and it is not impossible that other cytoplasmic motor systems are also microtubule dependent, but far too much of the evidence is circumstantial. Microtubule-dependent motors may be rather poorly expressed in cells because the movements associated with them are rather slow – but this should not be taken as an excuse for neglecting these systems.

# References

Roberts, K. and J. S. Hyams (eds) 1979. *Microtubules.* London: Academic. Contains extensive reviews of all the topics of this chapter.

Roberts, K. 1974. Cytoplasmic microtubules and their function. *Prog. Biophys. Mol. Biol.* **28**, 373–420. A good general review.

Goldman, R., T. Pollard and J. Rosenbaum (eds) 1976. *Cold Spring Harbor Conferences on Cell Proliferation.* **3**. Book C is mostly concerned with microtubules, several papers have been referenced in the appropriate place in the text.

*Cold Spring Harbor Sympos.* **46**. A more up-to-date source of semi-review papers.

*Symposium of the Society for Experimental Biology* **35**. Prokaryotic and eukaryotic flagella. A good source for details on cilia and flagella particularly for mechanisms.

# 4

# MOTORS
# OF OTHER
# SORTS

## 4.1 Introduction

In the two previous chapters we have seen how conformational change in a protein can be utilized to slide filaments past one another and, in discussing microtubules, have considered the use of assembly–disassembly reactions. Several other motor systems have been invented by Nature and will be considered in this chapter. The most important of these, and in many ways the most novel, is the mechanism by which flagellated bacteria propel themselves. We will also consider a 'pure' conformational motor, a rapid-acting assembly-driven motor and, somewhat superficially, some examples of hydraulic systems.

## 4.2 The bacterial flagellar motor

The flagella of prokaryotes have a totally different structure and mechanism of action from eukaryotic cilia and flagella and are not related, either by homology or analogy. Several groups of bacteria, both Gram-negative and Gram-positive are motile by means of flagella, including such pathogenic organisms as *Salmonella typhimurium* and *Vibrio cholerae*. Gram-negative organisms such as *Escherichia coli* have been the most common material for study and it is assumed, although the evidence is not available, that the mechanism is basically the same in most or all flagellated bacteria. The flagellum, unlike that of eukaryotes, is driven only at the base and may be considered as a sort of corkscrew-like propeller: it is a semi-rigid structure which may be several microns long and is composed of a protein, **flagellin**. Flagellin, which has a molecular weight of around 30–60 kd, is packed into the flagellar filament in an helical surface lattice, rather like tubulin in a microtubule; the whole structure having a diameter of approximately 20 nm. There are approximately 5500 sub-units in each complete turn of the flagellar helix. The filament of the flagellum self-assembles and a variety of slightly different packing patterns can occur, giving a family of different waveforms to the filament, the pattern being characteristic of the particular bacterial strain (Fig 4.1). Flagellin assembled into one form can be reversibly transformed into a flagellar helix of different pitch by altering the pH or ionic strength, which alters the interaction between the sub-units. Thus in *Salmonella* SJ670 the flagellum is a left-handed helix with a pitch of 2.3 μm and a diameter of around 0.4 μm at pH 6–8; at pH 4.7 flagella of this form are converted to tight coils with a pitch of 0.5 μm and diameter 1.2 μm; at pH 4.2 they transform to the 'curly' type, a right-handed helix with a pitch of 1.1 μm and diameter 0.3 μm. These transitions are reversible and are interesting because the curly form is unsuitable for generating translational movement, the pH of the environment may therefore affect the locomotory behaviour by altering the propulsive machinery directly. Transition between the various forms may also be brought about as a consequence of externally applied mechanical stress, which may have implications for the behaviour of the cells in flow conditions or when the viscosity of the fluid changes.

Interesting though the transformations of flagellar structure are as visible manifestations of an allosteric change, the more important question is the nature of the drive mechanism which causes the flagellar filament to rotate. That rotation, rather than a propagated wave, is the mechanism of action is shown by the experiments of Silverman and Simon (1974) who tethered bacteria to a clean glass slide by their flagella and showed that the cell bodies

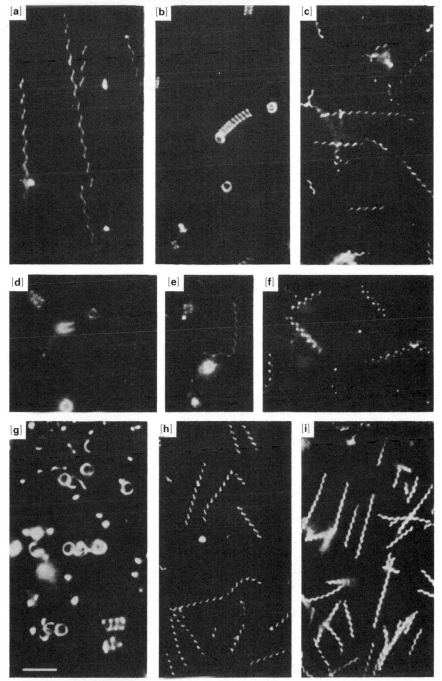

**Figure 4.1 Bacterial flagella** The morphology of *Salmonella* flagella under various conditions, seen using dark-field microscopy; scale bar represents 5 μm. Except in (f) the flagella are reconstituted from filaments of the wild-type strain SJ670. (a) Normal form, pH 7.0; (b) coiled form, pH 4.7; (c) curly form, pH 4.2; (d) and (e) mixed forms, pH 5.0 and 4.7 respectively; (f) semi-coiled form of SJ25 strain at pH 4.3; (g) coiled form, pH 11.0; (h) curly I form, pH 12.0; (i) curly II form, pH 12.5 (from Kamiya *et al.* 1982).

rotated alternately clockwise and anticlockwise (clockwise and counterclockwise in America). (The convention is to describe the rotation as though the observer is standing at the end of the filament, looking towards the bacterial body.) The bacterium, because of the propulsive characteristics of an helical propeller, has both a forward and reverse gear depending upon the direction of rotation of the propeller. In practice the reverse gear tends to cause changes in direction because the bacterium 'tumbles' during reversal and only moves effectively in forward gear, which drives the cell in a straight line. The alternation of straight runs and abrupt random changes in direction is controlled by the cell and is modified by the environment, as will be discussed in more detail in Chapter 9.

The base of the flagellum is the site at which the rotatory motion is generated. The flagellin filament is linked through a short proximal hook, which constitutes a flexible coupling, to a rod which is embedded in the cell wall and plasma membrane (Fig. 4.2). The hook polypeptide is a different gene product from flagellin, and about ten other proteins are involved in the construction of the rod and the rings or discs which anchor it in the wall and membrane. The rod and its associated ring structures constitute the motor proper (and may even serve as the whole motor in some gliding bacteria, see § 4.3). In Gram-negative organisms the rod has two outer (distal) rings, which anchor it in the lipopolysaccharide and peptidoglycan layers of the cell wall (L- and P-rings, respectively) and serve as bushes or bearings for the rotor (it is not clear whether they are fixed or whether they rotate with the hook). Gram-positive bacteria have a simpler wall structure and do not have these rings. The crucial rings for generating movement are the M (motor)-ring, which is embedded in the plasma membrane, and the S (stator)-ring, which is linked to the wall facing the M-ring. Interaction of the M- and S-rings generates the rotatory torque which drives the propeller. The problem, faced also by marine engineers, of preventing fluid from seeping around the propeller shaft is solved by having the M-ring in the plasma membrane which, being fluid, permits rotation without breaking the seal and allowing leakage. This arrangment does, however, prevent the easy access of cytoplasmic material to the active parts of the motor: it would be difficult to supply ATP to a motor of this sort. This problem does not, however, arise because ATP is not the fuel source; the energy comes directly from the so-called 'proton motive force' (**PMF**), a transmembrane electrochemical gradient of pH (proton concentration) and electrical potential. Artificially applied pH gradients (with the interior alkaline) will in fact serve to drive the motor (Matsuura *et al.* 1979), as will gradients of electrical potential (with the interior negative). Rotation can also be generated in starved cells by the opposite gradients, and the important thing to emerge from these experiments is that the direction of rotation is not affected by the sign of the gradient which is driving the motor. Calculating the resistive torque due to viscous drag allows a calculation of the work done in unit time: from this it has been estimated that approximately 300 protons suffice to drive the rotor through one complete revolution (if the motor is 100 per cent efficient, which seems unlikely).

An hypothesis for generating rotation by utilizing the proton motive force has been advanced by Berg *et al.* (1982), who propose that the M- and S-rings are linked by a bond which is 'broken' by reduction and 'made' by oxidation, the redox loop being directly linked to opposed hydrogen and electron carriers (a proton pump; Fig. 4.3). These authors further propose that the number of potential bond-forming residues on the S-ring is slightly greater or smaller than on the M-ring (so the M-ring would have (n) residues, the S-ring

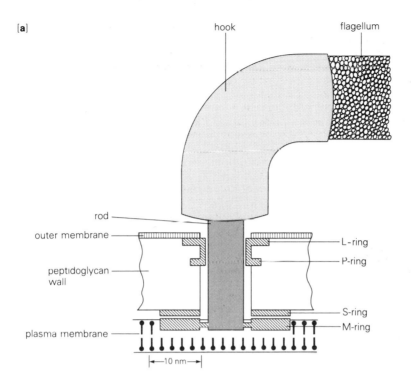

[a]

hook

flagellum

rod

outer membrane

L-ring

P-ring

peptidoglycan wall

plasma membrane

S-ring

M-ring

|←—10 nm—→|

[b]

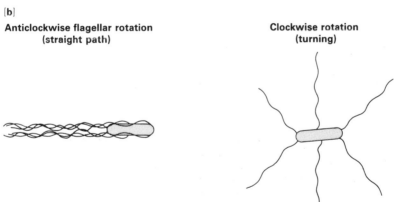

**Anticlockwise flagellar rotation (straight path)**

**Clockwise rotation (turning)**

**Figure 4.2  The bacterial flagellar motor**   (a) The business end of the flagellum is the basal region, the M-ring which rotates because of its interaction with the S-ring. The M-ring is embedded in the plasma membrane, and the S-ring in the wall. The L- and P-rings probably act just as bushes or bearings and are restricted to Gram-negative organisms, which have a more complex wall structure than Gram-positive bacteria. The hook is a flexible coupling between the rotating rod and the rigid filament. (b) When the flagella of a multiflagellate bacterium such as *E. coli* are rotating anticlockwise they form a bundle which rotates and drives the bacterium along a straight path. If the direction of flagellar rotation reverses then the bundle flies apart and the cell 'tumbles'.

(n + 1) or (n − 1)), and that there is a vernier which steps m(n ± 1) times per revolution as each bond is in turn 'made' and 'broken'. If 16 residues on the M-ring are oxidized and reduced, which would be in keeping with the rotational symmetry of the ultrastructure, then 15 or 17 residues on the S-ring would be involved. The dimensions of the M-ring are such that the 16 contributing residues would be 4.4 nm apart and those on the S-ring would be 4.7 or 4.2 nm apart. Each step would then be of around 0.29–0.26 nm and 240–272 steps would constitute a rotation. This number is interestingly close to the number of protons (300) which are calculated, on totally different grounds, to be required for a single revolution of the rotor.

Other models for the mechanism of rotation have been proposed but will not be discussed here: the model described illustrates a general principle of generating rotational movement by arranging a series of small stepping movements in series on a circular path.

If the motor does operate as described above then control of the direction of

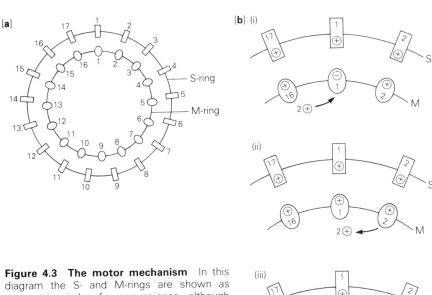

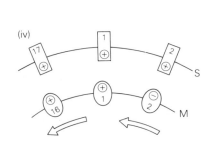

**Figure 4.3  The motor mechanism**  In this diagram the S- and M-rings are shown as concentric circles for convenience, although they are really apposed discs. The S-ring has 17 sub-units, one more than the M-ring and the motor steps around, taking 16 × 17 steps in a complete turn. The mechanochemical cycle is shown in (b): at the start S1 and M1 are aligned, (i). The addition of two protons to M1 alters its charge from negative to positive, (ii), and the removal of two protons from M2 makes M2 negative, (iii). Unlike charges attract, so M2 now swings into alignment with S2, causing the M-ring to rotate through 1.32° (360/16 × 17). The process can now be repeated. Although I have shown proton addition and removal as different stages, it is likely that they are tightly coupled. Even if the S-ring has only 15 sub-units, the basic cycle remains the same.

rotation would be exerted by determining the order in which the changes occurred on the sub-units that contribute to the linkage. Thus, by altering the bond-forming sequence from 1–1, 2–2, 3–3 to 1–1, 17–16, 16–15 and so on, the motor would shift from clockwise to anticlockwise rotation. Multi-flagellate bacteria exert a coordinated control over the direction of rotation of their flagella: when rotation is anticlockwise the effect is cooperative and the flagella tend to bundle together but clockwise rotation leads to centrifugal chaos and the bacterium spins on its axis. Knowing that all the flagella must be switched simultaneously does not, however, assist us in understanding the mechanism by which it is achieved.

## 4.3 Other bacterial motors

Not all bacteria swim by means of flagella; spirochaetes swim by rotating bodily (Canale-Parola 1978, Berg et al. 1978), others such as the Myxobacteria do not swim but move over solid surfaces in a fluid film (Clarke 1981). In the latter case it seems likely that the basic mechanism is the same as described above but that the flagellum is missing and only the hook is present (Pate & Chang 1979). This is based on the finding that rotor-like assemblies can be isolated from a cell wall preparation, and that motility requires these structures. Mutants lacking the motor-like structures are immotile and it seems fairly reasonable to suppose that these bacteria, which show some fascinating aggregation behaviour, very reminiscent of miniaturized Myxamoebae (Clarke 1981), use the 'normal' motor. For the spirochaetes, however, the situation is much less clear. It seems that the rotation of the body, which resembles a large bacterial flagellum in some ways, depends upon filaments which lie between the plasma membrane and the outer membrane. Exactly how these filaments, which are anchored at the ends of the body, bring about the movement is unclear and requires further attention. It is perhaps salutory to realize that even groups which contain important pathogenic bacteria, which attract more attention for obvious reasons, may have motor systems that have barely been studied and that are certainly not understood.

There are also movement systems in bacteria which are only vaguely described as yet: one such is the twitching movement of some Gram-negative bacteria, which seems to be driven by surface hydrophobicity (Henrichsen 1983). **Pili** on the surface have very hydrophobic properties and the bacteria therefore lie preferentially at an air–water interface, more a partitioning process than a means of locomotion as it turns out.

## 4.4 The spasmoneme of vorticellids

The peritrich ciliates, of which *Vorticella* is an example, look like miniature unicellular versions of hydroids; the cell body, which captures food particles by means of cilia, is supported on a stalk which attaches it to the substratum. Although it is advantageous to the organism to protrude into richer waters further away from the substratum, this position has its perils and an escape reaction, shortening the stalk and presenting a lower profile by withdrawing to safer regions, is likely to be of selective value. The contraction of the stalk is brought about by a single filamentous fibre, the **spasmoneme**, which extends from the basal region to the cell body, a distance of some hundreds of microns (Amos et al. 1976). The contraction of the spasmoneme may either cause coiling of the stalk sheath or, as in the case of *Zoothamnium*

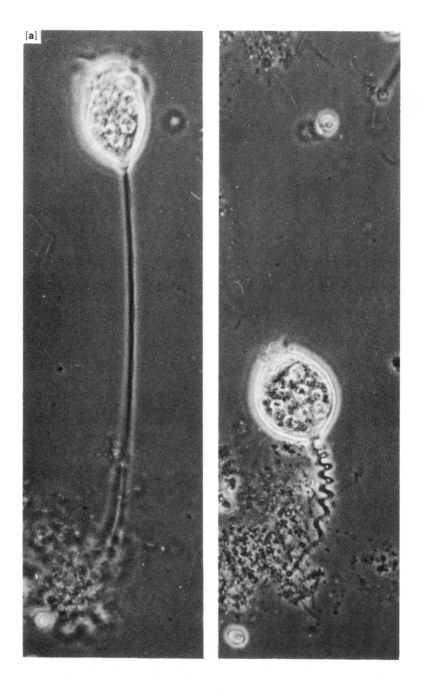

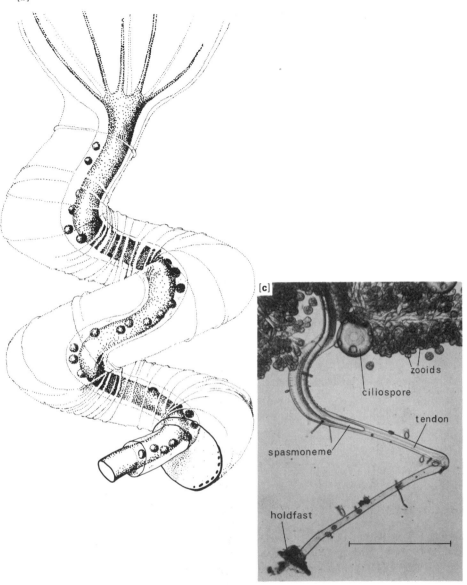

**Figure 4.4  The spasmoneme**  (a) The contraction of *Vorticella* is shown in these two phase-contrast micrographs. The fully extended stalk, approximately 150 μm long, contracts in about 4 ms and then actively re-extends (photographs courtesy of Dr W. B. Amos, Zoology Dept, Cambridge University). (b) A diagram of the stalk showing the helically coiled spasmoneme and the stiffening fibres which run beneath the sheath (from Amos 1975). (c) The arrangement of the spasmoneme in the giant colonial vorticellid, *Zoothamnium geniculatum* in which the knee-like arrangement amplifies the movement; bar represents 0.5 mm (from Weis-Fogh & Amos 1972).

*geniculatum*, the movement may be amplified through a knee-like hinge (Fig. 4.4).

Contraction of the spasmoneme is extremely rapid, much faster than the striated muscle of the putative predator from which the ciliate is escaping (15× faster than the fastest known striated muscle). Active expansion of the spasmoneme will also occur, rather more slowly. The isolated 'giant' spasmoneme of *Zoothamnium*, which can be dissected out by hand, will contract rapidly to 45 per cent of its initial length within 8 ms if the calcium concentration rises above $4 \times 10^{-7}$ M. It will expand, and exert a force as it does so, if the calcium level is reduced to $10^{-8}$ M by the subsequent addition of EGTA, a chelator of $Ca^{2+}$. In the isolated system ATP hydrolysis is not required, although work is being done by the experimenter in changing the calcium concentration. *In vivo* ATP hydrolysis is required to move calcium ions back into the reservoir, but the expansion and contraction phases are spatially separated from the ATPase (Weis-Fogh & Amos 1972).

The spasmoneme is built of 2–3 nm diameter fibres, arranged longitudinally, which have a periodicity of around 3.5 nm: the periodicity is probably derived from the dimensions of the sub-unit protein, **spasmin**, which has a molecular weight of around 20 000 (Routledge 1978). Interspersed with the fibres are longitudinally arranged tubules, which may run for several microns and which may be involved in the sequestration of calcium, acting in an analogous manner to the sarcoplasmic reticulum of striated muscle. Neither actin nor tubulin has been found in the spasmoneme, and it is thought that the basis for contraction is distinctly different from myosin- or dynein-based sliding filament systems, involving a simple conformational change in the sub-unit protein which changes the length of the multimolecular aggregate. The conformational change is actually brought about by the calcium-binding, rather than being a response to a signal which just happens to be a change in calcium ion concentration. Thus the spasmoneme may be an example of a series-linked conformational motor; such a motor would, of course, be capable of acting in contraction and, if packed in parallel to prevent bending, would also be capable of active extension.

Interestingly, the calcium sensitivity of this system is very similar to that of actomyosin, presumably because of constraints in the design of the calcium-sequestering system and the necessity for calcium levels to be compatible with those involved in the more conventional system which the ciliate uses for mitosis and for ciliary movement. Spasmin, the sub-unit, does not appear to resemble any of the calmodulin-type calcium-binding proteins and seems to be peculiar to this group.

## 4.5 Assembly–disassembly motors

The fusion of sperm and egg in fertilization is made more difficult in some cases by the need for a protective layer of gelatinous material around the egg. This gelatinous capsule is especially common in eggs which are released from the female for external fertilization. Penetration of the egg coat is facilitated by lysosomal enzymes derived from the **acrosome** of the sperm (a modified lysosome) and also, in some cases, by the production of a long thin acrosomal process which drives a channel through the protective coat as it is extruded. Two different mechanisms for production of this acrosomal process have been studied by Tilney and his co-workers; both mechanisms are interesting methods of producing movement and both use actin, a protein we are familiar with from previous discussion (Ch. 2), in a novel way.

**Figure 4.5  The acrosomal process of *Limulus* sperm**   An electron micrograph of an unreacted sperm showing the core filament bundle which extends from the acrosomal vesicle (v), through the doughnut-shaped nucleus (N), and is wrapped around the base of the flagellum. The basal body of the flagellum (B) can be seen (from Tilney 1975a).

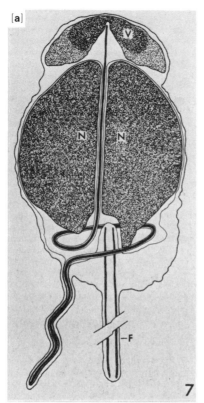

Figure 4.6 Discharge of the acrosomal process in *Limulus* (a) False discharge of the acrosomal process. This diagram, from Tilney (1975a), has been simplified by the omission of most of the coils of the process, only one of which is shown. The flagellum (F), the nucleus (N) and the acrosomal vesicle (V) are shown.

(b) Normal discharge of the acrosomal process. The remnants of the acrosomal vesicle (M) remain near the base of the acrosomal process, only a small part of which has been caught by the section. The flagellum has retracted and the axoneme is within the cytoplasmic mass (A) (from Tilney 1975a).

## 4.5.1 The acrosomal process of Limulus polyphemus

The horseshoe- or king-crab, *Limulus*, sole survivor of the ancient Mero-stomata, has spermatozoa of fairly complex structure, as shown in Figure 4.5. The sperm tail is a straightforward flagellum of conventional eukaryotic design and function, but the toroidal nucleus is penetrated by a long rod which is coiled several times at the base of the flagellum, distal to the acrosome. If the ionic composition of the bathing medium is altered a 'false discharge' of the acrosomal process will occur (Fig. 4.6a) but in normal activation the coiled rod straightens out and almost explosively projects the acrosomal process forward (Fig. 4.6b). The membrane of the acrosomal process is derived, in part at least, from the acrosomal vesicle, which is effectively lysosomal in character. As the rod straightens and the basal coils unwind, the rod rotates; a process which Tilney (1975a, b) has graphically described as one in which the acrosomal process screws itself into the egg coat, ultimately to fuse with the plasma membrane of the egg.

The ultrastructural appearance of the rod suggests that it is built of parallel filaments of 5–7 nm diameter, and the major protein constituent seems to be actin. When the rod is in the coiled form the actin is packed in such a way that it cannot be decorated with heavy meromyosin; when the rod straightens out the packing of the actin changes in such a way as to render decoration possible. At this stage it also becomes clear that the microfilaments are all

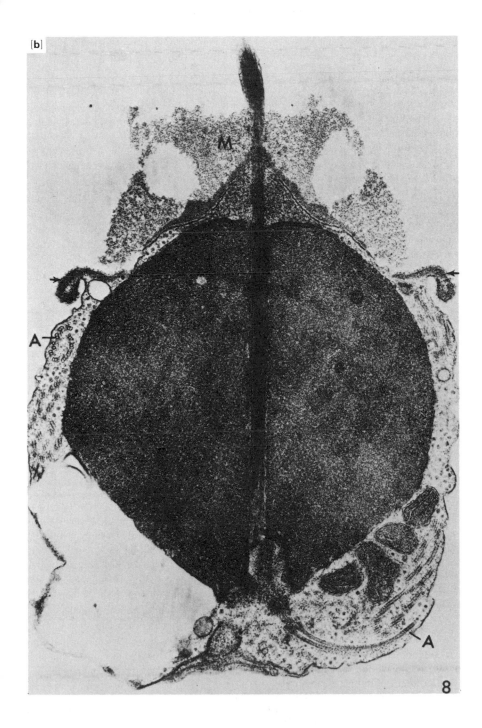

arranged in parallel with their 'barbed' ends at the membrane-associated distal end. It should be noted that the membrane-associated end of the bundle has fewer microfilaments than the coiled end (15 as opposed to 85), which can be taken as evidence for the polymerization of microfilaments at the barbed, membrane-associated end (Tilney et al. 1981). The failure of the microfilaments to bind actin when in the resting, coiled form is presumably because of the presence of other proteins which serve to stabilize the packing and which mask the heavy meromyosin binding-site. Inevitably, a fairly major secondary constituent of the rod, a protein of 230 kd, has been christened 'scruin'.

The movement of the acrosomal process, very much a once-in-a-lifetime event, seems to be brought about by a conformational change, not in the constituent protein but rather in the sub-unit packing: although the altered interaction of scruin and actin presumably involves a conformational change, it is not the shape change of the actin itself which generates the movement. If the spasmoneme of vorticellids (see § 4.4) is a pure conformational motor, involving a change in the shape of the sub-units, then this motor, which involves a change in the packing of the sub-units of a polymer, is a secondary design. An important implication is that actin in the acrosomal process is being used as a structural component, myosin is not involved. The possibility therefore exists that not all the actin molecules in a cell are for use in the actin–myosin motor.

## 4.5.2 The acrosomal process of Thyone

The echinoderm Thyone also exhibits an unusual acrosomal process, as do the other echinoderm species, Pisaster and Asterias. As with Limulus the mechanism of protrusion of the acrosomal process depends upon actin, but in the echinoderms the motor is of the 'assembly' type. The process of protrusion is shown schematically in Figure 4.7. In the non-activated spermatozoon there is an electron-dense mass immediately between the acrosome and the cup-shaped nucleus, as shown in Figure 4.8a (Tilney 1976a & b). Upon activation of the sperm there is an extremely rapid polymerization of this material leading to the formation of a microfilament bundle which serves to push the acrosomal process forward some 90 μm in only 10 s (Fig. 4.8b). A locally specialized region (the actomere), adjacent to the nuclear envelope (Tilney 1978), acts as a launching pad against which the process pushes itself forward to penetrate the jelly layers of the egg – like an arrow from Cupid's bow (if one wanted a fanciful analogy!).

Several aspects of this process are of interest. The local concentration of non-polymerized actin in the dense body (160 mg ml$^{-1}$) far exceeds that which would polymerize spontaneously in vitro, and the sub-unit mass is localized very precisely in the sperm head, starting its development near the base of the flagellum and later moving to the anterior position which it occupies in the mature sperm. As polymerization proceeds the micro-filaments are organized into a parallel array and not into a random meshwork; the normal rules for microfilament elongation are followed: G-actin adds on at the barbed (membrane-associated) ends, even though the delivery of protomer to the distal part of the process seems a silly way to do things (Tilney & Inoué 1982). (The actin concentration is, however, so high that even at the pointed end the equilibrium will favour assembly rather than disassembly, and some extension probably takes place here too.) Some of the problems have been resolved, at least in part. The G-actin of the dense body seems to be associated with the protein, profilin, which prevents polymeriza-

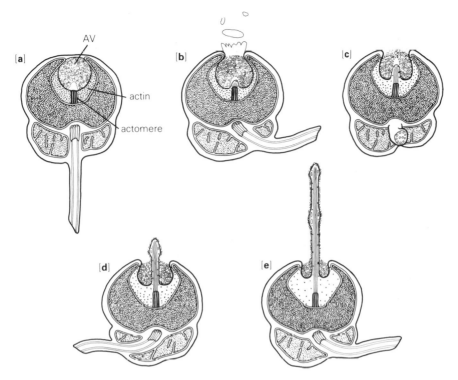

**Figure 4.7 Extension of the acrosomal process of *Thyone* sperm** A series of schematic diagrams illustrating the events of activation. The unreacted sperm, (a), has an approximately spherical acrosomal vesicle (AV) which is partly enclosed within the horse-shoe shaped nucleus. Between the nucleus and the acrosomal vesicle is an electron-dense region containing profilin-stabilized G-actin, and a specialized structure, the actomere. Following activation the acrosomal region swells slightly by the uptake of water, and the anterior membranes of the vesicle and the plasma membrane are shed, (b). Actin begins to polymerize onto the actomere, which is anchored basally to the nuclear envelope, and the process begins to develop, (c). As actin sub-units are added to the tip of the acrosomal process it extends, (d & e), and pushes the posterior part of the acrosomal vesicle membrane forward, eventually to fuse with the membrane of the egg. The contents of the acrosomal vesicle remain associated with the extending process and facilitate lysis of the egg jelly and binding via carbohydrate-binding proteins (bindins). The whole process takes only a few seconds, and the fully extended process can be 90 μm long (an extension rate of almost 10 μm s$^{-1}$). This series of diagrams is from Inoué and Tilney (1982) who base the sequence on their observations using differential interference contrast video microscopy and ultrastructural observations.

tion of the actin much as does the homonymous (and probably homologous) protein found in mammalian cells (see § 2.6.2). The integrity of the periacrosomal mass of G-actin, which is sufficient to permit its isolation by relatively low-speed centrifugation from disrupted DNAase-treated sperm, apparently depends upon an actin-binding protein which is unusual in that it binds G- and not F-actin. Treatment of the isolated periacrosomal material with protease *in vitro* solubilizes the G-actin and renders it assembly-competent (as will an increase in pH from 6.5 to 7.2). The localization of the G-actin seems to depend upon the localization of the actin-binding protein,

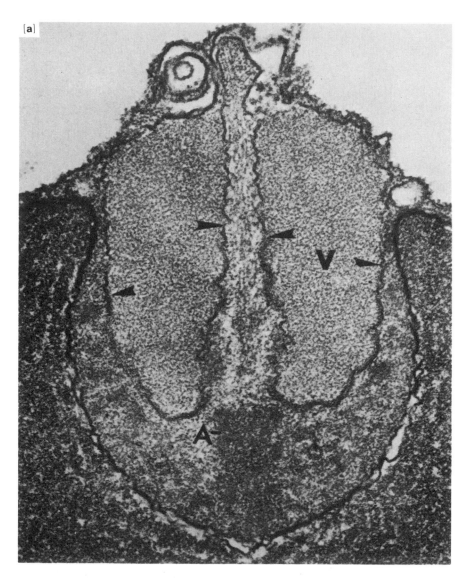

**Figure 4.8  The ultrastructure of the acrosomal process of *Thyone*** (a) A thin section through the apical end of a sperm which was fixed 1.5 min after activation by the addition of the ionophore X537A. Extending from the actomere (A) is the bundle of microfilaments which is responsible for the extension (from Tilney 1978). (b) Section through a sperm which was activated by ionophore, glycerinated, and then the actin filaments decorated with S1. The nucleus (N) is at the bottom of the picture, and the filaments are inserted into the actomere (A) (only part of which is visible) by their pointed ends. Extension of the process occurs by the distal addition of sub-units at the barbed ends of the filaments (from Tilney & Kallenbach 1979).

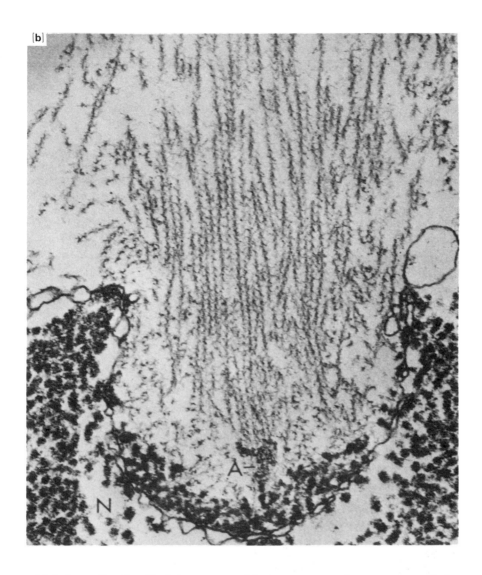

which traps the G-actin into an assembly-incompetent dense mass until the ionic (proton) influx associated with activation triggers polymerization (Tilney *et al.* 1978). Localization of the actin-binding protein remains to be explained, although there is some specialization of a region adjacent to the nuclear envelope, which can be recognized ultrastructurally (Tilney 1978). Notice that this explanation has just moved the problem one step further backwards, leaving the question of how the position of this region is specified and how this area actually localizes the actin-binding protein.

Assembly of the filaments is triggered by the influx of protons (this can be induced experimentally with ionophores), which leads to the activation of actin nucleation sites on the cytoplasmic face of the acrosomal vesicle opposite to the actomere (another problem in localization). The gradual build-up and relocation of the G-actin and its associated proteins during spermatogenesis has been documented extensively by Tilney (1976a & b) and will not be considered here.

Both the examples described above are of actin being used independently of myosin and might be dismissed as rather bizarre examples in obscure invertebrate animals. This would be to neglect an important message: having seen actin used in this way, and knowing the highly conserved structure of the molecule, we should at least be prepared to see actin as a structural protein in its own right and should bear this in mind when considering other, less specialized, systems. It is also an object lesson in the diversity of natural systems and the benefits of a broad zoological background for the researcher who wants model systems!

## 4.6 Hydraulic systems

Most biologists are aware of the important role of turgor pressure in maintaining the 'stiffness' of plant tissues, of some specialized mammalian tissues, and of the hydrostatic skeleton of nematodes, annelids and so on. Few of these examples are of direct interest to the cell biologist, although the activity of membrane-located ion pumps plays an important part in regulating water movement both across epithelia and also into intracellular vacuoles. Only two examples of the use of hydraulic systems by cells will be described, although it is perhaps worth commenting that cells, which are bounded by membranes of limited permeability, are susceptible to changes in the osmotic pressure of the environment. The blebbing sometimes observed on fibroblasts can be suppressed by increasing the tonicity of the medium, and the normal activity of most cells probably depends on the constancy of volume of the cytoplasm.

### 4.6.1 Nematocysts of coelenterates

As every schoolboy knows . . . some of the epithelial cells of organisms such as *Hydra* are specialized for catching prey and for defensive purposes. These cells are the nematocysts which will, when stimulated, shoot out a long, thin process. The everted process may inject toxins, or be barbed and thus serve to spear the hapless prey, or may be a sticky thread which entraps the prey. The eversion of the process occurs rapidly and may be triggered by chemical or mechanical means. Different versions of the nematocyst have evolved and they are 'borrowed' by certain nudibranch molluscs which graze on hydroids and use the nematocysts for their own defence. The mechanism has long been thought to be hydraulic and to involve an abrupt increase in volume of the nematocyst which can only be accommodated by everting the thread. The increase in volume comes about because of an influx of fluid from outside; not because of an alteration of permeability of the membrane, but because of a sudden increase in the osmotic strength of the cytoplasm. This conclusion is based on the work of Lubbock and Amos (1981) who showed that the increased osmotic strength arose as a consequence of sequestration of intracellular calcium ions, although the molecular events which lead to the altered osmolarity are unclear. Using the large nematocysts of the anemone *Rhodactis rhodostoma*, these authors were able to show that the permeability of the plasma membrane remained unaltered during the discharge, and that the process could be initiated by chelation of calcium with citrate or with EGTA. Whether the same mechanism is used in all cases and exactly how the calcium sequestration is brought about remain unresolved problems – which might enjoyably be pursued at marine field stations in sunny climates.

At a subcellular level a similar extrusion of material for defensive purposes is the release of mucocyst contents from *Tetrahymena*, a system which has been exploited in investigating the membrane specializations associated with the secretory process (B. Satir 1974).

## 4.6.2 *Hydraulic mechanisms in plants*

Many of the movements of plants, such as the curling of tendrils and the turning of leaves to the sun, depend upon hydraulic motors. In general, the movement is brought about by the swelling of groups of cells located at the leaf base, or as a central band in the tendril (Satter & Galston 1981). A slightly better known example is the opening and closing of stomata on the leaves of vascular plants, in order to regulate transpiration, by expansion or contraction of the stomatal guard cells (Raschke 1979). Most people are also aware of the 'sensitive plant', *Mimosa pudica*, which folds up its leaves in response to mechanical stimuli. The cellular basis of these movements has been investigated and seems to depend upon transmembrane fluxes of ions or small molecules (sugars or tannin-like compounds) which bring about a transmembrane movement of water. Specialized cells may be involved in active mechanical systems; such cells have thin cell walls with a high protein content and are capable of considerable extension. Most of the volume change is in the cell vacuole, and the commonest ionic species involved are potassium and chloride. The presence of fibrils 20–30 nm thick in the vacuole has led to some speculation about the involvement of contractile systems but the evidence for this is unsatisfactory: it seems likely that the motor system is primarily hydraulic. Plant cells are well suited to this sort of system, their walls can be modified to determine the shape change that will follow dilation and the presence of a wall reduces the risks of overexpansion and lysis. The topic of turgor movements in plants is reviewed by Raschke (1979) and by Satter (1979) and will not be discussed further.

## 4.7  Miscellaneous motor systems

It seems unlikely that the motor systems discussed in the preceding chapters and sections are the only ones which exist, and there are various systems in which the mechanism of movement is not understood. Most of the theoretical methods for generating movement have already been illustrated with real examples but this does not mean that all the possibilities have been explored. In some cases these will probably involve applications of known motors but with unusual geometry or bizarre coupling of the motor to propulsion, in other cases the motor mechanism will be entirely different. The sections which follow are designed to draw attention to a range of systems where, in general, our ignorance exceeds our knowledge.

## 4.7.1  *Spermatozoa of* Caenorhabditis elegans

The amoeboid spermatozoa of the nematode *Caenorhabditis elegans* move over a solid substratum at rates comparable to those of many other cells ($5-7\ \mu m\ min^{-1}$) but are very unusual in that they do not seem to contain appreciable amounts of the proteins normally associated with motor systems. The actin content is less than 0.02 per cent of the total protein, and there are no detectable microtubules; movement is insensitive to cytochalasin B, phalloidin and colchicine, as might be expected. It has been suggested that movement in this unicellular organism depends on the controlled insertion and recycling of membrane components (Roberts & Ward 1982), much as is required in some models of crawling movement (see Ch. 6). In its more familiar multicellular diploid phase the organism utilizes conventional actomyosin and tubulin–dynein motor systems.

Within cells the **saltatory** movement of large particles is poorly understood and may, or may not, be part of the broader problem of intracellular relocation of vesicles (Rebhun 1972, Wang & Goldman 1974). The selective routing of vesicles in the secretory and endocytotic processes is not well understood, although some of the complexities of what might be called the 'life-cycles' of various vesicles are beginning to be described (Besterman & Low 1983). The rearrangement of nuclei in virus-induced syncytia (Wang *et al.* 1979), or in natural syncytia, such as vertebrate **myotubes**, has a clear pattern and may be microtubule associated.

## 4.7.2   Shape changes in erythrocytes

Shape change in erythrocytes has received much attention. Immediately below the plasma membrane is a complex meshwork which involves the association of actin with spectrin and the linkage of these proteins to the membrane by other proteins such as ankyrin. These studies have contributed greatly to our understanding of the properties of the plasma membrane and of the cytoplasm immediately below. The relevance of the plasma-membrane-associated cytoplasmic proteins of the erythrocyte to the organization of 'normal' cell membranes is enhanced by the finding of spectrin-like and ankyrin-like proteins in other cells (Bennett & Davis, 1982). A recent review of erythrocyte shape change and associated phenomena is given by Branton (1982). The techniques developed for investigating the mechanical properties of the erythrocyte are now being applied to **leucocytes** (see Meiselman *et al.* 1984) and potentially can be applied to a range of cells. Unlike erythrocytes, many cells will alter their properties fairly rapidly in response to mechanical deformation, either attempting to resist further deformation or being stimulated to change shape even more. The active response of the test cell makes measurement more difficult – but it is the ability of cells to react that makes them so interesting.

## 4.7.3   Gliding movements

The gliding movement of gregarines has long been known and described but the mechanism is still unclear. The surface of the cell has a complex structure, with multiple layers of membrane and a pattern of ridges which is maintained by some cytoskeletal element (Schrevel *et al.* 1983). The cells, which are quite large (200–300 $\mu$m long in the case of *Gregarina* from the gut of the mealworm), move at speeds of 2–8 $\mu$m s$^{-1}$; movement can be inhibited by cytochalasin B and by trifluoperazine (TFP), an inhibitor of calmodulin (King & Lee 1982). Little or no actin is detectable (nor has calmodulin been detected, so the TFP experiment has to be interpreted with caution); it is clear that our understanding of the motor mechanism is incomplete. There are some similarities between the gliding of gregarines and the phenomenon of flagellar surface motility (reviewed by Bloodgood 1982), but this is not really a help because the movement of particles along the flagellum is not understood either. The strongest similarity is in the uncertainty which surrounds the two systems! The gliding of various eukaryotic algae (Yeh &

Gibor 1970, Nultsch 1974) and blue-green algae (more properly cyano-bacteria) (Halfen & Castenholz 1971a & b), remains poorly understood, although in the cyanobacterium *Oscillatoria princeps* helically arranged fibrils (6–9 nm diam.) seem to play some part in generating the gliding motion.

## 4.8 Summary

A wide variety of motor systems have been described in living organisms of which three are commonly encountered, the actin–myosin and tubulin–dynein motors of eukaryotes and the rotatory flagella of prokaryotes. To some extent it is possible to predict the means by which cells could use molecules to generate movements, the range is not enormous, and most methods are exploited by cells of one sort or another. It may be, of course, that we are only 'aware' of the possible solutions because examples exist and that other possibilities should have been listed: we should beware the dangers inherent in thinking we understand how living systems operate.

In the next two chapters we will address the question of how cells use their motor systems to move from place to place. Having a motor is of no particular value unless there is some way of using it to achieve a selective advantage.

## References

Symp. Soc. Exp. Biol. **35**, *Prokaryotic and Eukaryotic Flagella*. Several reviews on bacterial flagella, though quite detailed.

Berg, H. 1975. Bacterial behaviour. *Nature*. (Lond.) **254**, 389–92. A useful mini-review on the bacterial motor.

Stebbings, H. and J. S. Hyams 1979. *Cell Motility*. London: Longman. Gives rather more detail on some of the unusual motor systems.

Inoué, S. and R. E. Stephens 1975. *Molecules and Cell Movement*. New York: Raven. Has useful reviews on unusual systems, particularly acrosomal processes and the spasmoneme. (The latter can also be found in Symp. Soc. Exp. Biol. **30**, 273–302.)

# 5

# SWIMMING

## 5.1 General

From details concerning the motor we now turn to the application of the motor system to moving the whole cell around. In describing various motor systems most of the movements at a subcellular level have been discussed already and, with a few exceptions, we will restrict ourselves from now on to locomotory activities. Although 'locomotion' may seem an overly pedantic term, it implies with less ambiguity than 'movement' the activity of moving from place-to-place, and as such is a convenient usage.

Two fundamentally different locomotory patterns will concern us: the movement of free-swimming cells through a fluid phase, and the crawling movement of cells over a rigid **substratum** or through a semi-solid matrix. We will need, of course, to make an arbitrary distinction between swimming in very viscous fluids and crawling through very deformable matrices; the boundary between swimming and tunnelling can become rather uncertain. There are also examples of cells swimming through a thin layer of fluid which covers a solid substratum. In moving through such a fluid film the swimmer does not rely upon contact with the solid surface, whereas the crawling cell does.

Much of the work which has been done on cell behaviour, and which will concern us in later chapters, has been with cells that crawl. For this reason we will discuss the problems of swimming first, placing the description of substratum-associated movement immediately before the chapters dealing with cell behaviour.

## 5.2 Swimming

### 5.2.1 General considerations

If we ignore the problems of obtaining lift, then movement through a gaseous medium is essentially similar to swimming in a fluid, but although encysted amoebae and a variety of spores and gametes are dispersed by the wind they do not fly, they are moved passively. For cells, only liquid media are important for active movement. In order to move through a fluid the swimming cell must use its motor system to push a portion of the fluid medium in the direction opposite to that in which the movement is to take place.

The forward thrust in swimming is achieved by accelerating fluid backwards, and a given thrust can be obtained either by accelerating a large volume of fluid by a small amount, or accelerating a small volume to a high velocity. Because the power needed to obtain the thrust depends on the rate at which kinetic energy is imparted to the fluid, and because kinetic energy depends upon the square of the velocity (KE = $0.5 \times m.u^2$, where $m$ = mass of fluid and $u$ = velocity of fluid), thrust can be obtained more economically by accelerating large masses to small velocities. This is not intuitively obvious to most of us, nor particularly memorable, but an analogy may help: it is more efficient to propel a canoe with a broad bladed paddle moved slowly than with a thin stick moved very fast. The rate of doing work (and therefore the power required) is less with the paddle, although the total amount of work that must be done to propel the canoe remains the same.

Forward movement of a swimmer is resisted by two things: the inertial resistance of the fluid which must be displaced, which depends upon the density of the fluid, and the viscous drag experienced by the moving object, which is in effect a frictional resistance. The relative contribution of inertial and viscous constraints depends upon the size of the cell and the speed at which it is attempting to move. Cell biologists whose background is non-zoological may be rather unfamiliar with the problems of 'scaling-down' and a brief discussion of this topic, as related to the problems faced by the microscopic swimmer, is included in the next section. The frictional resistance in a fluid becomes a viscous drag because the fluid immediately adjacent to the solid body remains stationary and slippage at a distant plane within the fluid is resisted by the interaction of the molecules of the fluid – the interaction which determines the bulk viscosity of the fluid.

In considering the problems faced by a cell which is attempting to swim through a fluid we must take into account both the method by which thrust is obtained and the resistance of the medium.

## 5.2.2  Viscous drag and inertial resistance

Although it would be possible to calculate separately the contribution of inertial and viscous resistances, it is more convenient to take the ratio of the two, a ratio given by the **Reynolds' number** ($R$), which depends upon the size (which is related to the linear dimension, $l$) and velocity ($v$) of the moving object, the density ($p$) and viscosity ($\eta$) of the liquid, according to the equation:

$$R = \frac{lvp}{\eta} \tag{5.1}$$

The larger the value of $R$, the greater the contribution of inertial resistance and the less the viscous drag, a consequence (very approximately) of the decreased surface : volume ratio of larger objects.

At a Reynolds' number of one, both inertial and viscous forces are of equal importance; but, fortunately for the innumerate biologist, most large organisms operate with $R > 100$, whereas most cells have $R$ in the range $10^{-3}$–$10^{-6}$. In other words, for a cell the major constraint is viscous drag; for the human swimmer the main problem is displacing fluid. The important message is that our own experiences in the swimming pool are very misleading when we start thinking about a flagellate. For a cell, inertial effects are vanishingly small and when the motor stops forward movement will cease immediately and there is no gradual deceleration, only an abrupt halt (calculated to occur within a millionth of a body length in the case of a bacterium). It also follows that large models of sticks, string and sealing wax, may be excellent entertainment for demonstration purposes but are useless for measuring the forces involved at a microscopic level. The hydrodynamics of small objects moving through a fluid are relatively complex and become much more difficult to analyze when an undulating or oscillating propulsive system is involved. Further difficulties of analysis also arise when the viscosity of the medium is non-Newtonian, that is when the viscosity decreases as the shear force increases (because non-spherical molecules dissolved in the fluid become progressively more aligned as the flow rate increases).

At very low Reynolds' number fluid flow over an object is smooth and non-turbulent and there is no necessity for streamlining to reduce the drag coefficient. Apparently little is known about the advantages of different shapes for movement at low Reynolds' number and, judging from the diversity of form seen in small motile organisms and cells, where there seems to be no adaptive convergence of shape, streamlining may be unimportant.

## 5.3   Methods of obtaining forward thrust

The propulsive methods actually exhibited by cells fall into three categories, all of which rely upon the movement of thin projections from the cell. This appears to confound the argument presented earlier, that it is more efficient to obtain forward thrust by using broad paddles. The use of thin, whip-like projections presumably derives from the constraints imposed by resistive forces. By moving a long cylindrical projection fairly fast the size term (*l*) in equation 5.1 remains large, the velocity term (v) is increased and the total contribution of viscous drag is reduced. The three methods of obtaining propulsive thrust are:

(a)   the rotating propeller, which is used by flagellated bacteria;
(b)   the undulation of linear or planar drive systems, as with single eukaryotic flagella and ciliary fields;
(c)   the reciprocating oar-like movement of widely separated cilia.

The distinctions between these three methods are in some respects rather small, but for convenience they will be considered under different headings. To my knowledge no cell utilizes jet propulsion, probably because of the inefficiency of the method when viscous drag is the major resistive force (the jetting effects seen in the **comb-plates** of ctenophores are a possible exception, although these are rather larger than most cells; see § 5.3.3).

### 5.3.1   Rotation: bacterial flagella

The motor system which rotates the flagellum has already been discussed (§ 4.2); the propeller-like action of bacterial flagella can be demonstrated experimentally by attaching polystyrene beads to the flagellum (as an aid to visualizing the movement in the light microscope, the flagellum itself being below the limit of resolution), or the rotation can be shown indirectly by attaching the flagellum to a solid substratum by means of antibodies directed against flagellin (Silverman & Simon 1974). In the latter case the body of the bacterium rotates rather than the flagellum, and it is possible to estimate the power output by varying the viscosity of the medium and thereby altering the viscous drag coefficient. Further support for the drive mechanism being propeller-like comes from the observation that mutant bacteria with straight flagella are immobile, even though the mutant bacterium will rotate when the flagellum is anchored, proving that the motor system is functional (Silverman & Simon 1974). The helical propeller will displace relatively large volumes of the medium rearward at low velocity, and should therefore be a more efficient method of propulsion than one that attempts to move small volumes rapidly. An essential part of the system is that the counter-rotation of the cell body should normally be much slower than the rotation of the flagellum. If the rotation of the flagellum is stopped by tethering it to a slide with anti-body then the body will rotate. Were the flagellum to be detached whilst continuing to rotate (which is impossible since the drive mechanism depends on the flagellum remaining attached), then it would remain fixed in position or would move through the medium in the opposite direction to the normal movement of the cell.

130

Multiflagellate bacteria use a compound propeller with the individual flagella arranged in a bundle in which the separate flagella must slide over one another during rotation. If the flagella in such a bundle are experimentally cross-linked with bivalent antibody directed against flagellin, then the rotation of the compound bundle is halted and the cell stops (Berg 1975). Reversal of the direction of rotation of the individual flagella, which presumably occurs simultaneously in all, causes the bundle to untwist and the individual flagella to act as separate entities, the smooth forward movement achieved by normal rotation being transformed into a 'tumbling' motion of the cell body. For maximal propulsive efficiency it has been calculated that the ratio of amplitude and wavelength of the bends in the flagellum should lie in the region of 0.16, and this does seem to be the case for many bacterial flagellar bundles (Holwill 1974). The abrupt reversal of rotation of the propeller may also cause a change in the pitch of the flagellar helix. Such a change can be achieved *in vitro* by mechanically stressing the flagellum, to produce a conformational change in the packing of flagellin (Kamiya *et al.* 1982). Individual flagella will be much less efficient at moving the cell forward than the bundle, since some will act in opposition, so the displacement of the cell will inevitably be smaller with the motor operating in reverse. The disposition of flagella over the bacterial surface may mean that when untangled and acting individually they will impart a rotational movement to the body of the bacterium, thus contributing to the reorientation of the direction of movement which follows the tumbling phase. The direction of movement following an episode of tumbling is more-or-less independent of the previous path and altering the frequency of the tumbling alters the pattern of movement (see § 9.3 for a discussion of the mechanism of the response of bacteria to gradients, which depends upon the regulation of the frequency with which flagellar rotation is reversed).

## 5.3.2 Undulating drive mechanisms: eukaryotic flagella

The most obvious example of an undulating whip-like projection on the surface of a cell is the eukaryotic flagellum, as seen in the spermatozoa of many organisms and in unicellular organisms such as *Euglena viridis*. It is necessary to subdivide flagella into two groups: those which are smooth, and those which have projections normal to the long axis (**hispid** flagella). The reason for this subdivision will later become clear (see also Fig. 5.1). The distinction between eukaryotic flagella and cilia is somewhat arbitrary, since flagella can behave as stiff oars in some organisms or for part of the time (for example in *Chlamydomonas* when it is moving forward, but not when moving in reverse), and cilia can produce undulating bends. Both cilia and flagella work on the same molecular principle (see Ch. 3) and the common method of classification is loosely based on size and number. Flagella are in general longer and occur singly or in small groups, whereas cilia are shorter and often numerous. It should be unnecessary to stress that bacterial flagella are completely different both in structure and in mode of action from eukaryotic flagella, and that a qualifying adjective is often essential when writing generally about 'flagella'.

SMOOTH FLAGELLA

The hydrodynamic aspect of propulsion by undulating smooth flagella has received considerable attention (Blake & Sleigh 1974, Sleigh 1976), but it is not my intention to go into great detail here. Analysis, even of the simple case in which the undulation is regular and in a plane, is far from straightforward,

## [a] Planar beat

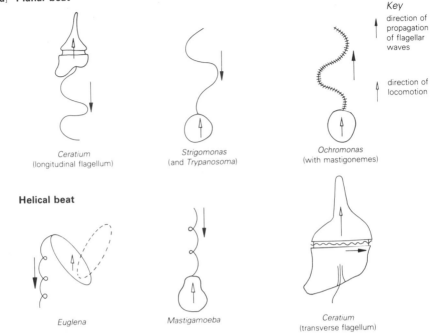

Ceratium
(longitudinal flagellum)

Strigomonas
(and *Trypanosoma*)

Ochromonas
(with mastigonemes)

**Key**

direction of
propagation
of flagellar
waves

direction of
locomotion

### Helical beat

Euglena

Mastigamoeba

Ceratium
(transverse flagellum)

[b]

**Figure 5.1  Swimming with flagella** (a) Some
of the ways in which flagella can be used to drive
a cell through a fluid medium are shown in this
diagram (from Sleigh 1974). (b) The beat pattern of
a flagellum can be illustrated nicely by the use of a
multiple-flash photograph on moving film. The
spermatozoan in this case is from a tunicate
(*Ciona intestinalis*); the sequence of images is
from top to bottom, and the bar represents 10 μm
(photograph from Omoto & Brokaw 1982).

and most flagellar beat patterns are really three dimensional.

One way of approaching the analysis is to use the approximation made by Gray and Hancock (1955), who express the forces in terms of surface coefficients of force. For a cylinder moving at velocity $v_L$ along its length and velocity $v_N$ at right angles to the axis of the cylinder, then the force along the axis is:

$$F_L = -C_L v_L l$$

(where $l$ is the length of the cylinder and $C_L$ is the axial force coefficient). The force normal to the axis is, similarly:

$$F_N = -C_N v_N l$$

For a long thin cylinder, like a flagellum, $C_N = 2C_L$ and

$$C_L = \frac{2\eta}{(\ln(2l/r) - 0.5)}$$

(where $r$ is the radius of the cylinder, $l$ is its length, and $\eta$ is the viscosity of the medium).

From these formulae it is possible to arrive at an expression for the relationship between forward velocity ($v_x$) and the velocity of wave propagation ($v_w$), which depends upon amplitude, wavelenth of the undulation, the number of wavelengths on the flagellum, and the ratio $C_L : C_N$. The sign of $v_x / v_w$ depends upon $(1 - C_L/C_N)$, and if it is positive the velocities of movement and of wave propagation will be in opposite directions. In other words, the direction in which the cell moves is opposite to the direction in which waves are propagated along the length of the flagellum, so that a wave must be initiated at the base of a smooth flagellum if the cell body is to be pushed along. Although basal initiation is more common than distal both are known; there may well be advantages to moving flagellum-first, as will be discussed later.

The velocity of forward movement is always a small fraction of the wave velocity, and the propulsive efficiency depends on the ratio of the amplitude and wavelength. Perhaps unsurprisingly, for those situations which have been examined in detail, the measured values of the ratio of amplitude to wavelength correspond with those predicted to give maximum efficiency. For helical waves it is also possible to predict that the optimum body size is 15–40 times the flagellar radius: most flagella have a radius of around 0.1 μm, so the optimum body size would be the hydrodynamic equivalent of a sphere of between 1.5 and 4.0 μm radius. Although this is indeed the case for many protozoa there are exceptions such as *Euglena viridis*, in which the body size is significantly larger than would be predicted for optimum propulsive efficiency. In the case of *Euglena*, however, the flagellum seems to be thickened by hair-like appendages, **mastigonemes** or flimmer filaments, of which there may be many thousands on each flagellum, and which may account for the apparent discrepancy (Bouck & Rogalski 1982). These filaments are around 5 nm in thickness, are composed of an heterogeneous group of glycoproteins, and are anchored to the axoneme. The flexible flimmer filaments seem to act only by modifying the effective radius of the flagellum and differ markedly from the stiff projections on hispid flagella.

## HISPID FLAGELLA

Unlike smooth flagella, the propulsive force generated by an hispid flagellum is in the same direction as the direction of wave propagation. If we were to add stiff projections to a smooth flagellum which was being used to push a cell (in a 'thought-experiment'), we would reverse the direction of movement and the cell would then be pulled along by its flagellum rather than being pushed head-first. For this to occur the ratio of $v_x : v_w$ must be negative in sign, which can only occur if $C_L / C_N$ is larger than unity. For the hispid flagella of two chrysomonad flagellates Holwill and Sleigh (1967) calculated that $C_L / C_N$ had a value of 1.8, so that $v_x / v_w$ would indeed be negative and the propulsive force would be in the same direction as that of wave propagation, as was observed to be the case. The projections involved, tubular mastigonemes, are about 20 nm thick and up to 1 μm long and are thought to be stiff, unlike the flimmer filaments mentioned in the previous section. The effect of lateral projections on the direction of the propulsive force generated by an undulating cylinder will be familiar to zoologists from the analogous comparison of eels (smooth) and ragworms (*Nereis*, for example) which have parapodia. It requires some thought to visualize the effective 'stroke' of the projection, which occurs on the crest of a wave, and Figure 5.2 may help. Essentially the projections act as oars which are brought back to the start of the stroke whilst 'hidden' on the inside of a bend. Because the mastigonemes are arranged as two rows (Fig. 5.3) it is essential for the above argument that the mastigonemes should lie in the plane of the flagellar beat and, although this has not been unequivocally demonstrated, it seems very probable.

One advantage which may accrue from decorating the flagellum is that the organism now swims flagellum-first, and water currents generated by the flagellum impinge upon the body near the base of the flagellum, where the gullet is normally located. An alternative, and possibly later development in the evolution of the ciliates, is to use smooth flagella but to reverse the direction of flagellar wave propagation (tip to base rather than the more obvious basally-derived initiation), or to change the relationship between body and flagellum so that the flagellum curves behind the body. A further advantage in leading with the flagellum is that avoidance reactions, which involve alterations in the direction of wave propagation (see § 5.4.2 & § 9.4),

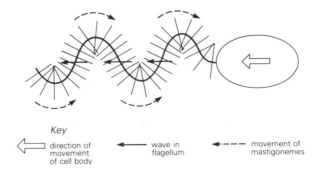

Key

⇐ direction of movement of cell body    ◀—— wave in flagellum    ◀--- movement of mastigonemes

**Figure 5.2 An hispid flagellum** The stiff mastigonemes are assumed to remain perpendicular to the axis of the flagellum as it bends. A wave moving away from the cell body will cause the mastigonemes to act as oars on the crests of waves, whereas in the inside of the wave they will be ineffective, and the overall effect will be to propel the cell flagellum-first.

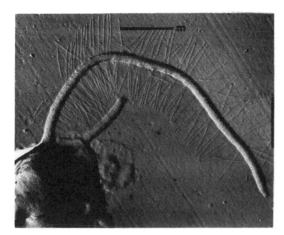

**Figure 5.3 The arrangement of mastigonemes on an hispid flagellum** A scanning electron micrograph of the flagellum of *Monas*. The specimen was shadowed with platinum–iridium, and the two rows of mastigonemes (m) can clearly be seen (photograph courtesy of Dr B. Nisbet, Zoology Dept, Aberdeen University).

can be controlled via receptor systems located right next to the motor. Putting the drive system at the back means that the message to go into reverse must be transmitted through the whole cell, or the cell body must penetrate the noxious area, before a response can be made.

UNDULATING SHEETS

Many ciliated organisms have such extensive fields of cilia that it is perhaps more realistic to consider the whole surface as one undulating sheet, rather that to consider the individual ciliary contributions. This approach, treating the 'surface envelope' which contains the tips of the cilia as the propulsive drive system, involves, however, a number of assumptions which may not be valid for all cases. The undulations are a consequence of the **metachronal** rhythm (Fig. 5.4, also § 5.3.3) which arises because cilia in different regions are at different stages of the beat cycle. Only for **symplectic metachronism**, in which the wave passes over the ciliary field in the same direction as that of the effective stroke of each cilium, is the model satisfactory. In the case of symplectic metachronism the propulsive force is in the opposite direction to that of wave propagation, just as for smooth flagella. *Opalina*, which has its cilia coordinated in this way, moves around at a speed of approximately $100 \, \mu m \, s^{-1}$, very close to the speed calculated for an undulating sheet with similar wave characteristics. A surface envelope model for *Paramoecium*, which has **antiplectic** metachronal waves, predicts a speed significantly lower than observed, suggesting that it is inappropriate to treat the system as an undulating sheet. Despite its quantitative inadequacies, the surface envelope model does successfully predict that for antiplectic metachronism the direction of movement of the cell should be the same as the direction of wave propagation. Alternative analytical approaches to considering the propulsive force generated by ciliary fields have been used, but will not be discussed. The problems involved in analyzing mathematically the hydrodynamics of ciliary fields are not ones which concern the cell, and it is probably sufficient for many purposes to realize that ciliates are quite common and that their method of movement is presumably well-adapted. They move impressively fast, particularly when compared with crawling cells!

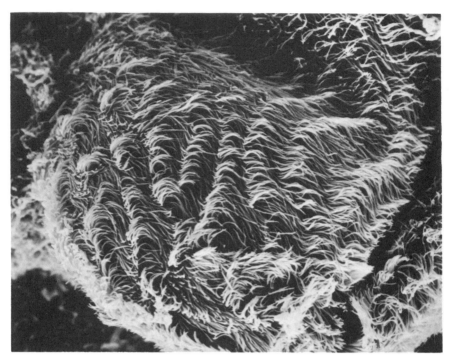

**Figure 5.4  Waves on the surface of a ciliate**  The metachronal beating of cilia gives the overall effect of an undulating surface. The ciliate shown here is *Opalina* (magnification, × 9500). The specimen was critical-point dried before shadowing (photograph courtesy of Lawrence Tetley, Zoology Dept, Glasgow University).

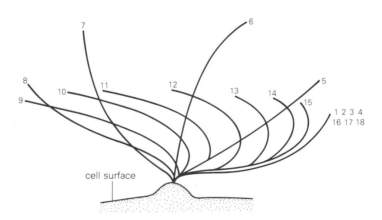

**Figure 5.5  Ciliary beating**  A simple diagram to show successive stages in the beat of a ctenophore comb-plate which is pushing fluid to the left. The beat pattern of an isolated cilium is essentially the same (from Sleigh & Barlow 1982, Fig. 2).

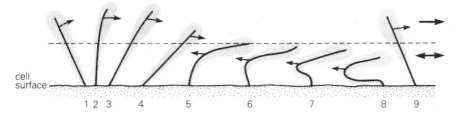

**Figure 5.6  Effective and recovery strokes**  The effect of a cilium on the local fluid environment can be separated into its effect on two layers: the layer nearest the cell surface is pushed one way during the effective stroke, and the other way during the recovery phase, so no useful work is done. The more distant layer is only acted upon by the cilium during the effective stroke (1–4) and not during the recovery (5–8) The cilium is not a simple rod, it is surrounded by a 'layer' of fluid which increases its apparent profile (shaded area).

## 5.3.3  Oars

### CILIARY BEATING

Most cilia beat in an oar-like fashion with a fast 'effective' stroke, in which the cilium is stiff, and a slower 'recovery' stroke in which the cilium is more flexible (Fig. 5.5, see also §§ 3.4 & 3.5). By far the easiest situation to consider is that in which cilia are isolated or, if in a small clump, beat synchronously. More commonly, however, cilia are distributed fairly evenly over the surface of a cell or as a broad band, and beat asynchronously giving rise to metachronal waves over the field of cilia. In these more realistic situations there is interference between adjacent cilia, and indeed hydrodynamic linkage (viscous–mechanical coupling) is probably the coordinating mechanism. For the simple planar beat patterns shown in Figure 5.5 it is obvious that fluid further from the surface of the cell is moved rearward, whereas during the recovery stroke only fluid nearer to the cell body is affected. Since the inner fluid layer is moved in the opposite direction in effective and recovery strokes it may well just oscillate and will not contribute to forward propulsion; the more peripheral layer of fluid is moved in one direction only and it is the acceleration of this fluid layer rearward that constitutes the propulsive force (Fig. 5.6). A similar mechanism can be used to move a viscoelastic sheet over a ciliary field, with the sheet 'engaged' only when the cilia are making their effective strokes. Such a situation might arise in mucus-covered ciliated epithelia, such as those of the vertebrate lung and on the gills of bivalve molluscs. The application of the ciliary beat cycle to movement of the cell over a surface is obviously also possible, with the effective stroke being equivalent to that of a limb while walking. The ciliate *Stylonychia mytilus* (Fig. 5.7) has 18 **cirri**, each composed of 8–22 cilia up to 50 μm long, which serve as 'legs' and allow the cell to move at speeds of up to 2–5 mm s$^{-1}$. The organism can also swim, and exhibits quite complex behavioural responses. The giant compound cilia which form the comb-plate of ctenophores such as *Pleurobrachia*, and which may contain many thousands of axonemes, are amongst the best examples of simple planar beat and, because of their size (600 μm × 800 μm, length × width) and the size of the organism, have received much attention. Another much studied example of simple ciliary beat pattern is the compound abfrontal cilium of the gill of *Mytilus edulis* (a tasty experimental animal).

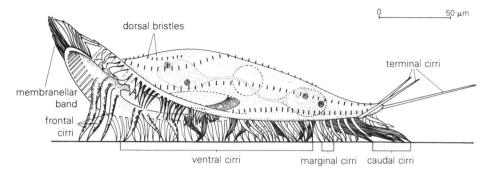

**Figure 5.7** *Stylonychia* **walking**  Side view of a stationary individual, which might set off to the left. The large compound frontal, ventral and caudal cirri are used as 'limbs'. The illustration also serves to indicate the complexity that can be achieved by the protozoa (from Grell 1973).

Ciliary beat patterns can be much more complex than a simple planar, oar-like, stroke – both at the level of the individual organelle and within ciliary fields. Some cilia, although having a more-or-less planar effective stroke, move out of this plane during the recovery stroke, bending either to right or left so that the tip describes a clockwise or anticlockwise path. Fascinating though such patterns may be, they are all just adaptations to local hydrodynamic circumstances, the need to avoid interfering with the effective stroke of adjacent cilia. We will ignore them.

METACHRONAL COORDINATION

Four types of metachronal coordination can be recognized in ciliary fields, and they are shown diagrammatically in Figure 5.8. Given the low Reynolds' number (around $10^{-3}$) at which ciliates operate, and the preponderant contribution of viscous drag, it is clear that cessation of motor activity would lead to an abrupt halting of forward motion. Were all the cilia synchronized then the organism would move forward in a saltatory fashion; by putting the ciliary beat at different parts of the surface out of phase, a smooth progression is achieved. Since the active cilia are then interacting with a fluid layer which is moving relative to the cell their effective rearward velocity is reduced, which may make the system more efficient. More importantly, the effect of arranging the ciliary beat in a metachronal fashion may be to minimize the mechanical interference between adjacent cilia, and allow each cilium to contribute to the greatest possible extent. In *Pleurobrachia*, interaction between adjacent comb-plates which are at different stages of the beat cycle, may enhance movement by a 'jetting' effect.

It is now generally accepted that, with the exception of the large ctenophores, the control of metachronal rhythm derives from viscous–mechanical coupling. By computer modelling the interactions of adjacent cilia, treating them as oscillators, it seems that a metachronal rhythm would tend to arise as a consequence of the activities of the cilia themselves, and that the wave pattern generated will be such as to provide least obstruction to each individual cilium.

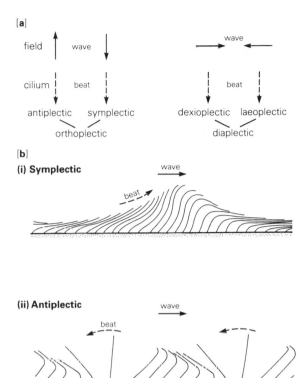

**Figure 5.8 Metachronal coordination** Metachronal beat patterns can be classified according to the relationship between the direction of ciliary beat and the direction in which the wave of activity passes (a). The planar possibilities, symplectic and antiplectic coordination, are illustrated diagrammatically in (b).

## 5.4  Control of the direction of ciliary beat

Most cilia and flagella beat in a characteristic fashion under normal circumstances, with the direction of the effective stroke being fixed with respect to the cell to which the cilium is attached. Under certain circumstances, however, the pattern of beat may change and cause an alteration in the direction of movement of the cell. This is of course particularly important in permitting a response to environmental cues, as will be described in Chapter 9. Two problems need to be considered, firstly the mechanism by which the direction of the normal effective stroke is defined and, secondly, the mechanism by which alterations are brought about.

### 5.4.1  The normal situation

Our knowledge of the mechanism of force production in cilia enables us to redefine the question of determination of stroke polarity in terms of the sub-sets of outer doublets of microtubules which are acting in each half of the beat cycle (§ 3.5). For a cilium with a simple planar beat, such as that illustrated in Figure 5.5, the doublets of the effective stroke act synchronously

139

and those of the recovery stroke are activated sequentially (metachronously). What determines the choice of effective-stroke doublets? Or, put another way, what determines the plane of symmetry of the cilium? Morphologically the cross-section of a cilium is bilaterally symmetrical, by virtue of the disposition of the inner pair of microtubules, and this morphological symmetry is reflected in the beat, which occurs at right-angles to the plane which passes through both central tubules. Thus the doublets closest to one of the two central microtubules act as one motor unit, with the two sets of outer doublets, which are equidistant from both inner tubules, being on the inside or outside of the curve, contributing least to the generation of motive force. Switching between the motors for effective and recovery strokes is poorly understood; one suggestion has been that the central pair rotates to reverse the direction of the beat – that the central pair are non-identical, one switching outer doublets on, one switching them off. This suggestion is not supported, however, by observations on the comb plates of ctenophores in which asymmetric morphological markers can be used to demonstrate that the central pair does not rotate (Tamm & Tamm 1981, see also Fig. 5.9). It seems unlikely that ctenophores are unusual in this respect.

Although the mechanism by which the polarity of beat is imposed is not clear, it does seem to depend upon the orientation of the central pair. If the central pair were helically disposed along the length of the cilium then we might, rightly, expect an helical beat pattern. Whether the outer doublets remain aligned parallel to the long axis, or whether they, too, are helically arranged remains unclear. A major unresolved question is, of course, how, in a field of cilia, the correct orientation of all central pairs is specified and how, on a ciliated epithelial sheet, the effective strokes of cilia on different cells are brought into alignment, not only with respect to the neighbouring cells but also with regard to the function of the epithelium in the organism. Once a majority of cilia have started beating with a particular orientation then it may well be that the remainder are forced to conform because of the

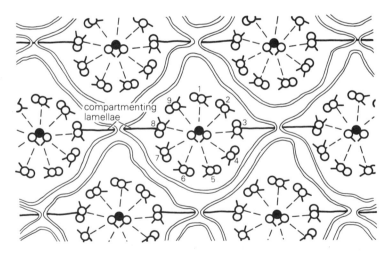

**Figure 5.9 Morphological markers for the axoneme** A diagrammatic cross section of a ctenophore comb-plate to show the structural features (compartmenting lamellae) which allowed Tamm and Tamm (1981) to exclude the possibility that the central pair of microtubules rotates in order to switch the system from forward to reverse.

viscous–mechanical coupling; that individuality is counter-productive. If this were the case then the central-pair microtubules of non-conforming cilia would have to be aligned secondarily; the mechanism for this hypothetical possibility is equally obscure. Such an explanation will not, however, suffice to generate an appropriate orientation of ciliary beat on an epithelial sheet without some means of specifying that the majority, which entrain the remainder, have their polarity of beating specified, thus returning to the original problem.

## 5.4.2  Reversal of ciliary beat

For a swimming cell to reverse its direction of movement the pattern of ciliary activity must be changed, although the alteration required will depend upon the mode of propulsion. For flagellates to move into reverse gear may require that the direction in which a wave is propagated along the flagellum is altered (see § 5.3.2); for ciliates the effective and recovery strokes must be interchanged. Figure 5.10 illustrates some of the different mechanisms by which the direction of movement can be reversed. All of these can be explained in terms of the dynein mechanochemistry as an alteration in the control mechanism which switches one set of outer doublets on and the other off, as a change in the pattern of dynein activation along the axoneme. Reversal may also require a change in the interaction of the outer doublets with the central core, through the radial spokes, although the half-axoneme and radial-spoke switches may turn out to be aspects of the same control system. Depending upon the characteristics of the beat cycle, the change in dynein activation required may vary, but a general feature seems to be that the alteration in the beat is mediated by calcium ions (Naitoh & Eckert 1974). Thus, in many cilia the normal beat (at low calcium concentration) involves synchronous dynein activation in the effective stroke and metachronal activation in the recovery stroke. Gradually increasing the concentration of calcium leads first to metachronal activation of the 'effective' dynein, then to synchronous activity in the recovery motor with metachronal activation in the effective motor and, at the highest calcium levels, synchronous activity in the recovery motor and total inhibition of the effective motor, thereby stopping the cilium from beating. This sequence of events will lead, at first, to a cessation of forward movement, then to a reversal (an escape response) and finally to paralysis of the motor system. Asynchronous beating of flagella caused by increased calcium levels probably arises by a shift from metachronal to synchronous dynein activation on one side of the axoneme. In demembranated (permeabilized) sea-urchin spermatozoa the symmetrical flagellar beat pattern can be restored, even at high calcium levels, by a brief trypsin treatment, which does not destroy the radial spokes. Clearly, calcium ion concentrations are of crucial importance in regulating the beat pattern, but it is not yet obvious how the synchronous activation of dynein is achieved: the activation is too rapid to be mechanical, it does not require an intact membrane and is therefore unlikely to be electrical. Reversal of the direction of bend propagation in the flagellum is also switched by a rise in the level of calcium, but the activity of the dynein remains metachronous (Sleigh & Barlow 1982).

Until the switching of activity from one half-axoneme to the other is understood the control of beat reversal is likely to remain obscure in molecular terms, although at a behavioural level it is sufficient to realize that an alteration in the membrane permeability to calcium is probably the common trigger. Hyperpolarization or depolarization of the membrane will

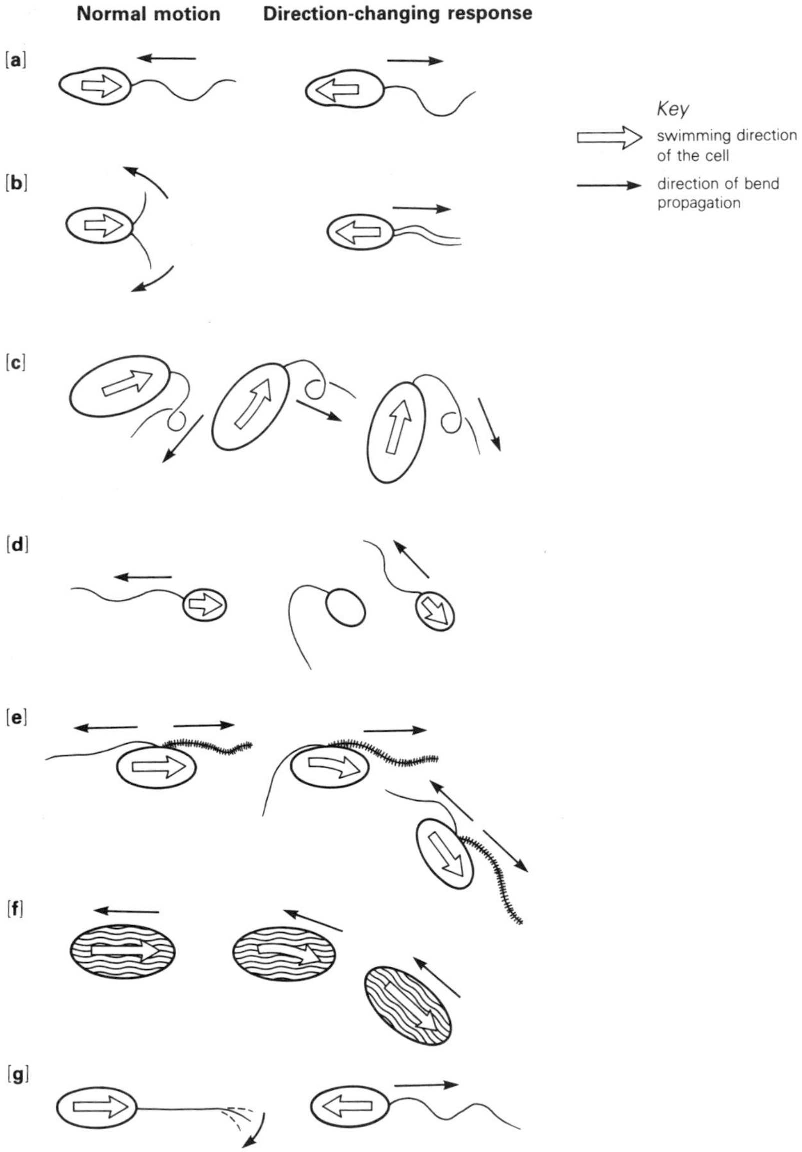

**Figure 5.10 Patterns of ciliary reversal** Various ways of engaging a reverse gear are illustrated in these diagrams from Holwill (1982). Typical organisms which exhibit these patterns are: (a) *Trypanosoma*; (b) *Chlamydomonas*; (c) *Euglena*; (d) *Blastocladiella emersonii*, spore; (e) *Phytopthora palmivora*, male gamete; (f) *Paramoecium*; (g) *Peranema*.

naturally affect the movement of calcium ions by affecting the electro-chemical potential gradient, even if the 'gate' itself is insensitive to membrane potential.

An alteration in the pattern of beating of individual cilia in a field may lead to a change in the pattern of metachronal coordination, and a shift from symplectic to antiplectic metachronism would suffice to reverse the direction of movement (see § 5.3.2). Reversal of the beat of individual cilia will also reverse the direction of propagation of metachronal waves, which will act directly to cause rearward movement. The changes of metachronal wave pattern as ciliary beat pattern alters, and the consequences for movement are too complex to enter into here; those who wish to pursue this topic should consult a specialist review, such as that by Machemer (1974).

## 5.5  Summary

Having briefly described some of the ways in which cells swim, and the viscous problems which face a minute swimmer, we will move to the simpler problems of crawling; simpler because they are restricted to two dimensions and because experimental material remains in focus while we watch it moving! Nevertheless, it is worth remembering that the ability to swim is important both for very simple organisms which live in fluid environments, and for half the successful mammalian gametes. It is perhaps also worth pointing out that the inability of mammalian phagocytes to swim makes fluid-filled lesions less accessible to the immune defences of the body and therefore more persistent.

## References

Sleigh, M. A. (ed.) 1974. *Cilia and flagella*. London: Academic. Probably the most comprehensive set of review articles, and still the best source.

*Symp. Soc. Exp. Biol.* 35 1982 *Prokaryotic and eukaryotic flagella*. More up-to-date than Sleigh, less biological and more mechanistic.

# 6

# CRAWLING MOVEMENT

---

## 6.1   Introduction

Those cells that cannot swim require a solid substratum in order to move around. In conventional tissue culture movement is restricted to a planar substratum which can be an optically flat surface, making the analysis of movement somewhat easier than might otherwise have been the case. More recently attention has been directed to movement of cells through rigid and deformable matrices but, as with crawling over a flat surface, there is an interaction with a solid substratum which marks the difference from swimming in a fluid phase. Three groups of cells have been studied extensively: the large free-living amoebae, mammalian leucocytes and 'tissue-culture' cells, the latter supposedly behaving in the same way as their counterparts *in vivo*. More attention has been directed towards those cells which grow easily in culture, fibroblasts and their transformants, though other cells have been used, including some epithelial cells, myoblasts and neural cells. Although a few studies of cell movement *in vivo* have been made, particularly of cells in early embryos, these studies have not been directed towards the question of mechanism so much as that of behaviour.

Essentially the question to be addressed in this chapter, as in the previous one, is that of coupling the motor system to the environment – the transmission and drive system of the vehicle in our analogy. The motors appear to be actomyosin-based, with perhaps one exception.

## 6.2   A simplistic analysis of the problem

For a cell to move over a surface the contractile motor system must operate to move the cell centre toward a distal attachment; new distal attachments must then be made, adhesions at the trailing edge broken, and the cycle repeated. An obvious problem is that the actomyosin motor system is basically a contractile one and it is therefore easier to pull rather than push, yet some form of protrusion (or pushing) is needed in movement. In addition there are three other problems, the formation of reversible attachments, the production of distal attachments, and the proximal connection of the distal attachments so that the motor does useful work on the cell as a whole. In the car of our analogy, the engine must be fastened to the bodywork, the wheels must gain traction (but the tyres must not adhere irreversibly to the road), and new attachments must be made in front of the old ones when moving forward, as will be the case provided the wheel rolls rather than sliding or spinning.

### 6.2.1   Attachment

To interact mechanically with the environment there must be some form of attachment either by adhesion or by anchorage. When we are dealing with attachment at the cellular scale, involving separations between plasma membrane and substratum of tens of nanometers, the interactions can be hard to visualize in terms of our own macroscopic experience. On a perfectly flat surface the attachment must be adhesive in nature, although this need not necessarily be by the formation of direct linkages – it would be sufficient to couple the surface of the cell to the substratum by a film of very viscous polymer. Alternatively, there could be binding of cell surface moieties to ligands that are irreversibly adsorbed on the surface or to integral ligands of the substratum, electrostatic interaction between the surfaces, or a combination of receptor–ligand and physicochemical mechanisms according to taste.

Our continuing ignorance about the mechanism(s) of cell adhesion makes it more difficult to predict the nature of the interaction: some details of what we know will be discussed in Chapter 7.

Once the surface becomes topographically more complex (once the curvatures of surface features approach cellular dimensions) the opportunity arises for more complex interactions. On a surface such as that shown in Figure 6.1, the cell could push by extending **pseudopodia** rearward or pull by producing a hook-like anchor. This sort of non-adhesive anchorage/interaction becomes easier to visualize if the cell is crawling through a three-dimensional meshwork or over a grid (Fig. 6.2). In a fibrillar matrix an expanded pseudopod could provide an anchorage (Fig. 6.3).

Irrespective of the mechanism used, the attachment must meet certain criteria if it is to be adequate for the purpose. First, the anchorage must resist the force exerted by the cell by remaining static or at least moving backwards less than the cell moves forward; if the attachment point is mobile then the cell will remain static. Secondly, the attachment must be reversible. If the cell cannot relinquish an attachment then it must abandon part of its substance, leaving a trail of cellular material behind, or must remain in place. Once we recognize the necessity for transience, the absolute strength of the attachment, provided it exceeds the minimum to resist the pull of the cell, becomes unimportant, but the duration of the attachment becomes an important parameter. The attachment must persist for long enough for the motor to have the opportunity to act upon it, and the time scale for the lifespan of adhesions will therefore be determined by the time needed to assemble or relocate components of the motor and, in the case of an actomyosin motor, the time required for the myosin cross-bridge cycle to be repeated on several occasions.

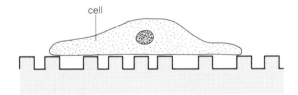

**Figure 6.1  Crawling on a rough surface**  The surface features are of such a size that the cell could potentially use hooks, or push against positive footholds, rather than relying completely on adhesive anchorage. Few real surfaces are likely to be as regular as the one illustrated, but the argument does not require symmetry in the surface features.

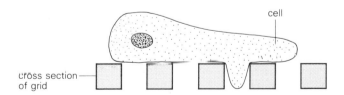

**Figure 6.2  Crawling over a grid**  The possibility of obtaining non-adhesive anchorage when crawling over the surface of a grating is obvious; it is the difference between crawling on a ladder or on a plank.

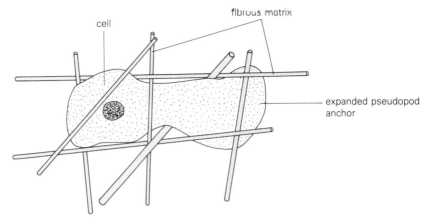

**Figure 6.3  Climbing through a meshwork**  In a matrix composed of fibres there are various ways of gaining an anchorage or reacting on the environment. The imaginary cell illustrated here is fist-jamming through a gap, and having to distort itself considerably.

## 6.2.2   Formation of distal attachments

Although we can consider this as a stepwise procedure, the formation of new attachments and the subsequent release of old ones, the process is likely to be a smooth progression – much more like a tracked or wheeled vehicle than a bipedal gait. If we visualize a cell as in Figure 6.4, moving left to right, then new attachments must be formed on the right-hand side: the cell must protrude processes to form distant attachments which are then overrun. Having drawn attention to the contractile nature of the actomyosin motor in Chapter 2 we must devise an indirect application of the motor or an alternative mechanism for this protrusion.

## 6.2.3   Linkage

If we attach a contractile element to a new distal site and cause it to shorten, we achieve nothing unless the proximal end of the contractile motor is 'attached' to the cell as a whole. An obvious requirement is for structural elements within the cell which give coherence to the contents.

## 6.2.4   Resistance to movement

In the case of free-swimming organisms viscous drag limited the rate of movement (Ch. 5): for many crawling cells this is unlikely to be a problem since the rate of movement is extremely slow. For a cell penetrating a dense matrix, however, there may well be a resistance to be overcome. Some of the resistance to forward movement will be a consequence of the work involved in deforming the matrix, and part may come from the necessity to deform the cell. The relative contribution of each resistance will depend largely upon the mechanical properties of the matrix, and a rigid matrix will require extensive cellular deformation. Once the motor stops working the cell will immediately come to a halt because of the viscous drag; there is no necessity for a braking system. Having outlined the basic requirements of the drive mechanism let us consider the movement of cells in more detail.

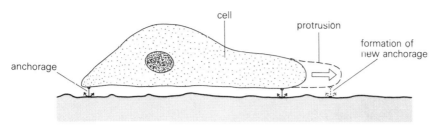

**Figure 6.4  Simple problems in crawling**  The cell shown here is meant to be moving from left to right. The basic problems are to protrude a portion of the cell to the right, to form a new adhesion, and to discard the old adhesions (or anchorages).

## 6.3    Amoeba

The large fresh-water amoebae such as *Chaos chaos*, *C. carolinensis*, and *Amoeba proteus*, are attractive cells in which to observe movement and have, in consequence, received considerable attention. These phagocytic organisms, which may be predatory or herbivorous, must move around in order to obtain their food and may attain speeds of 2–5 μm s$^{-1}$; unimpressive to the student of ciliates, perhaps, but faster than most of the crawling cells we will consider later. It is simpler to consider the movement of the **monopodial** form shown in Figure 6.5, although **polypodial** amoebae are just as capable of movement. Large, roughly cylindrical, pseudopods can be protruded from almost any part of the cell body except the tail (uroid) region, and constitute the major locomotory organelle. Protoplasmic streaming within the cell is marked and accompanies forward movement. This streaming is the most conspicuous difference between the movement of the giant amoebae and the smaller free-living amoebae, for example cellular slime-moulds such as *Dictyostelium* (see Ch. 9) which have also been studied. The smaller amoebae resemble metazoan cells much more in failing to exhibit a massive streaming, possibly because the flow is not concentrated in a central channel, or possibly because the flow in giant amoebae serves also to circulate the cytoplasm, something which is likely to be more important in such large cells. The central cytoplasm of the amoeba, the **endoplasm**, flows forward to the region immediately behind the **hyaline** cap where it transforms to apparently more granular **ectoplasm**. A frequent assumption is that the forward streaming of cytoplasm is associated with the forward movement of the cell: the basis for this assumption may not be very sound.

### 6.3.1   Evidence for an actomyosin motor

There seems little doubt that the motor is actomyosin based: partly because cytoplasm extracted from amoebae or glycerol-permeabilized amoebae will contract if ATP and $>10^{-7}$M Ca$^{2+}$ are supplied, and partly because of the well-documented existence of filaments composed of actin and of myosin in the cytoplasm (Pollard & Ito 1970, Pollard & Korn 1971). Thin filaments are found both in the central streaming cytoplasm and in the more rigid ectoplasm, where they are intimately associated with the plasma membrane. The thick filaments are remarkably similar to myosin filaments found in other non-muscle contractile systems.

Because amoebae live in fresh water their plasma membrane is extremely

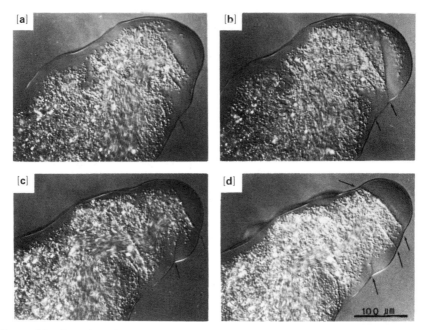

**Figure 6.5 Amoeba** Sequential photographs of a monopodial amoeba which is moving from left to right. Notice the differentiation of the cytoplasm into granule-rich and granule-poor regions and the hyaline cap; bar represents 100 μm (photograph courtesy of Dr D. L. Taylor, Carnegie-Mellon University).

impermeable both to water and to ions, and it is therefore difficult to achieve good fixation of deep cytoplasmic areas and almost impossible to vary the intracellular ionic environment except by very drastic permeabilization procedures. The cytoplasm can, however, be isolated after puncturing the plasma membrane with a micropipette, and this naked cytoplasm proves a useful model (Taylor *et al.* 1973). Varying the composition of the bathing medium causes the cytoplasmic droplet to change consistency; this can be gauged by stretching the cytoplasmic drop while observing with polarizing optics. Stretching induces birefringence because linear elements become ordered: if the cytoplasm is merely viscous then the birefringence decays with time, whereas if the cytoplasm is viscoelastic the birefringence should remain stable provided tension is maintained.

The most remarkable property of these isolated cytoplasmic droplets is their behaviour in solutions containing both $Mg^{2+}$-ATP and calcium at $7 \times 10^{-7}$ M (at or above the threshold for the calcium control). (The various

**Figure 6.6 Flare-streaming in isolated cytoplasm** (a) Sunburst-type flares from an ▶ isolated cytoplasmic droplet from *Chaos carolinensis* when 'flare' medium is added; bar represents 100 μm (see Table 6.1 for details; photograph from Taylor *et al.* 1973). (b) and (c) A loop streaming out from a droplet such as that shown in (a). The two photographs were taken a few seconds apart and show marked changes, particularly in the left hand portion of the loop which is being pulled by a contraction at the bend; bar represents 10 μm (from Taylor *et al.* 1973).

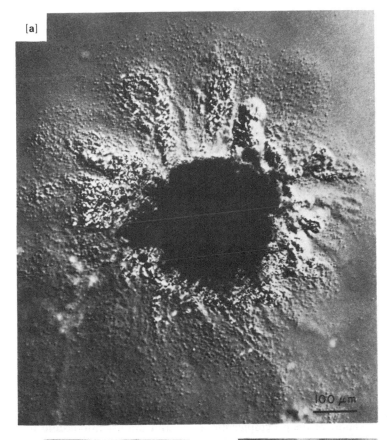

[a]

100 μm

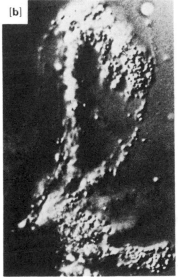

[b]

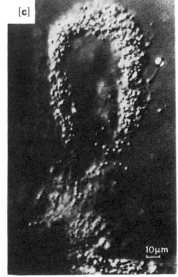

[c]

10μm

**Table 6.1** Solutions used to modify the properties of isolated cytoplasm from *Amoeba*.

|  | Stabilizing | Contracting | Relaxing | Flare-streaming |
|---|---|---|---|---|
| free-$Ca^{2+}$ (M) | $<10^{-7}$ | c. $10^{-6}$ | $<10^{-7}$ | c. $7 \times 10^{-7}$ |
| $MgCl_2$ (mM) | 0 | 0 | 0 | 0.5 |
| ATP (mM) | 0 | 0 | 1.0 | 0.5 |

All contain Pipes buffer (5 mM; pH 7.0), dipotassium EGTA (5 mM), KCl (27 mM), NaCl (3 mM), and are of the same osmotic and ionic strength. More detail can be found in Taylor *et al.* (1973).

solutions which have been used are described in Table 6.1). The behaviour of these isolated cytoplasmic droplets in $Mg^{2+}$-ATP + high calcium solutions has been described as a 'sunburst' or 'flare' pattern in which loops of cytoplasm are protruded, rather like pseudopods (Fig. 6.6).

Two important points emerge from these studies. First, the system behaves in manner very reminiscent of striated muscle contraction in terms of the requirements for calcium ions and ATP. Secondly, motile behaviour (streaming) occurs in an 'open' system, although the droplet is anchored to the substratum. Despite this additional information about the motor it is, however, still not clear how the cytoplasmic streaming is linked to forward movement, nor how the pattern of cytoplasmic movement is organized within the intact amoeba. Forward movement of the cytoplasm in the centre of the pseudopod is associated with protrusion, how is the flow generated? Two hypotheses have been advanced, and vigorously defended by their proponents. One, the ectoplasmic tube contraction model, depends upon contraction in the gel-like ectoplasm of the lateral and posterior regions squeezing the more fluid endoplasm forward like toothpaste from a tube (Fig. 6.7). The alternative model (Fig. 6.8) involves the fluid viscoelastic endoplasm being pulled forward and assembled into gel-like ectoplasm in the frontal contraction zone. Although my prejudice is in favour of the latter hypothesis there is, despite much work, no unequivocal evidence to distinguish the two models. Both will be described, together with the major arguments.

### 6.3.2 Ectoplasmic tube contraction

In this model the gel-like ectoplasm would, by contraction, cause a small positive pressure within the cytoplasm which would force more fluid endoplasm forward along the line of least resistance. Local weakening of the cortical microfilament meshwork, which resists the deformation at the leading edge, would then permit blebbing and thus the necessary forward protrusion (Komnick *et al.* 1973). A small pressure gradient does seem to exist from front to back, the evidence coming from experiments in which the amoeba crawls through a small hole between two chambers, the pressure within each chamber being recorded. Also in favour of this model is the more conspicuous filament meshwork in the tail region, the greater folding of plasma membrane at the rear and the apparent contraction of cytoplasm in this region. Contraction has been assessed by looking at the separation of particles within the cytoplasm, which diminishes in the tail region (Rinaldi & Opas 1976). On this model the birefringence in the forward moving endoplasm would be a consequence of flow.

At first sight the flare-streaming of isolated cytoplasmic droplets (Taylor *et al.* 1973) would seem strong evidence against such a model since their isolated cytoplasmic droplets are not a closed system. However, the plasma

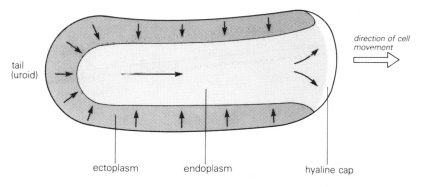

**Figure 6.7 Ectoplasmic tube contraction** On this model for amoeboid locomotion, contraction in the lateral and posterior ectoplasm squeezes the more fluid endoplasm forward, inflating a protrusion at the front. Birefringence in the endoplasm arises from flowing (oriented) molecules on this scheme.

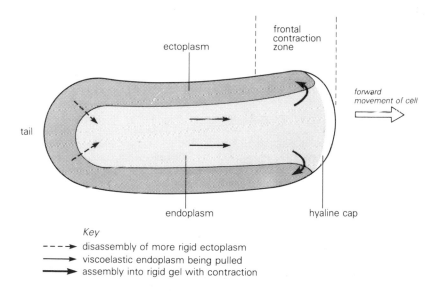

**Figure 6.8 Frontal-zone contraction** The opponents of ectoplasmic tube contraction prefer to think of the endoplasm being transformed into more rigid ectoplasm in the frontal zone, causing protrusion by an assembly process. Contraction of the ectoplasm is proposed to generate the particle-free hyaline region, and the birefringence in the viscoelastic endoplasm arises because it is being pulled forward (stress birefringence).

membrane of the intact amoeba is unlikely to contribute to the mechanical properties of the cell, it is the sub-plasmalemmal microfilament meshwork which gives rigidity: there seems no reason to suppose that the centre of the gel should not be squeezed whether or not the gel is surrounded by a phospholipid permeability barrier.

### 6.3.3 Frontal zone contraction

Assembly of endoplasm into more gel-like peripheral ectoplasm at the advancing tip of the pseudopod, accompanied by contraction which would generate the hyaline cap, would, on this model, pull the endoplasm forward and push the pseudopod tip outward by an assembly process. Birefringence in the endoplasm would be tension-generated, and would be lost as the viscoelastic endoplasm contracted into an isotropic gel to form ectoplasm. Relaxation of the ectoplasm in the tail region would be accompanied by transformation to endoplasm. The various states of cytoplasm are shown in Figure 6.9, and the mechanical properties ascribed to cytoplasm in different regions are consistent with the properties observed in isolated cytoplasmic droplets.

This model can also be described in terms of actin gel transformations in the light of what has been discovered since the model was first proposed. The endoplasm might well be a cross-linked actin gel associated with calcium-sensitive binding proteins of the gelsolin type, so that raising the calcium concentration would weaken the gel matrix. Weakening the cross-linking of the gel would also permit myosin to interact effectively with the micro-filaments to cause a contraction of the gel, provided ATP was available, producing a gel with reduced volume. Relaxation of the system would then require ATP to release the myosin from actin *and* sequestration of calcium ions to prevent further myosin–actin linkage: as the calcium concentration dropped the rigidity would be restored as a consequence of the reformation of cross-links composed of the calcium-sensitive actin-binding proteins. This is essentially the solation–contraction coupling model proposed by Hellewell and Taylor (1979) on the basis of their work on isolated cytoplasm from amoebae.

In support of the model it has been shown that microinjection of calcium into the cell leads to a local contraction and the beginnings of pseudopod formation, no matter where the injection is made. The injection of EDTA to chelate calcium ions leads to a local relaxation. Thus a gradient of calcium concentration in the cytoplasm could be an important control mechanism, and mitochondria, which are found predominantly in the tail region, might

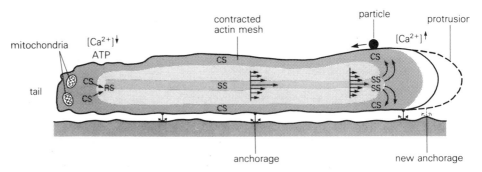

**Figure 6.9 Cytoplasmic events in a moving amoeba** This diagrammatic side view of an amoeba moving from left to right shows some of the activities that might be involved if the frontal zone model is correct, and accommodates current views on actin gel formation. Although anchorages are shown, and must exist, we know little about their disposition. Particles on the dorsal surface will move rearward on most (but not all) species.

well be the calcium-sequestering organelle. There is, of course, an element of circularity in the latter argument since the tail region containing the mitochondria is defined on the basis of the locomotory activity of the cell.

If, however, control of movement were exercised through a cytoplasmic calcium ion gradient then the 'free' calcium levels should exceed $7 \times 10^{-7}$M in the ectoplasmic region – some 10 per cent of the cell volume. Studies in which the calcium-sensitive light-emitting protein, **aequorin**, was micro-injected into *Amoeba proteus* indicate, within the limits of this technique, that less than 0.1 per cent of the cytoplasm could have such a calcium level (Cobbold 1980). This observation would then imply that the contracted state of ectoplasm, other than at the extreme frontal region, could only be maintained if the myosin–actin interaction could not be broken. The semi-permanent myosin–actin cross-linking might arise because no ATP is available, or because the rapid reformation of actin cross-links after contraction locks the gel into its 'contracted' configuration (Taylor *et al.* 1980), and relaxation would require 'high' concentrations of calcium to release the cross-linking. Insufficient data about the levels of ATP and calcium ions in different regions of the cell leaves this an unresolved question.

All these models do is to explain the forward flow of cytoplasm, which is in itself insufficient to explain forward movement. Protrusion of a pseudopod must be followed by the formation of distal attachments, and there is evidence for the existence of small 'leg-like' pseudopodia which support the cell and which are the sites of attachment. The nature of the adhesive interaction with the substratum, and its decay in the uroid region, have not been investigated in detail in amoebae. We must turn to the consideration of another cell type, the fibroblast, to approach this problem.

## 6.4 Fibroblast locomotion

### 6.4.1 General

If a small piece of embryonic chick heart is placed on a coverslip and a suitable nutrient medium is supplied, the explant becomes anchored to the coverslip and within a few hours cells are seen moving out over the coverslip surface (Fig. 6.10). These fibroblastic cells have (perhaps unfortunately) been a favourite cell type for studies on cell movement and behaviour. Their emigration from the explant is a wound-healing response (futile in this case) that *in vivo* would knit together the edges of a wound and restore the connective tissue matrix. Many of the cells grown in tissue culture are considered to be fibroblastic, although frequently this means only that their morphological appearance resembles that of the primary outgrowth from an explant as described above. Strictly speaking a fibroblast is a mesodermally-derived cell of the connective tissue that has the capacity both to degrade and to synthesize fibrillar collagen and the other components of the interstitial connective tissue matrix. Fibroblasts move as individuals and do not form permanent **desmosome**-like junctions with one another. In a wound healing response the wound area, typically involving a fibrin clot, must be infiltrated by fibroblasts and the edges of the wound must be drawn together before new connective tissue is laid down. There are, in consequence, two phases to the response, the early stage in which cells move into the area and the later stage when the cells have to pull the wound edges together (Jennings & Florey 1970, Harris 1982). Neglect of the differences in cell behaviour which will be required at the two stages has led to considerable confusion.

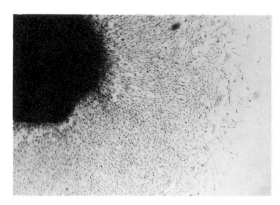

**Figure 6.10 Fibroblasts migrating from an explant** A micrograph to show the radial outwandering of fibroblasts from a fragment of embryonic chick heart (the dark mass), which has been placed in nutrient medium on a protein-coated glass cover-slip. This culture was set up by undergraduates during a practical class in the Cell Biology Department and, after being allowed to grow for 48 h, has been fixed and stained with Kenacid blue.

In the early stages of emigration from the explant the fibroblast moves rapidly, in the later stages it slows down and exerts traction without relinquishing its attachment sites, or exchanging them only slowly. Fibroblasts in the latter phase have well developed stress fibres (Couchman & Rees 1979, see Ch. 2) and can be obtained in greater quantity in culture. Stress fibres have been assumed to be the primary motor system, and much of the discussion of fibroblast movement has centred around the contractile bundles. The rapidly moving fibroblast of the early infiltration phase does not have conspicuous microfilament bundles and, because the cell is changing position, we know that the lifespan of each adhesive interaction with the substratum is relatively short. As the cell slows down, the duration of the adhesive interaction may well increase and stress fibres become conspicuous (Fig. 6.11). An increased lifetime of the adhesions at this stage may be associated with the increased production of the extracellular matrix protein fibronectin (which is known to occur), or might be regulated intracellularly in response to some environmental cue. It is therefore possible to argue that the 'early' fibroblast has adhesive interactions with the substratum which are dispersed broadly over the leading edge and that as the lifetime of these interactions increases, a lateral component of the contractile force causes clustering of the adhesion sites to form the focal adhesions more typical of the slow moving fibroblast. These focal adhesions would therefore be 'patches' of integral membrane proteins into which are inserted the microfilaments of the contractile bundle. The stability of the bundle is further enhanced by lateral interactions between microfilaments, both at the insertion sites (vinculin) (Geiger 1979, Burridge & Feramisco 1980, Geiger *et al.* 1981) and along the bundle (filamin, α-actinin). Once the anchorage is localized into a focal adhesion it can be visualized by light and electron microscopy (Fig. 6.12) (Izzard & Lochner 1976, Heath & Dunn 1978) and the *local* stress on the substratum will be much greater. The implications of this are perhaps better seen by analogy: if we want to anchor a tent we distribute guy-ropes to several points, if we wanted to use the tent as a sail to move a boulder we would fasten all the ropes to the single object to be moved.

Although the idea that the microfilament bundle is the primary motor is almost certainly misleading, the rearrangement of the motor into a more organized linear contractile element provides an opportunity to examine the components and to visualize the attachment site. The assumption is that only a rearrangement has been made and that the basic mechanism for movement remains the same whether attachments are dispersed or clustered. On this argument the contractile system which moves the cell centre toward the

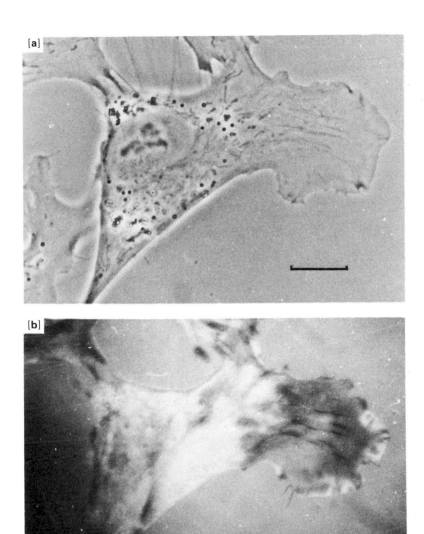

**Figure 6.11 Contact with the substratum in the early phase of emigration** Phase contrast (a) and interference reflection (b) micrographs of a primary chick heart fibroblast shortly after it has emigrated from an explant culture of the type shown in Figure 6.10. Both types of cell–substratum contact can be seen in the interference reflection picture. The dark lines are focal contacts (10–15 nm separation), relatively few in number here (contrast with Fig. 6.12); the grey area is of more diffuse adhesive contact (30–50 nm separation). The light areas are ones in which the separation between the cell and the substratum is such that there is probably no adhesion; bar represents 10 μm (from Abercrombie 1980).

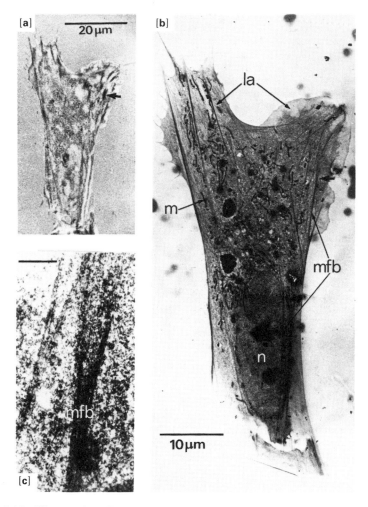

**Figure 6.12 Filament bundles and focal adhesions** The association between micro-filament bundles and focal adhesions in chick heart embryo fibroblasts which have been grown in culture can be demonstrated by correlated high-voltage electron microscopy (HVEM) and interference reflection microscopy (Heath & Dunn 1978). The HVEM picture (b) shows the focal contacts, one of which is arrowed. The interference reflection micrograph (a) shows the same cell with the same focal contact arrowed. The lamellar region (la), a mitochondrion (m), the nucleus (n), and microfilament bundles (mfb) can all be seen. The focal contact and its associated microfilament bundle are shown in greater detail in (c). Scale bars: (a) = 20 μm, (b) = 10 μm, (c) = 0.5 μm.

attachment sites is of microfilaments anchored distally to integral membrane proteins interacting, through myosin, with microfilaments associated with a proximal anchorage within the cell. The proximal anchorage seems to be perinuclear (Izzard & Lochner 1980) and may involve the pericentriolar microtubule organizing centre, which is in turn linked to the intermediate filament meshwork of the cell.

Once the adhesions have been localized to form focal adhesions their distribution can be seen using interference reflection optics, and this technique coupled with high-voltage electron microscopy of the whole leading lamella region showed the invariable association of microfilament bundles with adhesion sites (Heath & Dunn 1978). A variety of other techniques support the idea that the attachment sites of a fibroblast are distributed peripherally, as shown in Figure 6.13. Sections cut vertically through attached fibroblasts show discrete areas of close approach to the substratum with considerable electron density on the cytoplasmic face and indications of the insertion of microfilaments (Heaysman & Pegrum 1973a, 1982). These electron-dense plaques are distributed in the same way as focal adhesions, and the close approach (15 nm) of the plasma membrane to the adsorbed protein on the substratum is consistent with these electron-dense plaques being adhesion sites. More direct evidence comes from the experiments of Harris (1973b) who showed that a glass microneedle could be moved freely between the fibroblast and the substratum except at the margins; moving the needle at the margin led to local detachment. Similarly, Armstrong and Lackie (1975) found that neutrophil leucocytes would crawl around underneath fibroblasts but were restricted in their access to the marginal areas. If a fibroblast is made to round-up, as it does just before mitosis or when treated with very small amounts of trypsin, then in scanning electron micrographs it is seen to be anchored by guy-ropes (**retraction fibres**) which remain attached to the substratum at sites corresponding to those of focal adhesion (Fig. 6.14) (Revel 1974, Abercrombie & Dunn 1975).

## 6.4.2 Forward protrusion

We now have a cell anchored distally either by dispersed or focal adhesion, and with these adhesions connected to a proximal anchorage within the cytoplasm by contractile elements. The cell centre becomes the focus of a competitive tug-of-war and will be moved toward the largest number of adhesion sites (assuming all sites have comparable contractile bundles). The competitive tensions cause distortion of the nucleus and when the attachments are in two areas the nucleus appears elliptical in plan view with the major axis of the elipse parallel to the axis of tension and the direction of movement. The area of the cell with the greatest number of adhesion sites, or the area exerting the greatest pull, will be the front because the rest of the cell will be pulled along by this region.

The next stage, if the cell is to change position, is for new attachments to be formed beyond the present distal sites. The protrusion of the leading lamella is an essential part of the process. If the fibroblast is moving over a solid substratum, but with the dorsal surface bathed in fluid medium, then the protrusive activity leads to prominent '**ruffles**' being formed, which can be seen in the light microscope and are shown in vertical section in Figure 6.15. If, as is a more realistic model of the situation *in vivo*, the cell is constrained on the dorsal surface by medium of high viscosity or viscoelasticity, the forward protrusion is more marked and the 'ruffling' activity is damped down. The protrusion takes the form of a thin (0.5 μm thick) **lamella** of organelle-free cytoplasm or, in some cases, a thin microspike (Fig. 6.16). Transmission electron micrographs of these protrusions (see Fig. 6.15) show the content to be an amorphous mass of microfilaments, which may or may not be cross-linked (Abercrombie *et al.* 1971, Tosney & Wessells 1983). Immunofluorescence studies with antibodies directed against motor elements, or with labelled actin injected into the cell, show this protrusive region to be rich in actin and actin-binding protein but apparently deficient in

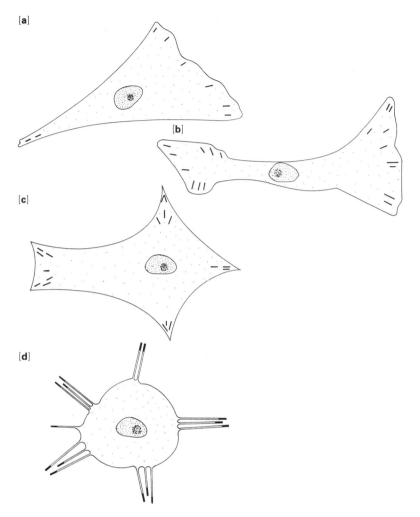

**Figure 6.13  The distribution of adhesive anchorages**  Various techniques have been used to show the distribution of the adhesions that a fibroblast might use for forward movement. An archetypal fibroblast as in (a) would have focal adhesions located under the leading edge and at the tail (although a cell without such focal adhesions would probably move faster). It is generally supposed that a bipolar fibroblast such as that in (b) would have an equal number of adhesions at each end, and a multipolar (and well-spread) fibroblast such as that in (c) would have adhesions at the vertices of the cell. If the cell shown in (c) were to round up then it would remain tethered by retraction fibres which are attached at the sites of focal adhesions (d). (See Fig. 6.14, also Figs 6.11 & 6.12 which demonstrate contacts by interference reflection microscopy.)

myosin (Herman *et al.* 1981). This suggests that protrusive activity derives from local assembly of actin gel meshwork, a conclusion which is supported by the observation that protrusive activity is blocked by the drug cyto-chalasin B, which interferes with actin polymerization by competing for nucleation sites (see Ch. 2).

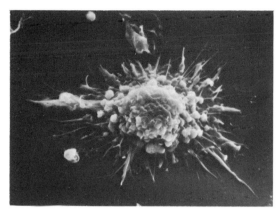

**Figure 6.14 Guy-ropes tethering a rounded fibroblast** This scanning micrograph of a partially rounded BHK fibroblast shows both the complexity of the cell surface topography and thin 'guy-ropes' tethering the cell to the substratum. Each guy-rope is fastened to a focal adhesion (photograph courtesy of Lawrence Tetley, Zoology Dept, Glasgow University). Compare the structure of this cell with the more flattened BHK cell shown in Figure 6.18.

Thus protrusion depends upon the formation of a mechanically resistive meshwork of microfilaments. The rate of assembly of the meshwork may be limited by the rate of delivery of sub-units, or by the availability of nucleation sites and cross-linking agents. That assembly of the leading lamella may be limited by sub-unit availability is suggested by the experiments, carried out independently at the same time by Dunn (1980) and by Chen (1981), who showed that mechanical detachment of the rear of a moving fibroblast led to a temporary increase in protrusive activity as the microfilaments of the tail were disassembled. Another observation that would seem to be consistent with this sort of explanation is the progressive increase in protrusive activity as the contractile ring of cytokinesis is disassembled and the final separation of the daughter cells approaches (this is based simply on watching films and has not, to my knowledge, been placed on an objective footing). Whether the sub-unit for the assembly of the meshwork is G-actin or microfilament fragments does not matter to the argument. As with streaming in amoeba, the delivery of sub-units could either be by rear (and lateral) contraction, forcing non-cross-linked cytoplasm into the area of least resistance, or by assembly and contraction at the frontal zone acting to pull viscoelastic cytoplasm forward. The rearward movement of denser areas of microfilaments as 'arcs', which can be seen in the light microscope (Heath 1981) (Fig. 6.17), does perhaps support the idea that the meshwork contracts once protrusion has occurred and would be consistent with a frontal assembly model. On either hypothesis the ability to form a resistant gel of microfilaments in the frontal zone is an essential part of the mechanism of forward movement. Frontal assembly with subsequent contraction of the meshwork has been advocated by Dunn (1980) and is probably the most plausible hypothesis at present.

### 6.4.3 Formation of distal adhesions

Not only must the cell project a leading lamella forward but this protrusion must form attachments to the substratum that are relinquished as the cell moves on. Various pieces of evidence point to the leading edge having adhesive properties which differ from those of the remainder of the cell surface: it is to this area that particles such as erythrocytes adhere best, for example, and the ventral surface of most of the cell is not adherent to the substratum; but the source of the difference in adhesive properties is unclear. Two main hypotheses have been advanced to explain the temporarily enhanced adhesiveness of the newly protruded leading edge, and also to

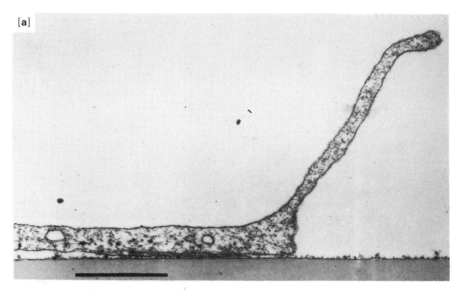

[a]

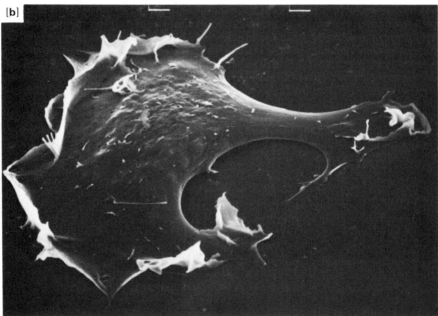

[b]

**Figure 6.15  Ultrastructure of a ruffle**  (a) A transmission micrograph of a vertical section through the leading edge of a chick heart fibroblast. The cell is in close contact with the substratum, and the cytoplasm is noticeably electron dense adjacent to the contact site, the probable site of insertion of a microfilament bundle (which is not conspicuous in this section). The rearing projection seems to contain little except a rather amorphous fibrillar matrix: almost certainly an actin gel; bar represents 1.0 μm (from Abercrombie *et al.* 1971). (b) A scanning micrograph to show ruffles on the surface of a human fibroblast; 10 μm between bars (photograph courtesy of Julian Heath, MRC Cell Biophysics Unit, Kings College, London).

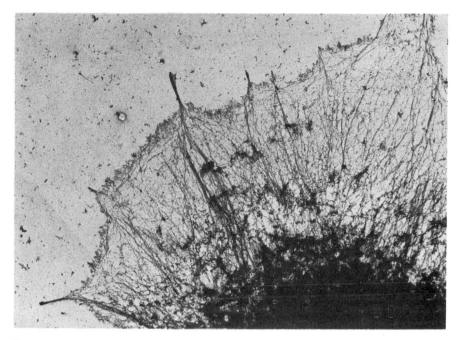

**Figure 6.16   Cytoskeletal architecture in the leading edge**   This high-voltage electron micrograph of a detergent-extracted cell shows the complexity of filaments in the leading edge. This cell, a plasmatocyte from the butterfly *Aglais urticae*, has microspikes, supported by bundled microfilaments, which form the first protrusion (photograph courtesy of T. M. Preston & D. H. Davies, University College, London).

explain the important observation that particles which attach to the dorsal surface of the cell are moved centripetally towards the perinuclear zone (Abercrombie *et al.* 1970, Harris & Dunn 1972). The rearward movement of particles (and of polymeric ligands such as the lectin **concanavalin A**) occurs at a rate which is almost exactly twice that of forward protrusion (75 μm h$^{-1}$ for rearward particle movement on fibroblasts, compared with a forward speed of 36 μm h$^{-1}$), so that a particle moves back relative to a fixed point on the substratum at the same time as the leading lamella is protruded further. The particle may well be an anchorage point which is mobile: instead of the cell moving to the particle, the particle moves towards the cell centre. This observation, it will be noticed, rules out the possibility that the cell *rolls* forward: the upper surface of a caterpillar track moves forward not back.

One hypothesis, which has now fallen rather into disfavour, is that the protrusion of the leading lamella is associated with the insertion of new membrane at the front, and that membrane is internalized in the perinuclear area and recycled through the cytoplasm. A continuous rearward flow of membrane would occur and would sweep particles towards the internalization area, much as patches of cross-linked membrane are formed into a cap. (Fibroblasts, it should be remembered, form caps as a perinuclear ring rather than at the posterior.) This hypothesis, advanced by Abercrombie *et al.* (1972) for fibroblasts, and by Bray (1973) for the nerve-growth-cone, was modified by Bretscher (1976), who proposed that there might be a rearward flow of lipid which would sweep cross-linked membrane proteins backwards

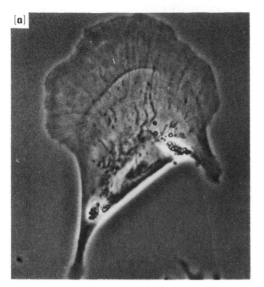

**Figure 6.17 Arcs in the leading edge** The leading edge of a moving fibroblast contains 'arcs' which move rearward, towards the nucleus, as the cell moves forward.

The arcs are shown in phase contrast, (a); in permeabilized cells, (b); using a fluorescence staining technique with anti-actin antibody, (c); and in vertical section by transmission electron microscopy, (d). The arc region clearly contains a dense microfilament array (photographs courtesy of Julian Heath; (b) and (c) are from Heath 1983).

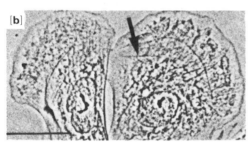

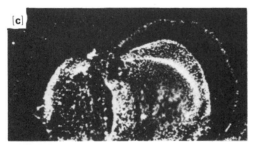

but which would permit maintenance of an isotropic distribution of non-cross-linked integral proteins simply by forward diffusion in the plane of the membrane. Such an idea is difficult to test, but the relative rates of diffusion required of individual and cross-linked membrane components are well within the limits of the observed values and the hypothesis would certainly explain some otherwise puzzling observations. Because the flow is of recycled membrane (Harris 1976), pulse-labelling experiments will not reveal this membrane traffic, although the rearward movement of monovalently-tagged proteins or labelled phospholipids would be convincing evidence in favour of the model. There is some evidence for rearward flow of non-cross-linked membrane components from the experiments of Middleton (1979) with monovalent Fab fragments of antibody directed against the surface antigen Thy-1, and the work of Roberts and Ward (1982) on the amoeboid sperm of *Caenorhabditis elegans* (see § 6.6.4). A further argument which is sometimes advanced is that since the capping process (§ 2.5.1) is insensitive to cytochalasin B and to colchicine when the drugs are used separately, but is inhibited when they are used together, there are two independent mechanisms, one relying upon membrane flow and the other upon a cytochalasin B-sensitive microfilament system. This argument is by no means conclusive.

A second hypothesis, advanced without any supporting evidence (Lackie 1980), is that the assembly of a microfilament meshwork in the leading lamella requires the membrane to flow forward to cover the protrusion, and that only those integral proteins which are not linked to cytoplasmic structures flow with the phospholipid so producing an area of membrane depleted of those proteins which inhibit adhesion. It would be necessary to suppose that some of the mobile proteins serve both as nucleation sites for microfilaments and as adhesion sites, and that once the meshwork begins to contract they move rearwards. Once surrounded by inhibitory proteins they lose their capacity to act as attachment sites and the attachment 'decays' once it moves proximally or is overrun by the cell. Clustered attachment and nucleation sites are more resistant to this decay process, possibly because the inhibitory proteins do not diffuse among them so readily, and so patched areas are moved rearward by the attached microfilaments: a particle is, of course, a multivalent ligand, as is the substratum. This hypothesis has not been disproven – but its predictions have not been tested.

## 6.4.4   Decay of adhesions

If a crawling cell is going to change its position it is obviously essential that the adhesions to which the motor system is anchored are transient, or else the cell will merely stretch out along the path it is attempting to follow. The rate at which adhesion anchorages are relinquished may have a number of important consequences for the behaviour of the cell. If isometric tension in the contractile system is maintained because the adhesions do not decay, then there may be a reorganization of the microfilaments to form a distinct bundle and a focal adhesion by lateral slippage of the adhesion sites. We might also predict that if disassembly of the microfilaments requires decay of the adhesion, or if bundles persist longer under isometric tension, then, as bundles form, the supply of sub-units becomes restricted and forward protrusion at the leading edge slows down. Thus cells with well developed stress fibres and focal adhesions are likely to be the least motile, and this does seem to be the case (Kolega *et al.* 1982). Since there appears to be an inverse correlation between the duration of adhesive interactions and the rate of movement, the control of the decay of adhesions is clearly of considerable interest, although we know very little about the mechanisms involved An obvious possibility, that proteolytic degradation of the adhesions is important, does not seem to be the case (Forrester *et al.* 1983). The stability of focal adhesions seems to depend upon lateral stabilization of the microfilament bundle, rather than the converse, and the protein involved may be vinculin. Vinculin is one of the cytoplasmic targets of the protein kinase that phosphorylates a tyrosine residue, a kinase first recognized as the product of the avian Rous sarcoma virus *src* gene and now known to be present in the non-transformed cell. Phosphorylation of vinculin is associated with loss of focal adhesions (Rohrschneider *et al.* 1982, Sefton *et al.* 1982) and of microfilament bundles, and the effect can be mimicked by the intra-cytoplasmic injection of antivinculin antibodies (Birchmeier *et al.* 1982). One problem with this attractive hypothesis is, however, that only a small proportion of the vinculin becomes phosphorylated, and it is not clear whether we are dealing with a primary cause or a secondary consequence of the altered microfilament distribution. The implications of the altered morphology of motile machinery in cells that do not form focal adhesions will be discussed in more detail in Chapter 10.

## 6.4.5   *Summary of fibroblast movement*

If, for a fibroblast moving over a rigid substratum, we arbitrarily consider a single step forward (arbitrarily because the process is really continuous), the sequence of events is as follows:

(1)   protrusion of the leading edge by assembly of actin meshwork;
(2)   formation of a new distal adhesion by virtue of the adhesive properties of the newly protruded process;
(3)   contraction of the meshwork by the interaction of microfilaments attached to integral membrane proteins with microfilaments linked proximally;
(4)   forward movement of the cell centre as the microfilament array contracts;
(5)   decay of the adhesion as it becomes centrally (proximally) located, the decay of the adhesion being associated with disassembly of the contractile system, providing recycled sub-units for protrusion.

From what we know of fibroblast and amoeboid locomotion we can build a composite model of the coupling of the actomyosin motor system to forward movement of a cell. An important part of the system is that the machinery can be rearranged on a relatively short time-scale, the components of a microfilament bundle or a meshwork are available, at a later stage, for the assembly of a new bundle attached to a more distal site or new meshwork for forward protrusion. Even though we can see ways in which the system might be controlled there is, as yet, no detailed information about the biochemistry of the control process. Some of the behavioural studies, which will be discussed in later chapters, give clues as to the controls required and there is no shortage of possible control systems; it remains to be seen which of the possible controls actually operate. It is not impossible that overall control can be exercised by several systems, and that cells are influenced by one control under one circumstance and another control when conditions change. Other reviews on fibroblast movement are to be found in Middleton and Sharp (1984) and in Bellairs *et al.* (1982).

## 6.5   Fibroblast spreading

The tendency of fibroblasts to flatten on the artificial substrata used in tissue culture depends upon the locomotory machinery of the cell. The transition from spherical to flattened occurs when cells are newly plated, and also when cytokinesis is complete. Freshly subcultured cells which have adopted a spherical shape in suspension are poor specimens for studying the flattening process because their surfaces have been modified in the suspension process, often by proteolytic insult, and the rate of renewal of normal surface characteristics is probably limiting. Most cells round up during mitosis, however, and the flattening of the daughter cells following cytokinesis is rapid, possibly facilitated by the availability of sub-units which are disassembled from the contractile ring.

That fibroblasts are actively pulled into a flat form is clear both from experiments in which some of the cell's adhesions are artificially dislodged, which leads to an elastic recoil of part of the cell, and also from culture systems in which the spreading fibroblasts can be seen to be exerting tension on the substratum. The fibroblast seems almost to be pegged out like an elastic sheet, and a cell such as that shown in Figure 6.18 exhibits a concave margin between apices with focal adhesions to the substratum at the points. Since adhesions used for gaining traction are transient, holding the cell into a flattened form must require continual replacement of the adhesions, and the strength of the attachment must never fall below that minimum required to resist the tension in the cell. Very brief adhesions would require more 'effort' by the cell in replacing the adhesion molecules (assuming that such molecules are involved) and the longer the adhesion lasts the lower the metabolic cost. Thus we might predict that cells which flatten extensively would have rather long lived adhesions, which in turn would lead to well-organized bundles of microfilaments or stress-fibres, and that these cells would be rather slow moving. On the whole this expectation is borne out although, since there is no good assay for the duration of individual adhesions, the evidence must be considered rather circumstantial at present. Proteins such as fibronectin, which enhance the spreading of fibroblasts, may therefore act to stabilize adhesions: an untested prediction would be that the kinetics of spreading on fibronectin would be different from those on other proteins because of the altered rate-constant of adhesion turnover. Now that

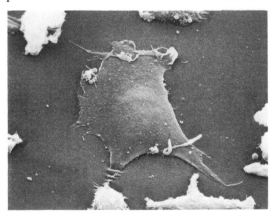

**Figure 6.18 A well-spread fibroblast** A scanning micro-graph of a well-spread BHK cell, and some rounded cells nearby. A portion of this cell is shown in greater detail in Figure 7.9b (photograph courtesy of Mike Lydon, Unilever Research). (See also Fig. 6.15b.)

detailed kinetic studies on spreading are being carried out (Bardsley & Aplin 1983a, b), this may be tested.

An interesting side issue relating to spreading behaviour is the interaction between proliferation rate and the projected area of cells. The experiments of Folkman and Moscona (1978), who altered the extent of spreading by varying the properties of the substratum, suggest very strongly that under conditions where growth factors are in restricted supply cells must spread beyond a certain minimum area if they are to enter S-phase and later divide. When growth factors are supplied in greater concentration, as in the experiments done by O'Neill *et al.* (1979) using serum concentrations of up to 60 per cent v/v, the necessity for spreading is lost, suggesting that the spreading behaviour is in some way linked to growth factor uptake. Since occupancy of the epidermal growth factor (EGF) receptor stimulates phosphorylation of vinculin, the bundle-stabilizing protein mentioned in the previous section, a biochemical basis for this feedback seems to exist. Notice that the argument here is that failure to stabilize the bundle–focal adhesion complex leads to a rounded morphology, and that this could plausibly be achieved by altering the duration of adhesions. Although the argument is very speculative it provides a neat connection between morphology and proliferative capacity, a connection that students of transformed cells have long desired.

## 6.6 Movement of other cell types

Studies on giant amoebae and tissue fibroblasts have provided much of the evidence for the mechanism of crawling movement and the 'amoeboid' and 'fibroblastic' modes of locomotion have become more obviously similar as the details have emerged. In general it is probably safe to assume that most cells crawl over a substratum in more-or-less the same way, using the actomyosin motor system to generate contractile forces and the actin gel-assembly process for protrusion of the front of the cell. Nevertheless there are some differences which will be mentioned below and at least one example of a crawling movement which does not come about in the same way.

### 6.6.1  *Neutrophil leucocytes*

Polymorphonuclear leucocytes, particularly the **neutrophil granulocytes** of mammals, are primary phagocytic cells which will move into tissues in response to inflammatory stimuli (Wilkinson & Lackie 1979, Lackie 1982; reviewed extensively by Wilkinson 1982). They move rapidly $(10–15 \ \mu m \ min^{-1})$ and are possibly the fastest-moving mammalian cells. Their pattern of movement has often been described as 'amoeboid' and they do not have the clearly differentiated leading lamella which characterizes the

**Figure 6.19  Tail fibres of moving neutrophils**  Neutrophils, when moving over rather ▶ adhesive surfaces, often have a complex tail which presumably represents old adhesions which are being slow to detach. The cytoplasm within these thin processes has bundled microfilaments, which are not found elsewhere in the cell. (a) Scanning micrograph of a neutrophil which is moving from right to left, dragging its tail (arrowed) behind it (from Armstrong & Lackie 1975). (b) and (c) Interference reflection and phase contrast pictures of neutrophils moving (rather poorly) over a surface; the branched tails can be seen, each branch ending presumably in an adhesion. Using the interference reflection system, the dark areas are regions which are in close contact with the substratum and are presumably areas of adhesion.

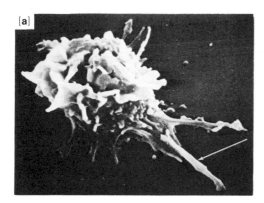

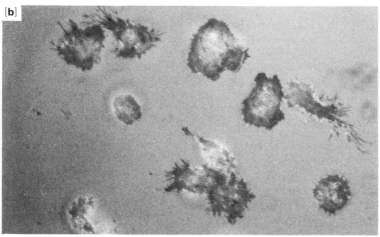

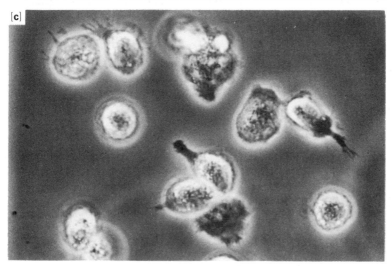

fibroblast. They contain, however, the same actomyosin machinery as fibroblasts; the differences lie mostly in the organization which is much more labile. Protrusions at the front of the cell are small, and microfilament bundles do not often form except perhaps in the tail region (Fig. 6.19) (Wilkinson & Lackie 1979). Contact with the substratum is much more diffuse when viewed with interference reflection microscopy (Armstrong & Lackie 1975), and the only examples of focal adhesions may be small contact areas at the ends of tail fibres. The tail 'fibres' seem to restrain forward movement and are seen most clearly when the cells are moving on a rather adhesive substratum. Tension in these fibres is maintained for longer than in the frontal regions and this may account for the development of bundled microfilaments associated with more distinctly focal adhesion sites.

The pattern of movement of neutrophil leucocytes could be accounted for if their adhesive contacts with the substratum were generally brief and if the bundle-stabilizing proteins were not 'active' or were expressed only poorly. The rapidity of movement implies that the half-life of contacts must be shorter than in a slow-moving cell, and the rapid turnover of attachments would give the cells a more rounded morphology (Lackie & Wilkinson 1984). In some respects the neutrophil leucocyte resembles malignant transformants of fibroblasts, and the invasive properties of leucocytes and of malignant cells may be associated with the organization of their motile machinery. Other leucocytes do move, particularly monocytes, which leave the bloodstream and move into tissues where they differentiate into tissue **macrophages**; and lymphocytes, which move through the tissues of the lymph node during the process of recirculation.

One interesting aspect of the locomotory behaviour exhibited by neutrophils which has not been fully explained is the formation of 'constriction rings' in the cell body which appear to move rearward whether the cell is in suspension, in a deformable matrix, or on a planar substratum (Keller 1983, Haston & Shields 1984, Shields & Haston 1985). Presumably these constriction rings represent a locally contracted region of cortical microfilament meshwork and they may be important when the cell is moving through the interstices of a fibrillar extracellular matrix, such as connective tissue. The surprising feature is that the constriction rings are produced when the cell is in suspension: the motor appears to be working even when the transmission system is disengaged.

## 6.6.2   The nerve growth cone

Newly differentiated nerve cells are approximately spherical: the formation of the long cellular processes known as axons takes time and depends upon the movement of the distal end of the axon, the nerve growth cone (Fig. 6.20), coordinated with the growth of the axon. The nerve growth cone apparently acts as a sort of tractor, unreeling the growing axon while the cell body remains stationary. In many ways it appears that the nerve growth cone behaves like a fibroblast in its locomotion (Bray 1973, 1982). Two aspects of the[?] movement are of interest: the mechanism of movement and the mechanism by which the direction of movement is controlled. The latter is of particular interest in morphogenesis because the pattern of innervation of a tissue depends upon the final position of nerve growth cones, and in the mapping of retinal neurons onto the optic tectum of the brain, for example, the problem is of great complexity. It is with the former problem, the mechanism of movement, that we are concerned here, the direction finding problem will be discussed in Chapter 8.

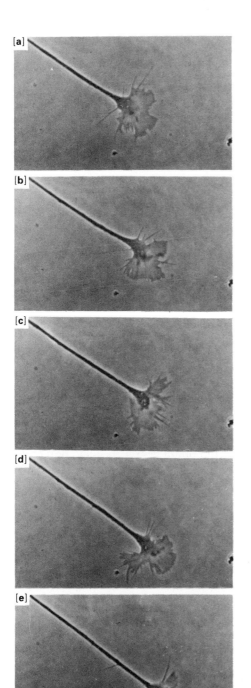

**Figure 6.20 Movement of a nerve growth cone** A series of photographs, taken at two minute intervals, of a nerve growth cone. Notice the microspikes, particularly in (c), which are obliterated as the leading edge spreads further, (d). The pictures have been aligned, using stationary particles as markers, to show the forward progress of the cone. The growth cone is approximately 15 μm wide (photographs courtesy of Dennis Bray, MRC Cell Biophysics Unit).

The growth cone itself has a leading lamella and moves at a rate of up to 50 µm h$^{-1}$. Attachment sites are formed in the front and decay posteriorly; the axon itself is not attached to the substratum and is therefore straight *in vitro* because of the tension generated by the moving tip. Posterior transport of particles on the dorsal surface of the nerve growth cone has been observed, as in fibroblasts, and particles accumulate near the junction of growth cone and axon. Forward protrusion is by microspike formation, the web between the anteriorly directed spikes being filled as forward movement proceeds, and protrusion is sensitive to cytochalasin.

The growth cone may be considered as an invasive portion of the neuron: innervation often occurs relatively late in tissue morphogenesis and it is therefore important that the tip should have the ability to move through territory which is already occupied by other cells. The ability of the cone to move over other cells may be partly due to the low shear-stress imposed upon the substratum by the diffuse attachment area, and certainly the cone does not exhibit **contact inhibition** of locomotion when it contacts a fibroblast (see Ch. 10).

### 6.6.3 Epithelial movement

Epithelial cells, unlike fibroblasts and leucocytes, do not normally operate individually but rather as part of a coherent population composing a sheet (Kolega 1981). Since epithelia cover or surround tissues, it might be predicted that their locomotory behaviour would tend to spread the cell population over the maximum area. This seems to be the case, only cells at the margin of the sheet are actively motile and the attachment of the sheet to a substratum seems to be mostly marginal, although some cells in the second row back from the edge and some central cells may also form attachments. Protrusive activity occurs at the margin and appears to be basically similar to protrusion in the systems already discussed (Middleton 1982). Epithelia seem to form hemispherical bulges at the leading edge rather than flat ruffled sheets or microspikes. The mechanical integrity of the epithelial sheet depends upon specialized cell junctions, particularly desmosomes (*maculae adhaerens*) which link the intermediate filaments, which in epithelia are of **cytokeratin**, but also microfilament-associated junctions (*zonulae adhaerens* or **adhaerens-type junctions**). Rather unfortunately, the majority of the work done on epithelial movement has been with pigmented retinal epithelium, a very atypical epithelium which is phagocytic and has neither cytokeratins nor desmosomes.

We will have occasion to refer back to epithelial cells in the context of locomotory polarity (Ch. 7) and cell–cell contact behaviour (Ch. 10).

### 6.6.4 An exception: locomotion of the spermatozoa of C. elegans

Not all spermatozoa have flagella and those of the nematode, *Caenorhabditis elegans* move in an amoeboid fashion (Fig. 6.21). The major peculiarity is, however, the virtual absence of actin in the sperm. It seems difficult to visualize an effective actomyosin motor system when less than 0.02 per cent of the total protein of the cell is actin, and the absence of tubulin renders the other major alternative motor-system equally unlikely. Various features of the sperm, including immotile mutants, make them an excellent experimental subject and there seems to be good evidence for a rearward flow of non-cross-linked membrane components in the frontal zone (Roberts & Ward 1982).

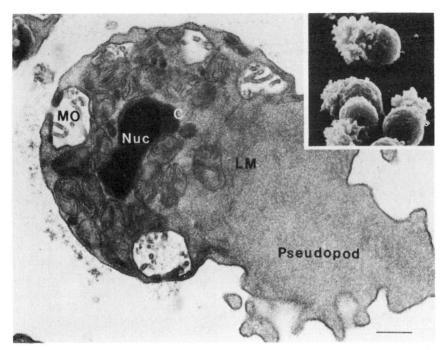

**Figure 6.21  The structure of the amoeboid sperm of *C. elegans***  The large picture shows the ultrastructure of a mature spermatozoon fixed in glutaraldehyde and tannic acid and post-fixed in osmium. The pseudopod contains neither microfilaments nor microtubules; the bar is 0.5 μm long. The inset shows four spermatozoa with extended pseudopodia (from Ward *et al.* 1981)

Centripetal membrane flow is restricted to the anteriorly-directed pseudopod and there is insertion of membrane at the pseudopod tip. Membrane flow is perturbed in immotile mutants. This suggests quite strongly that models of crawling movement that depend upon membrane flow cannot be neglected and should provide a stimulus for the re-evaluation of our theories of movement.

## 6.7  Summary

In this chapter we have looked at the problems faced by a cell crawling over a substratum, and by drawing on observations from several cell types it has been possible to provide a fairly detailed explanation of the way in which movement is brought about. Even though it is now possible to visualize how the operation of microfilament assembly and contraction systems bring about movement there are still areas of considerable ignorance, particularly concerning the adhesion anchorages. Neither the molecular structure nor the control of the lifespan of these attachments is understood at present. Having now described the motor(s) and the way in which motor systems are coupled to forward displacement, we will turn to problems concerning the 'traffic rules', the ways in which cells respond to environmental cues.

# References

Middleton, C. A. and J. A. Sharp 1984. *Cell locomotion in vitro: techniques and observations*. London: Croom Helm.

Bellairs, R., A. Curtis and G. Dunn (eds) 1982. *Cell Behaviour* (memorial volume for Michael Abercrombie). Cambridge: CUP. Currently *the* source of reviews in this area. (Abercrombie's Croonian Lecture is also reprinted in this volume.)

Abercrombie, M. 1980. The crawling movement of metazoan cells. (The Croonian Lecture 1978.) *Proc. R. Soc. Lond.* **B 207**, 129–47. Well worth reading both for content and style.

Curtis, A. S. G. and J. D. Pitts (eds) 1980. *Cell Adhesion and Motility*. Cambridge: CUP. (Symp. Brit. Soc. Cell Biol. **3**). More on adhesion than on motility, but a useful collection of reviews.

Porter, R. and D. W. Fitzsimons (eds) 1973. *Locomotion of Tissue Cells*. (Ciba Foundation Symp. **14**, New Series.) Amsterdam: Elsevier.

# 7

# MOVING IN A UNIFORM ENVIRONMENT

## 7.1   Introduction

Having described various motor systems and the ways in which they can be used to drive cells around, we must now consider factors which control and regulate movement, for if self-propelled cells are not controlled by rules of behaviour then biological systems would be worse than large cities in the rush hour. Our vehicular analogy 'asks' for a driver and a set of traffic rules, and the remaining chapters attempt to describe these aspects of cell behaviour.

The simplest form of movement is that of molecules in liquid or gaseous phases, and the behaviour of such systems is rigorously described by diffusion laws. Particles moving in a random fashion collide and change their direction of motion: they follow a path best described as a random walk in which movement is as likely to be in one direction as in any other.

In an infinitely large volume a single particle will continue to move in a straight line unless acted upon by some external force (the basis of Newton's second law of motion) but as the number of particles increases collisions lead to changes of direction. Once the packing reaches a critical level collision follows collision and, although the direction of movement is constantly altering, the particle remains in place – the formation of a solid phase. Of course there is slightly more to the story, for example there are interactions between particles which restrict the escape probability at a boundary, but given enough kinetic energy all solids will change to liquid or gas.

We can use the insights of physics in considering the simplest form of behaviour, the movement of cells in uniform environments, and can ask how the properties of the environment influence the behaviour of cells. Under this heading we can consider the paths taken by cells in such isotropic environments, the influence of external factors (such as adhesion to non-cellular components or viscosity of the medium), and internal factors (which may derive from the design of the motor system). The effect of interactions with other cells, the formation of solid tissues and the escape of cells from such tissues, will be left until the final chapter.

## 7.2   Random walks and internal bias

It is considerably easier to consider and illustrate the problem of random movement if we restrict the environment to two dimensions. Fortunately this restriction is not unrealistic because most tissue cells require a solid substratum upon which to move, and much of the work on this topic has been with cells moving on isotropic planar substrata, protein-coated coverslips in most cases. Unlike the molecules which physical scientists describe, cells have their own source of kinetic energy, the ability to move, and they can turn without collisions being necessary.

Cells have finite speed, there is a limit to the rate at which the motor can run; if they are to achieve a net displacement then there must also be some restriction on the frequency or extent of turning. This is clearer if we take an example, based upon that given by Dunn (1981). A cell which moves at $10~\mu m~min^{-1}$ and makes a turn every minute might follow a path such as that shown in Figure 7.1a. If turns are made every 30 s then the path might be as in Figure 7.1b, and as the interval between turns decreases the path degenerates into that shown in 7.1c where the cell is 'jittering' on the spot, as would a molecule in a solid. This pattern of change can also be achieved if we keep the turning frequency constant, but progressively decrease the

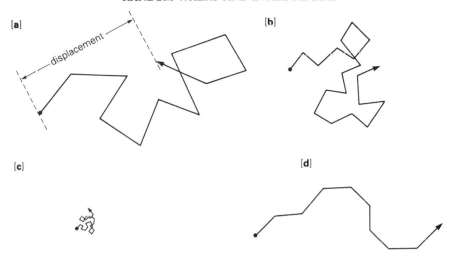

**Figure 7.1  Random walks**  In a simple random walk turns occur at random intervals and the probability of turning through a large angle is the same as the probability of turning through a very small angle. In (a) the cell is making 'steps' of around ten units and turning fairly randomly. If the turning frequency increases, or if the speed drops, then the path might look more like that shown in (b) where the steps are on average five units long. If the steps become very short, either because the frequency of turning increases or because the speed is very low, then the overall displacement is very small, (c). If, however, turns never exceed ± 45°, as in (d), then a rather different path is seen in the short term; in the long term it will not be dissimilar to (a). (Note that the use of 'steps' to describe a sinuous path has dangers: see Fig. 7.3.)

speed. In most cases we are interested in net displacement of the cell, the distance the cell centre has moved after a period of time, and it is clear that if the cell is to achieve a significant displacement in a realistic time then the frequency of turning must be restricted (or the speed must be infinitely high). By 'a significant displacement' we mean a displacement which is some integer multiple of the cell's dimensions, and a realistic time might be the time between mitotic divisions: we must use a time scale which is biological and not geological.

An alternative to restricting the frequency of turning is to restrict the angle through which turns are made. If the maximum turn is 90°, with an equal probability of making a turn of any angle between +90° and −90°, then the path looks very different, as in Figure 7.1d, and the net displacement is considerably greater. As the maximum permitted angle of turn decreases, the path becomes straighter and the displacement increases to a limit set by the speed of the cell. Another way of affecting the path would be to impose restrictions on the direction of the turns, so that the cell tends to turn in the chosen direction. This would require that the cell 'knew' where it was trying to go and, since we do not believe cells to be sentient, requires that there is an external cue or bias: the environment would have to be anisotropic, and behaviour in anisotropic environments will form the subject matter of later chapters.

Since cells do make significant net displacements in uniform environments but have an upper limit to the speed at which they can move, there must clearly be some constraint upon turning behaviour which has its source within the cell. From observation we know that cells tend to be polarized to some extent, the active protrusions of the leading lamella of a fibroblast, for example, are restricted to a small number of sites, and a moving neutrophil or amoeba has a 'front' and 'back' which can be recognized on morphological grounds alone. The implication is that the process which gives rise to a protrusion is restricted to a particular area of cytoplasm and, recognizing this, it becomes obvious that the turning behaviour is likely to be restricted by the necessity of rearranging the motile machinery in order to make a turn. If forward protrusion requires the delivery of sub-units for assembly into actin gel (see Ch. 6) then it will be easier to maintain movement if there is a well defined site for delivery, so that constant re-routing is unnecessary. Thus we might expect, from a consideration of the motile machinery itself, that the turning behaviour would be limited by internal constraints and that, in the short term, the path of a moving cell would show the phenomenon of **persistence**. An objective way of demonstrating this persistence is to look at the displacement of cells as a function of time: we need to obtain a measure of the time-base upon which the reorganization of the machinery takes place. For a small particle behaving in classical way, the displacement changes with time in a predictable fashion, the mean square displacement increases linearly with time, i.e.

$$<\mathbf{d}^2> = 4Dt$$

where $\mathbf{d}$ is displacement in unit time; $D$ is the diffusion coefficient, a measure, effectively, of population mobility; and $t$ is time.

If we then apply this equation to the displacement of cells we find that at 'short' times the displacement deviates from linearity because of the internal bias in turning behaviour, as shown in Figure. 7.2. This method was used by Gail and Boone (1972) to determine the persistence behaviour of 3T3 cells, and a derivation of the method has been used by Dunn and others to investigate the movement of chick heart fibroblasts and neutrophil leucocytes (Dunn 1983, Allan & Wilkinson 1978).

From these sorts of studies it is possible to obtain a measure of the persistence and an accurate measurement of the magnitude of the instantaneous velocity of a cell. The latter measure is to eliminate a problem in the method by which cell movement is usually recorded, the analysis of time-lapse film in which the position of the cell centre is marked at fixed time intervals. Because the cell follows a tortuous path the distance actually travelled by the cell will be greater than the sum of the distances between successive marked positions (see Fig. 7.3) and an estimate of speed based on the track length will underestimate the true speed over the real path.

These analyses suggest that the path followed by a cell in a uniform environment can be described by an equation which treats the path as a random walk but takes cognizance of the tendency of the cell to persist over short distances: it is then possible to describe the behaviour in terms of two parameters, the speed ($S$) and the persistence time ($P$). A practical way of estimating these parameters is shown in Figure 7.4, and has been used by Wilkinson et al. (1984).

Another aspect of behaviour in an isotropic environment is the behaviour of the cell population. If the environment is truly devoid of directional cues then there should only be a random and rather small **displacement** of the population, the random displacements of individual cells lead to exchange of position but, at equilibrium, a random distribution of cells. The magnitude of the probable displacement of the centre of the population can be predicted by

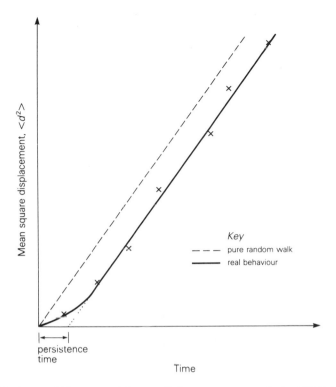

**Figure 7.2 Internal bias – persistence** If the mean square displacement of a random walker is plotted against time, the line should pass through the origin (broken line). For cells, however, the line shows a characteristic deviation at short times because of the tendency for the cell to persist. The intercept on the x-axis is a measure of the persistence time. The method was first applied by Gail and Boone (1972).

treating the individual displacements of the cells as steps in a random walk (see Dunn 1981). If the cells are not randomly distributed then the environment is non-uniform, but provided the only anisotropy is in **distribution** then the diffusion of cells will lead to a random distribution in the same way as gas molecules will distribute themselves through a container. The diffusion coefficient for the cell population can be derived from the speed and persistence parameters:

$$R = 2\ S^2P$$

where $R$ = population diffusion coefficient, $S$ = speed, $P$ = persistence time. (If $S$ is in $\mu$m and $P$ is in seconds, then $R$ is in $\mu m^2\ s^{-1}$.) It is, of course, necessary to consider the boundaries of the area or volume occupied and to make the assumption that collisions with the wall have no effect upon the subsequent behaviour. Once we are dealing with populations of cells then the distribution becomes important and the cell behaviourist can draw upon analytical methods from other disciplines for predicting or measuring distribution. Perhaps the only important point to bear in mind is that a random distribution is not a uniform distribution: cells will not be evenly spaced.

179

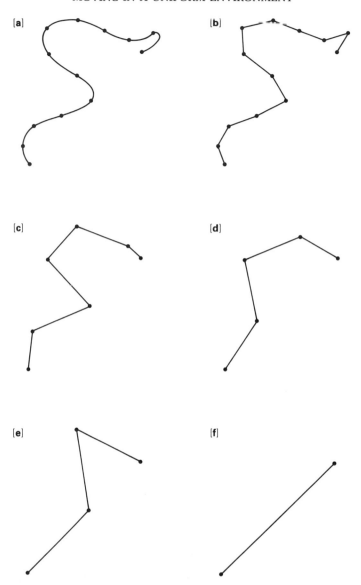

**Figure 7.3  Steps and curved paths**  Depending upon the time-lapse between recording the successive positions of a cell which is moving along a smooth path, the distance travelled by the cell, and therefore the speed of the cell, will be underestimated. The measurement can be improved by choosing a shorter time-interval, but the effort required increases too.

Probably the most important aspect of this analysis of the random movement of cells is that their behaviour can be described by two parameters, their speed and their characteristic persistence time, which can be combined to give a measure of the rate at which the population will expand to occupy a territory, the population diffusion coefficient. This means

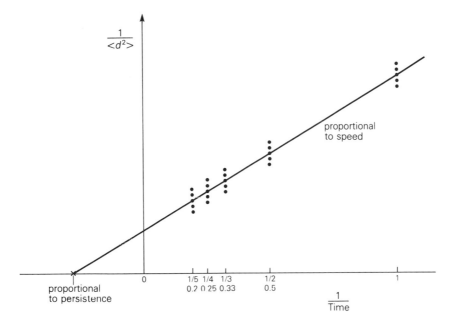

**Figure 7.4 Graphical estimation of limit speed and persistence** An escape from the dilemma posed in Figure 7.3 (underestimating cell speed as a result of choosing too long an interval between plotted positions) is to plot the reciprocal of mean square displacement against the reciprocal of time. The gradient of the line gives the speed. Essentially we are determining what happens to speed as shorter steps are used to approximate the path, by considering the effect of using multiples of the shortest steps actually used. The intercept on the x-axis is related to the persistence behaviour – the line would go through the origin if there were no persistence.

that in considering the behaviour of cells in uniform environments we need to ask how these two parameters are affected, and in terms of understanding the motile machinery we need to consider these two features of behaviour. Shifting from two dimensions to three does not affect any of the arguments about the parameters we need to measure.

## 7.2.1 Speed

One limit to the speed at which a cell can crawl is undoubtedly the time course of the myosin cross-bridge cycle, but for most motile cells this is as likely to be limiting as the speed of the petrol–air explosion in the cylinder of an internal combustion engine is likely to limit the speed of a car. More realistically, the limits on speed are probably set by the assembly constants for the reorganization of the machinery, the rate of forward protrusion and the decay characteristics of posterior attachments. No doubt the relative importance of these different limitations will vary from one environment to another. Some motors, particularly the highly organized ones such as striated muscle and cilia, probably operate at only one speed under normal conditions. For the more complex systems where transient assembly and linkage are involved it is not clear which is the rate-limiting factor or indeed whether there is a single limiting factor. At present, therefore, there is little to say about speed control except to notice that there does seem to be an upper limit for all motile cells. Free-swimming cells will probably be limited by viscosity more than by any other constraint, and crawling cells may be limited by the adhesion–detachment cycle. Variation in the speed of a particular cell type is frequently observed but the source of the variation remains obscure. For a crawling cell such as a neutrophil there are two main possibilities: that the motor is operated at different rates or that the motor runs at a constant rate but that the efficiency of coupling to forward movement is varied. In other words we do not know whether cells use an accelerator, a gearbox or variable-slip wheels.

Factors which alter the speed of movement are referred to by some authors as **orthokinetic** factors and, since experimentally it is easier to alter the fluid phase over cells which move on a solid substratum or in which cells are swimming, the tendency has been to describe **chemokinetic** factors, although the mechanism by which the variation in speed is brought about is made no clearer by this usage. A problem in many studies is to determine whether the soluble factor which has been added changes the speed by altering the substratum, by binding to specific receptors and triggering changes within the cell, or even by masking charged groups on the surface, thus altering the interaction of the cell with the substratum.

## 7.2.2  Persistence, turning and polarity

In some cases it is possible to define a turning event in unambiguous terms. The reversal of ciliary beat and the reversal of the direction of turning of a bacterial flagellum are both discrete and observable phenomena: for a crawling cell defining a turn is a much more subjective business and obtaining an objective measurement of turning behaviour has been far more difficult. In the case of bacteria the shift from anticlockwise (counter-clockwise) to clockwise rotation of the flagellum causes a marked change in the pattern of movement – anticlockwise rotation gives a relatively straight path or 'run' whereas clockwise rotation induces a 'tumble', the bacterium spins on its axis and sets off again in a random direction when anticlockwise rotation resumes. Control of reversal frequency in bacterial flagellar rotation is the key to their 'directed' locomotory behaviour and is a good example of a **klinokinesis**, changes in movement pattern being achieved by altering the frequency of turning behaviour. Confusingly, the phenomenon is usually discussed as a bacterial **chemotaxis**, and their altered turning behaviour in gradients of diffusible substances will be described more fully in Chapter 9.

Reversal of ciliary beat in *Paramoecium* causes the cell to shift from forward to reverse movement and, since the cell tends to reverse somewhat erratically, when forward movement recommences it is likely to be in a different direction. This is the basis of the (so-called) 'avoidance reaction' and, again, is usually discussed as a component of a directional (**tactic**) response. Flagellates can alter their direction of movement by altering the flagellar wave pattern and *Chlamydomonas*, which normally swims with the two flagella directed anteriorly, can switch into reverse as an avoidance mechanism (see Ch. 5).

It is when we consider the turning behaviour of crawling cells that we encounter significant difficulty. The path followed by a cell is generally sinuous and we cannot define turns without introducing a subjective element. Very small changes in direction, of a cell moving in a path with a large radius of curvature for example, may or may not be considered as involving turns depending upon the resolution of our angle measurement and, in practice, upon the time interval between marking the positions of the cell centre. Objective methods, hinted at in a previous section, have only rarely been applied to the problem. Two different phenomena may be involved in the turning behaviour of a crawling cell such as a fibroblast: one is the gradual shift in dominance between the competitive 'fronts' and the other is a more subtle change in the organization of a single frontal region. Fibroblasts are generally drawn as, and frequently actually are, bipolar cells in which one region is more active than the other: the front or leading lamella competes with a smaller region of ruffling activity at the rear. Immediately after cytokinesis a fibroblast is approximately spherical (although the single MTOC of the pericentriolar region does define a possible axis), but once moving there must, by definition, be a front and a back. There is some evidence that the bipolar organization depends upon the integrity of the microtubular system; treating flattened cells with colchicine apparently abolishes this polarity and active protrusion begins to occur over the whole perimeter (Vasiliev *et al.* 1970). The inability to maintain polarity leads to a cessation of displacement, the dominance of the leading edge is lost and the cell attempts to move in all directions at once. It is, however, also worth noticing that the differentiation of the cytoplasm of a fibroblast into central (granule-rich) and peripheral (granule-poor, flattened) lamellar regions is lost upon colchicine treatment.

Under some circumstances dominance may be exchanged between front and rear of a normal fibroblast and the cell begins to move backwards; if the two competing locomotor regions remain equally matched then the cell will adopt a markedly bipolar shape.

Epithelial cells seem to lack the internal polarity determination shown by fibroblasts, and when isolated an epithelial cell behaves rather like a colchicine-treated fibroblast. Isolated individual fragments of epithelial cells will, however, demonstrate some polarity, in that they move in a single direction, yet these fragments have no demonstrable microtubules (Euteneuer & Schliwa 1984). The differences between fibroblasts and epithelial cells may well reflect the difference in social behaviour, an isolated epithelial cell is a laboratory curiosity and *in vivo* epithelial cells have their polarity defined by their interaction with the other epithelial cells of the sheet (Kolega 1981). The active cells of the margin have a front which is defined in that it is the only free margin.

The cell type in which polarity is most marked is probably the neutrophil leucocyte. In order to change direction a neutrophil does not, generally, engage a reverse gear but moves around in a tight turn, one side of the leading edge becomes more active than the other. Under certain circumstances neutrophils may develop a new 'front' in an area previously quiescent but this is exceptional (Gerisch & Keller 1981, Zigmond *et al.* 1981). Even though the polarity is strongly determined, the cell does respond to microtubule disruption, not by losing polarity completely but by increasing the magnitude or frequency of turns. The path becomes more tortuous although the speed remains unaltered. An important implication is that the motor and steering systems are distinct, the latter apparently microtubule associated. Another observation that tends to support this view and that gives us another piece of information is the work of Englander and Malech (1981) on the movement of neutrophils from patients with immotile cilia syndrome. This rare condition is associated with a defect in the motile machinery of cilia, which in some cases lack dynein arms on the outer doublets. The ciliary defect leads to male sterility and to recurrent respiratory tract infections because lung clearance is impaired: more interestingly the syndrome is also associated, in approximately half the cases, with dextrocardia (Kartagener's syndrome). This suggests, though by no means proves, that the asymmetry in the disposition of internal organs, which leads to the normal coiling of the intestine and to the left ventricle delivering blood to the systemic circulation, is in some way dynein dependent. One could visualize early morphogenetic movements which might be involved in the generation of the asymmetry of organs and which could be affected by the loss of internal microtubule-derived polarity. Neutrophils from these patients move at normal speed but their tracks are far more tortuous, much as with colchicine-treated neutrophils, and the turning behaviour seems to be altered. Englander and Malech estimated turning by scoring the frequency of recrossing of the path, a simple objective measure. If our understanding of the syndrome is correct then the implication is that the microtubule-associated polarity of movement depends upon an interaction between dynein and cytoplasmic microtubules, possibly in the delivery of material to the anterior part of the moving cell, although it should be stressed that this is only a speculation and we do not understand the basis of polarity determination. The position of the pericentriolar MTOC and the nucleus, and the relationship between the axis defined by these two and the direction of movement is disputed; the clear association which was thought to exist may have broken down (Anderson *et al.* 1982, Schliwa *et al.* 1982).

A change in the turning behaviour, whether an alteration in frequency of turning (a decrease in persistence time) or the extent of turning, would be a klinokinetic effect in the terminology sometimes used. For crawling cells there are no clear examples of such klinokinetic effects, apart from the colchicine-induced perturbation, and most alterations in movement pattern come from changes in speed, orthokinetic effects. Since the population diffusion coefficient is much more sensitive to a change in speed (since $R = 2S^2P$) this may not be surprising.

## 7.3 Effects of altering the physical properties of the environment

In order to move forward the cell must exert a rearward force on its surroundings, and it is therefore obvious that the mechanical properties of the environment will influence the rate of movement. Of course there will be situations in which the linkage of the motor to forward movement is not a limiting factor, but this does not invalidate the argument. A real-life example of this would be a comparison of wheeled and tracked vehicles. Both might move at equal speed on a good road surface but when the terrain changes to marshy ground or becomes snow-covered the wheeled vehicle soon stops while the tracked vehicle continues at constant speed. Potentially the wheeled vehicle could perhaps have travelled much faster on the road surface; the limiting factor was the accelerator rather than the traction. For cells propelled by cilia or flagella the only significant factor is likely to be the viscosity of the medium, and the effects of altered viscosity have been discussed in Chapter 5. A viscoelastic medium will make swimming very difficult, if not impossible. Most of the discussion which follows relates to cells which crawl either over a two-dimensional substratum or through a three-dimensional matrix. We will assume throughout that the nature of the environment approximates to physiological: that the temperature, pH, ionic composition and tonicity are such that the cell is viable and its metabolism is functional.

Two major environmental factors must be considered, the adhesive interaction and the rigidity or deformability of the substratum or matrix.

## 7.3.1 Adhesion

For a cell moving over a planar substratum, that is over a surface which has only smooth curves of radii much greater than the dimensions of the cell, then the attachment must be 'adhesive', using the term in a broad sense. Although in principle the interaction could be through a viscous coupling-medium between the cell surface and the substratum, rather than an adhesion in the strict sense, this will operationally seem the same. Before entering into a discussion of the effect of varying the adhesive interaction, we will digress and briefly summarize the mechanisms of cell adhesion. Surprisingly, there is no clear answer to the question of how cells adhere, but there are several hypotheses about the mechanisms which may be involved. The interactions can be subdivided into two main categories: those which depend upon the surface properties, the physicochemical interactions, and those which depend upon more specific interactions of the receptor–ligand type. Reviews of cell adhesion, which may be helpful in expanding this area, are those of Grinnell (1978), Hubbe (1981), Bongrand et al. (1982), Edwards (1983) and Trinkaus (1984). A simple introduction is given by Curtis (1984).

PHYSICOCHEMICAL FORCES IN ADHESION

These theories are mostly based upon the relatively well-characterized interactions of colloidal particles which flocculate when the forces of attraction exceed those of repulsion. The theoretical work on colloid flocculation was done by Derjaguin and Landau, Verwey and Overbeek, and is therefore often referred to as the DLVO theory. The repulsion forces for colloids are electrostatic, the tendency of similarly charged particles to repel one another; similar forces may well operate between cells, which have a negative surface charge by virtue of charged residues on integral membrane components. The cell differs from a colloidal particle in that the surface negativity derives from point charges, but the electrostatic interaction should still occur. The force of attraction in colloids is from electrodynamic forces, mostly the London–Van der Waals dispersion forces, which act between materials of similar composition. Since plasma membranes all have the same basic composition these forces could operate for cells. The relative strengths of the attraction and the repulsion forces fall off with distance but follow different power laws in each case: the balance between the two forces will therefore vary as a function of the separation, and there seem to be two separation zones in which attraction exceeds repulsion (Fig. 7.5). These two separation distances are calculated for cells, on the basis of several assumptions, to be less than 2 nm for adhesion in the primary minimum and around 15–25 nm for adhesion in the secondary minimum; distances which correspond well with the separations seen in tight junctions and in non-specialized apposition of surfaces (Fig. 7.6).

Many of the observed properties of cell–cell adhesions could be accounted for by such theories, the effect of divalent cations in shielding point charges would account for the divalent-cation dependence for example (although not perhaps the differences seen in divalent ion specificity), and there is a considerable amount of evidence to suggest that these interactions may play a part in cell adhesion. The extent to which biological adhesion processes can be explained on the basis of physicochemical theories, and a more detailed exposition of the application of colloid flocculation theories, can be found in Curtis (1967). The metabolic dependence of cell adhesion does not rule out such interactions, although it is necessary to suppose that active redistribution of surface components or their continual replacement is involved, or that

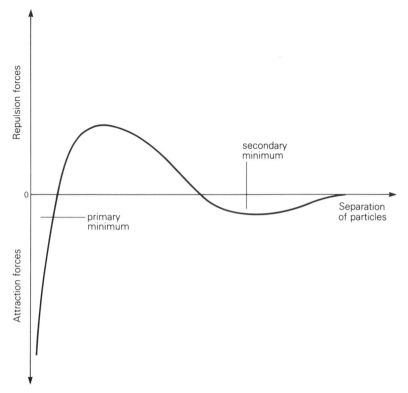

**Figure 7.5 Physicochemical forces of adhesion at different separations** The balance of adhesion and repulsion alters as the separation of two identical charged particles increases. The main interest lies in the observation that there are two 'minima' (where attraction exceeds repulsion), one rather small and therefore likely to be easy to escape, and one very deep. For cells these minima are calculated to occur at separations of 15–25 nm and 1–2 nm respectively. The primary (deep) minimum might give a rather irreversible adhesion, whereas the secondary minimum might only hold the cells together rather weakly.

the local ionic composition in the medium immediately next to the surface is maintained by metabolically-dependent mechanisms. None of these is inherently improbable.

Another physicochemical interaction which may be important is to do with the relative wettability of the surfaces which are in contact. If two surfaces are hydrophobic then the apposition of the surfaces and the exclusion of aqueous medium will be thermodynamically more favourable. If two phospholipid bilayers are allowed to come into intimate contact then they will fuse, and it has often been supposed that one rôle of the hydrophilic carbohydrate residues on integral membrane components is to prevent such fusion from taking place when cell contacts cell. The relative ease with which hydrophobic particles are phagocytosed, and the relative immunity of hydrophilic particles such as capsulated bacteria have been explained in terms of hydrophobic interactions (Van Oss *et al.* 1975), and the hypothesis has considerable appeal in explaining the behaviour of phagocytes.

An interaction, which is always important when dealing with biological

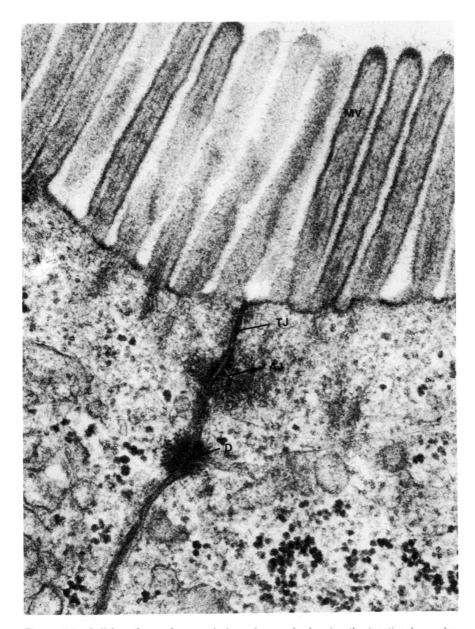

**Figure 7.6 Cell junctions** A transmission micrograph showing the junctional complex between two intestinal epithelial cells. The apical (luminal) surfaces have microvilli (MV) and three junctions can be seen. The apical tight junction (TJ: *zonula occludens*) serves as a gasket, the adhaerens junction (AJ) is the site of insertion of microfilaments, and the desmosome (D: this is a rather poor specimen!) is linked to intermediate filaments. Below the junctional complex the plasma membranes are apposed without specialization (photograph courtesy of the Anatomy Dept, Glasgow University).

fluids, is the adsorption of protein or glycoprotein molecules to surfaces. A clean glass or plastic surface will avidly adsorb protein, and when protein-containing medium is used the whole surface has an adsorbed film of protein that is thick enough to be seen in the transmission electron microscope. In the absence of fluid-phase protein the cell itself will adsorb to the glass; an interaction which is independent of temperature, divalent-cations and metabolic activity, and an interaction which tends to be irreversible. Selective adsorption and desorption could have considerable influence on adhesion, and the study of cell adhesion through adsorption studies has recently been attempted (Pethica 1980). Molecules such as fibronectin (reviewed by Edwards 1983, Yamada 1983), which adsorb readily to surfaces, have a marked effect on fibroblast spreading and, since these are important components of the extracellular matrix, their adsorption may well be of biological importance.

RECEPTOR–LIGAND TYPE ADHESIONS

An alternative mechanism for adhesion is that there is some form of molecular cross-bridging between adherent cells or between the cell and material adsorbed onto an inert surface. The interaction could be direct or indirect, with **ligands** on the cells and bifunctional **receptors** acting as cross-links or with mutual interactions between receptors on one cell, ligands on the other and *vice versa* (Fig. 7.7). Such theories offer ample scope for introducing specificity in the interactions and for explaining differences in adhesive strength by varying the affinity of the receptor. In this context 'receptors' are taken to be discrete sites with affinity for a particular ligand: the receptor could conceivably be its own ligand, but there would have to be a finite number of sites, the standard rules of saturation kinetics should apply. The nature of such receptor–ligand interactions has attracted much bio-chemical attention and there was a vogue for carbohydrate-binding receptors, or 'cell surface lectins' as they are sometimes known. A more recent enthusiasm is for tissue-specific cell adhesion molecules, for example the neural cell adhesion molecule (NCAM) (Edelman *et al.* 1983, Edelman 1984, Grumet *et al.* 1984). Some adhesive interactions may involve such protein–carbohydrate links and there are other receptor-mediated interactions, such as the interaction between the Fc portion of the IgG molecule and the Fc-receptor of mammalian phagocytes that may in some circumstances be important. As with the physicochemical theories it is necessary to suppose that, for example, there is turnover of receptors or ligands, in order to accommodate the dependence of cell adhesion on continued metabolic activity by the cell. Divalent-cation sensitivity of adhesion might easily be explained by alteration in receptor conformation when divalent ions are bound, and there is some evidence for this actually occurring (Hyafil *et al.* 1980).

Despite the wealth of hypotheses, the mechanism of cell adhesion remains obscure. It seems unlikely that the various models are totally incompatible, and probably cell–cell interactions depend both upon physicochemical and receptor-mediated mechanisms – although some cells may rely more heavily upon one rather than the other.

## 7.3.2   The influence of adhesion on movement

Even though we cannot explain the mechanism of adhesion we know that cells do adhere and if, for example, we allow a cell suspension to flow through a chamber, a proportion of cells will become trapped on the walls

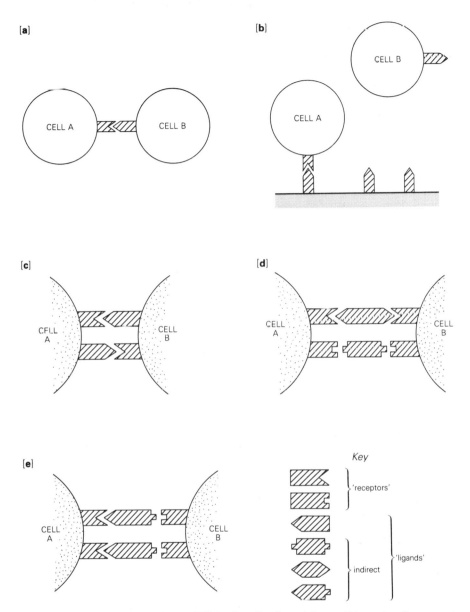

**Figure 7.7 Receptor–ligand possibilities in adhesion** It is possible to visualize several different ways in which receptor–ligand interactions might be involved in cell–cell adhesion. In (a) cell A has a receptor for a ligand found on cells of type B; if the ligand is not cell-bound, but is a soluble protein (for example) which will adsorb onto surfaces, then cell–substratum adhesion might be mediated in this fashion, (b). Alternatively, both cell types may have receptors and ligands (c); although the receptors and ligands are shown as identical, this need not necessarily be the case. The ligand could be a bivalent soluble molecule, with receptors on both cells; the receptors could be identical and the ligand symmetrical, as in (d), or the situation might be as shown in (e), with the two receptors of different specificity. The examples illustrated do not exhaust the possibilities, but the basic ideas remain the same even if variations are introduced in the detail.

(provided the adhesive interaction between the cells and the wall of the chamber exceeds the fluid shear force which will tend to dislodge them). We can also predict that if a cell cannot form an adhesion it will be unable to move over a flat surface (except under the influence of external forces), and that an irreversible adhesion means that the cell is irreversibly held in place. The adhesion which is formed between the cell and the surface over which it is attempting to move must be sufficiently strong to resist the force exerted by the contractile motor system, and must last for long enough for the machinery to have an opportunity to act. The forces required to move a cell are not enormous, but this does not remove the necessity to react upon the environment. The question of transience in adhesion has already been discussed in the context of forward movement (Ch. 6), and if the adhesion is transient then, provided the adhesive interaction exceeds the minimum requirement, the strength of the adhesion becomes irrelevant. Whether strong adhesions, that is adhesions which require a greater force to break, are likely to be of longer duration than weak ones is not obvious. In order to achieve the minimum necessary adhesion the cell can increase the strength of the adhesion per unit area, or can increase the area over which the adhesion takes place. The latter is, of course, limited by the surface area of the cell, the former by the properties of both the cell and the substratum over which it is moving.

In order to introduce transience into the adhesion, an essential requirement if the cell is to move from place to place, the cell must either alter the cell surface in the contact area or must discard the adhesion together with a portion of itself. There is evidence that on occasion cells do adopt the apparently drastic expedient of leaving a trail of discarded material behind them, but this does not seem a very economical solution and a more elegant solution would be for the cell to give each adhesion a finite lifespan.

Relatively few studies of the effect of altering the strength of the cell–substratum adhesion on the rate of movement are interpretable. Adhesion measurements are imprecise and are usually made on a population of cells, whereas the measures of speed are made on individual cells. Several studies show clearly that an irreversible attachment prevents movement, and that too low an adhesive interaction leaves the cells rolling passively. With neutrophil leucocytes, substituting a strong receptor–ligand interaction (Fc-receptor – Fc moiety) for a weak interaction with an albumin-coated surface (the binding of albumin is weak and non-saturable) reduced cell speed; an increasing proportion of the cell's time was spent stationary, the cells became flatter and many cells became irreversibly anchored (Lackie & Wilkinson 1984). By varying the charge on a substratum coated with poly-L-histidine (by altering bulk pH), Sugimoto (1981, Sugimoto & Hagiwara 1979) was able to show an alteration in the speed of movement of mouse fibroblasts, which decreased as the area of contact increased. There does appear to be a decrease in cell speed as the adhesiveness of the substratum increases, but the details of the association are not fully described as yet. Gradients of adhesiveness do affect cell behaviour, with trapping of cells on the more adhesive regions, as will be described in Chapter 8.

### 7.3.3  Adhesion requirements in matrices

Attachment to a flat surface depends upon an adhesive interaction but there are other means of obtaining anchorage in three-dimensional matrices. The cell might produce hooks which are mechanically entrapped, or might protrude a narrow pseudopod and then expand the pseudopod distally

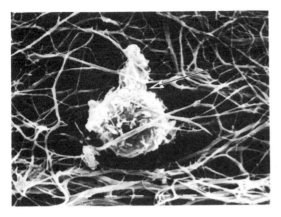

**Figure 7.8 A neutrophil 'fist-jamming'** A scanning micrograph of a neutrophil leucocyte moving through a collagen gel made from Type I collagen which has been purified and allowed to polymerize from the soluble form. The suggestion is that the cell gains anchorage by pushing a narrow protrusion through a gap in the gel meshwork and then expanding the protrusion so that it cannot easily be withdrawn. The cell illustrated has (perhaps) just expanded the process which is arrowed (photograph courtesy of Alastair Brown, MRC Cell Biophysics Unit).

(Fig. 7.8). Alternatively, the cell might rely upon a rearward protrusion to drive the cell body forward. In principle, therefore, a cell might move through a meshwork without forming adhesive contacts provided the meshwork had dimensions that permitted the cell to squeeze through and that allowed the use of 'mechanical' anchorage. The problems of movement in a matrix are different from those of movement over a surface, and yet many cells in multicellular organisms move almost exclusively in such matrices. Adhesion measurements cannot be made on a cell that is surrounded by a solid matrix and it is therefore difficult to predict what the effect of cell–matrix adhesion will be. An indirect indication comes from studies on the behaviour of neutrophil leucocytes in a matrix composed of hydrated collagen (Brown 1982). Neutrophil leucocytes do not adhere to collagen-coated glass yet they manage to move well in a fibrillar matrix of collagen. Increasing their adhesion to the collagen by adding manganese ions to the buffer solution enables them to move over a collagen coat but seems to slow down their movement in the three-dimensional matrix. Lymphocytes, which are apparently too poorly adhesive to move over planar substrata, move well in three-dimensional matrices: a further indication that adhesion is not a requirement for movement in a matrix. These studies, and others, also indicate that a rearward protrusion is unlikely to generate forward movement by pushing the cell, since the cells penetrate the gel matrix from a fluid-gel interface where there is nothing above to resist the protrusion.

If the cell moves through the matrix using non-adhesive anchorage mechanisms, then any adhesions formed will be superfluous and cells may become immobilized if their adhesion to the matrix, or to other cells within the matrix, is increased sufficiently. This might be important in morphogenesis, trapping primordial germ cells within the developing gonads for example.

## 7.4   Roughness

A planar surface may be relatively smooth or may have surface features which are well below the dimension of cell processes. A rough surface provides better friction at a macroscopic level but it is rather harder to predict what the effect would be for a cell. Resident peritoneal macrophages appear to accumulate on a rougher polystyrene surface, whereas both normal and transformed fibroblasts and some macrophage lines (P388D1) show the opposite preference (Rich & Harris 1981), but the cause is not clear.

## 7.5 Rigidity and deformability

Whether the attachment of the cell to components of the environment be adhesive or mechanical the contractile forces from the motor system must meet some resistance if the cell is to move forward. As mentioned previously (Ch. 6) small particles are transported centripetally on the upper surface of a moving fibroblast, and it is essential that the anchorage point used for locomotion does not behave in this way. Thus, for cells moving over a planar substratum the two-dimensional rigidity of the surface is of importance. Two sets of experiments illustrate this very nicely. In the first, cell movement over the interface between silicone oil and aqueous medium was examined: by varying the bulk viscosity of the silicone oil the rigidity of the interface could be varied. Marked differences were found when different cell types were tested (Table 7.1), neutrophil leucocytes translocating on oil of $<10^{-5}m^2s^{-1}$, whereas chick heart fibroblasts needed oil of $>10^2m^2s^{-1}$. Similar findings also come from experiments using elastic substrata made of silicone rubber (Harris et al. 1980). Here deformation of the substratum can be visualized as wrinkling in the elastic sheet (Fig. 7.9), and again some cells clearly apply greater stress to the substratum. The effects of cells in causing contraction of suspended collagen gels are much the same; fibroblasts cause massive shrinkage whereas macrophages and neutrophils cause no damage.

An interesting correlation is that the surface viscosity required for spreading, or the elastic deformation induced in a deformable substratum, is directly related to the extent to which attachments are localized under the cell. Those cells which form focal adhesions seem to put much more tension on the substratum than do those with broad contact areas, and generally the latter have invasive capacities (see Ch. 10). There are two aspects to this: the cells which form focal adhesions have well-developed linear contractile elements, which may generate forces that are greater in magnitude, and the force is concentrated on a small area of the substratum. The latter explanation for the cell's inability to spread will suffice when the surface is plastic (the oil interface) but is less satisfactory for the elastic substrata where the local contact area is coupled to the remainder of the surface.

Possibly the most interesting non-rigid surface is the surface of another cell, although, since we do not know how the moving cell forms attachments on

**Table 7.1** The ability of different cells to spread at the interface of medium and silicone fluid.

| Cell type | Viscosity required (Stokes) to permit spreading |
|---|---|
| rabbit peritoneal neutrophil | 0.03 |
| mouse peritoneal macrophage | 0.1–0.5 |
| Hep-2 (human carcinoma line) | 125–500 |
| chick heart fibroblast | 2500 |
| 3T3 (mouse line) | 10 000 |

Unless the silicone fluid has a bulk viscosity greater than the value shown the cell remains rounded (based on data from Harris 1973c).

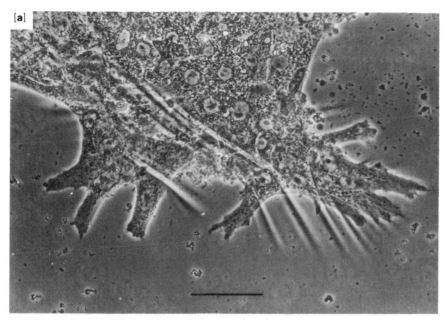

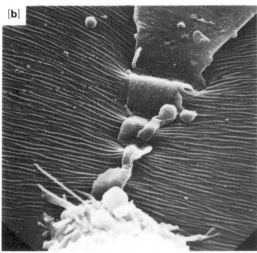

**Figure 7.9  Tension exerted by cells on deformable substrata** (a) Wrinkles in a silicone rubber sheet. The effects of cells exerting traction on their substratum are rather splendidly demonstrated when fibroblasts crawl over a silicone rubber sheet which becomes wrinkled. Invasive cells do not cause the same distortion. The cell sheet (pigmented retinal epithelium from the embryonic chick eye) illustrated here has created both tension and compression wrinkles in the sheet of silicone rubber; bar represents 50 μm (photograph courtesy of Albert Harris, University of North Carolina, Chapel Hill). (b) The BHK cell in this scanning micrograph has been grown on a plastic substratum which has an adsorbed protein film. Tension exerted by the cell in the process of spreading has wrinkled the protein film (photograph courtesy of Mike Lydon, Unilever Research). (The whole cell is shown in Fig. 6.18.)

another cell, it is hard to predict whether the surface will have the 'viscosity' of the phospholipid bilayer or whether the attachment sites are less mobile because of interactions with the cytoskeleton of the cell forming the substratum. Only a few cell types have the ability to move over other cells and some of the implications will be discussed in Chapter 10.

Agar or agarose gels form rather complex surfaces which may change their properties if annealed (Maroudas 1973); the inability of cells to spread on these surfaces may be because of the lack of rigidity or it may be that such surfaces are non-adhesive because of their hydrophilicity. Partly because the properties of such polymers are not well understood, they have not been used very extensively for studies on movement, although tissue cells are frequently plated onto such surfaces for other reasons.

Similar problems of the stability of the anchorage point arise for cells moving in fibrillar matrices, such as collagen or fibrin gels, and an indication of the low resistance to deformation (of these gels) is that fibroblasts within these gels do not have stress fibres. Fibroblasts plated on the surface of a collagen gel rapidly cause shrinkage and collapse of the gel.

If, however, the matrix is rigid or has high resistance to deformation then a second problem arises for a cell moving in such an environment: the cell must deform if the matrix will not. The ability of a cell to deform is probably related to the nature and composition of its cytoskeletal elements, and there seem to be considerable differences between cells, possibly because of differences in the composition of the cytoskeletal elements. One limiting factor may be the size of the nucleus which, being approximately spherical, is much less deformable than the lamellar cytoplasm which has a greater reservoir of membrane. Although we may speculate about such problems it remains an area of cell behaviour which has received little experimental attention.

## 7.6 Summary

In an environment that is uniform with respect to those properties which affect locomotion, cells show a pattern of movement which can be described fairly well as a random walk with internal bias; cells are less likely to make turns through very large angles than make small deviations from a straight path. In the absence of an external cue, the cell, if it is to make a significant displacement, must commit itself to movement in one direction rather than in all directions simultaneously – it must have some internal polarity. Polarity may derive from the microtubular cytoskeleton, but there is little known about the determination of internal asymmetry in the organization of the locomotory machinery. On a flat substratum the anchorages (which are essential if the action of the motor is to be translated into a reaction on the surroundings and move the cell forward) are probably adhesive, but the problems of moving through a matrix are rather different and adhesion may not be so important. The interaction between locomotion and adhesion is complex and further complexities arise when it becomes necessary to consider the rigidity of the substratum, its ability to resist the stresses put upon it by the cell.

Uniform environments are probably very rare, most cells will be operating in anisotropic environments and the asymmetries of the environment may well be important in determining the non-random distribution of cells. In the following chapters we will consider non-uniform environments, situations in which there are vectorial or axial cues which direct the movement of cells.

# References

Berg, H. C. 1983. *Random walks in biology*. Princeton, NJ: Princeton University Press. A valuable and thought provoking book; an approachable treatment of the more theoretical aspects of random walks for the committed biologist.

Dunn, G. A. 1981. Chemotaxis as a form of directed cell behaviour: some theoretical considerations. In *Biology of the chemotactic response*, J. M. Lackie and P. C. Wilkinson (eds), 1–26. Cambridge: CUP. Despite its title this article sets out the problems of isotropic environments in a very clear form. Much easier for the general biological reader than Berg.

Dunn, G. A. 1983. Characterising a kinesis response: time averaged measures of cell speed and directional persistence. In *Leucocyte locomotion and chemotaxis*, H.-U. Keller and G. O. Till (eds) 14–33. Basel: Birkhauser.

# 8

# ANISOTROPIC ENVIRONMENTS

## 8.1 General

One of the major problems facing the student of cell behaviour is in accounting for the non-random distribution of cells: random distributions require little explanation but are of far less interest. For motile cells to become non-randomly distributed their behaviour must be more complex than that of 'diffusing particles' and their pattern of movement must account for the observed distribution. Non-randomness may arise either because of cell–cell interactions, which will be the subject of Chapter 10, or because the environment is non-uniform and the movement of the cell is affected by the anisotropy of the environment. It is essential to realize, in the discussion which follows, that the cell responds to the environment and is constrained to behave in a particular fashion; the cell does not 'perceive' the characteristics of the environment and then 'choose' to respond in a particular way, as might a more complex sentient animal. Although the study of animal behaviour may provide useful insights which we can apply to the study of cell behaviour, the difference between purely reflex behaviour by the cell and stimulus–response coupling in animal behaviour should always be remembered.

Sources of non-uniformity in the environment can be analyzed in terms of the information that they could provide to the cell: the major subdivision is between features which give vectorial cues, that is those which could impose a directionality upon movement, and those which give axial cues, that would constrain movement to an axis or to a plane (Dunn 1981). The differences between these two sorts of non-uniformity are illustrated in Figure 8.1. Alterations in the properties of the environment that are scalar or uniform throughout the area of interest were discussed in the previous chapter and are usually referred to as **kineses** or as kinetic responses: the effect may be to cause a change in the speed of the cell or in the frequency or the extent of turning, but such kinetic responses do not determine or influence the direction in which a cell will turn. Those properties of the environment that give vectorial or directional information are generically referred to as taxes, and those that impose axial or planar constraints are **guidances**. A tropism differs from a taxis in being a movement of part of a sedentary organism (the stem or leaf of a plant, for example) usually due to growth, rather than movement of the whole organism: tactic responses are likely to be more common in motile cells. The term 'tropism' was used interchangeably with taxis in the older zoological literature but this usage is obsolete and it is useful to keep the terms separate.

A consideration of the possible sources of non-uniformity provides a series of subdivisions for discussing the problem. The variables which are most likely to be important are listed below together with the relevant tactic response:

(a)  adhesion, **haptotaxis**;
(b)  diffusible substances, chemotaxis;
(c)  light, **phototaxis**;
(d)  flow, **rheotaxis**;
(e)  magnetic fields, **magnetotaxis**;
(f)  electric fields, **galvanotaxis**;

(g)  gravity, **geotaxis**;
(h)  oxygen tension, **aerotaxis**;
(i)  temperature, thermotaxis;
(j)  shape, (guidance);
(k)  rigidity, (guidance).

Some of these variables have received much more attention than others, in particular the behaviour of motile cells in gradients of diffusible substances

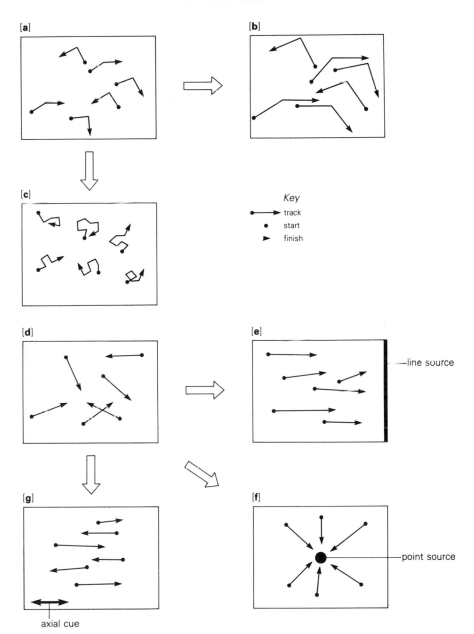

**Figure 8.1  Kinesis, taxis and guidance**  Kineses alter the 'random walk parameters' of a moving cell without imposing a directional cue. In (a) the starting and finishing positions of cells are shown, together with an intermediate position (these can be thought of as two consecutive steps); in (b) there is a positive orthokinesis, the steps are longer; in (c) there is a positive klinokinesis, more turns are made and the paths are more tortuous. In the remaining boxes only single steps are indicated (these could be displacements over considerable time periods). The random behaviour in (d) can be checked with a vector scatter analysis (see Fig. 8.9); in (e) there is a net displacement of the population towards a line source, in (f) the cells are moving towards a point source. A guidance cue, as in (g) does not cause a net population displacement but the displacements are restricted to an axis.

(chemokinesis and chemotaxis) which will be the subject matter of a separate chapter (Ch. 9). In some cases the distinction between kinetic and tactic responses has not been made, and before embarking on a discussion of each of the variables we will digress to show how the distribution of cells can be altered without directional cues being available.

## 8.2 Trapping and avoidance

An old-fashioned way of solving the housefly nuisance is to hang up a fly-paper, a piece of paper coated with glue to which flies stick irreversibly. It is not necessary to suppose that flies hurl themselves deliberately at the trap, but a local accumulation of flies will nevertheless occur. We could say that the fly-paper constitutes a local area in which the speed of movement is reduced to zero, and the trap only works if the flies are moving in the remainder of the environment. This is, of course, an extreme example of non-uniform adhesiveness in the environment, but a local increase in population density will occur if the speed of movement is lower there than elsewhere; the speed need not be zero in the trap. This effect can be shown using a modelling technique (Fig. 8.2) in which the rate of movement (displacement) is diminished in one half of the chamber. Since the rate of movement of a cell that follows a random walk depends on the magnitude of the instantaneous velocity and upon the frequency and extent (average angle) of turning (see Ch. 7), the accumulation of cells in the 'slow' part of the chamber may arise either because of an alteration in speed (an orthokinesis, negative in this case) or in turning (a klinokinetic response). The result of changing one or both of these parameters is not always intuitively obvious. A positive orthokinesis will lead to a greater dispersion of the cells (analogous to the effect of warming up a gas), a positive klinokinesis will lead to decreased dispersion as the frequency of turning rises – the cells will tend to 'jitter' on the spot rather than achieving a significant displacement.

Thus, in order to modify the dispersion of cells it is only necessary to alter the motile behaviour in a quantitative fashion; it is not strictly necessary that the cell should perceive the direction in which to move in order to enter the trap. The redistribution of cells which arises in response to local changes in

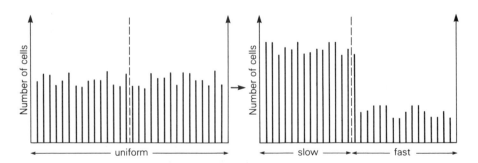

**Figure 8.2 Consequences of altered kinetic behaviour** In (a) cells are randomly distributed through a closed compartment; in (b) the speed of the cells is greater in the right-hand half of the compartment, and slower in the left. With time the distribution of cells alters, so that the cell density is greater in the 'slow' compartment. Based on computer simulations described by Dunn (1981).

kinetic factors may, however, be slow, and will become slower as the new equilibrium position is approached. For this reason it may be advantageous to the cell to have a mechanism which directs the movement in response to gradients of environmental information, but the 'expense' or difficulty involved in building the sensory mechanism must be taken into account.

## 8.3 Gradients

### 8.3.1 Gradients of kinetic factors

In the previous section we considered the simple case of two adjacent compartments with different properties which altered the parameters of movement of cells in those compartments. A more complicated situation, but a logical extension of the two-compartment case, is the existence of a gradient in environmental properties from one side of the compartment to the other. For simplicity we could consider the case of a linear gradient and a linear response of the cells to the change in the environmental parameter which is varying, and model the effect of introducing pure orthokinetic or klinokinetic factors. This sort of modelling has been done, and some of the consequences of introducing such information into the environment are shown in Figure 8.3. A common complication in analyzing real problems of behaviour is that the gradient is rarely a simple one, and the response of the cell is not a straightforward concentration-dependent alteration in behaviour. Gradients may sometimes be purely spatial, that is stable with respect to time, but more frequently they have a temporal component as well, so that the cell experiences waves of altered environmental properties. Even in a stable linear gradient we need to assume that the cell's behaviour in response to the change is a predictable and consistent one, that there is no change in the response with time or at critical threshold concentrations, for example. Biological systems rarely respond in a linear fashion, and cells often show responses such as adaptation, they respond to changes in the environment with a behavioural alteration which is transient. The enhanced motile activity on first exposure to a stimulus may decline to the base-line level within minutes, for example.

The important point to realize about the response to anisotropic cues is that complex distributions can be brought about by varying simple parameters. It is often unnecessary to assume that the cell actually 'perceives' a gradient and makes a directed response. A true tactic response is rare and should be invoked only as a last resort: proving that a tactic response is involved can be extremely difficult.

### 8.3.2 Gradients of adhesion

Since the adhesive interaction of a cell with the substratum or matrix influences the speed of movement, the local level of adhesiveness will influence the distribution of cells, by acting as a kinetic factor. A secondary consequence of an alteration in adhesiveness arises from the anchorage-dependence of many tissue cells *in vitro*. If cells can only survive and proliferate when they have a solid substratum upon which to attach and spread, then cells will only be found where the adhesive interaction with the surface permits the motile machinery to flatten the cell.

An elegant demonstration of the influence of adhesion on movement comes from the experiment done originally by Carter (1965), in which a gradient of

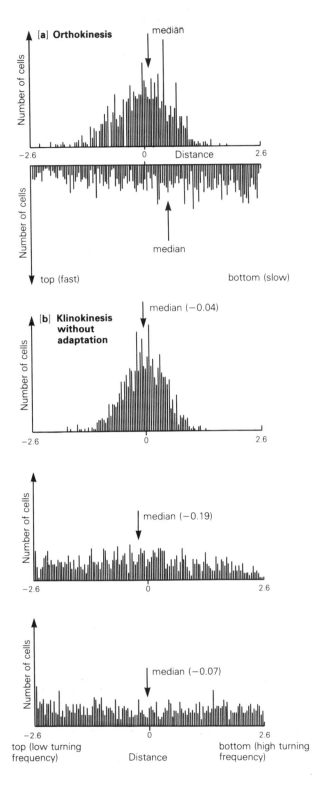

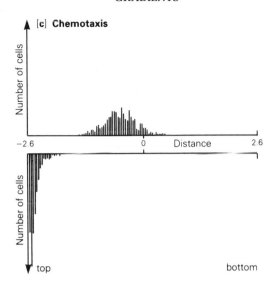

**Figure 8.3** **Distribution as speed or turning behaviour changes** Computer simulations of the effects of altering speed, persistence, or the probability of turning one way rather than the other. The number of cells at different points in the linear environment (the x-axis) is shown as a vertical bar (from Vicker *et al.* 1984). (a) In this simulation the cells move faster as the concentration of the substance increases. The gradient decreases linearly from X = −2.6: the speed of the cells at the highest concentration (−2.6) is three times greater than at the lowest concentration. 1000 cells start in the centre of the field (at 0) and the distribution is shown after 5000 steps in the upper panel and after 60 000 steps in the lower panel. Although slight shifts up-gradient occur at first, after a longer period the shift is down-gradient with the mean at +0.39 (median +0.51). In other words the cells accumulate more in the region where they move more slowly. (b) Here the persistence of the cells increases as concentration increases (a negative klinokinesis, the cells tend to turn less frequently). The probability of the cell turning at the next step is 0.05 per cent at the top of the gradient and 50 per cent at the bottom of the gradient; there is no adaptation. All cells start at 0 and the distributions after 5000, 60 000 and 260 000 steps are shown in the upper, middle, and lower panels respectively. The gradient makes remarkably little difference! (c) Simulation of the effect of a directional cue. A cell moving up-gradient (to the left) has a 55 per cent probability of making its next turn up-gradient (and only 45 per cent probability of moving down-gradient), whereas a cell which is moving down-gradient turns randomly. All cells started at the centre (0); the top panel is the distribution after 5000 steps, the lower panel the distribution after 60 000 steps. Despite the fact that only a 5 per cent 'bias' has been introduced, the accumulation at the top of the gradient is remarkable.

adhesiveness was provided by shadowing palladium onto a cellulose acetate-coated glass slide. Cellulose acetate is non-adhesive (and non-wettable), but the metal-coated surface is suitable both for adhesion and movement. Fibroblasts and other tissue cells on such a haplotactic gradient move towards the more adhesive region where they accumulate (Fig. 8.4). The accumulation of cells at the top of the gradient could arise simply through a trapping phenomenon, or there could be a directional response by the cells. If the direction of movement is influenced by competition between contractile elements then those adhesions which resist contractile stresses most effectively, or which last longest, will be at the front of the cell. The front of the cell has in this case been determined by the properties of the environment. Originally, the movement of cells on such a gradient of wettability was proposed to depend upon the differential wettability of the palladium-shadowed and naked substrata; a water droplet placed on such a gradient would spread more towards the top of the gradient (Carter 1967a). The subsequent discovery of the complex biochemistry behind the motile machinery of the cell makes it improbable that the driving force for movement actually comes from the hydrophilic interaction of the cell with the substratum; it is nevertheless clear that the environment acts upon the internal motor machinery of the cell and imposes the direction in which movement will occur. The cell is not a free agent.

Islands of hydrophilic surface within a sea of hydrophobicity have also been used experimentally to study the behaviour of cells and as 'clonal' territories (Carter 1967b). Using a copper grid (of the sort used to support thin sections in the electron microscope) as a mask, metal can be shadowed onto a bacteriological-grade Petri dish (a polystyrene surface which is of low adhesiveness for many cells), or onto any surface of chosen adhesive properties. With this sort of approach Harris (1973a) was able to show that cells exhibited an hierarchy of adhesive preference (glass > palladium >

[a]

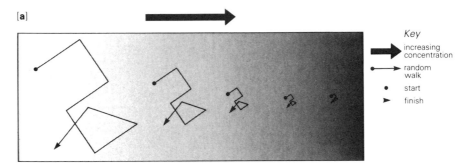

*Key*

increasing concentration

random walk

start

finish

[b]

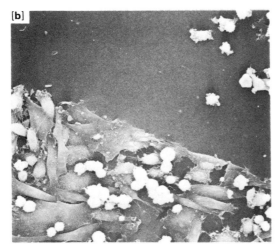

**Figure 8.4 Haptotactic gradient** (a) A gradient of adhesiveness can be set up by shadowing palladium onto a cellulose acetate-coated slide. The gradient is indicated by the increased shading, and the effect on random cell tracks is shown. The top of the gradient is an adhesive trap. (b) A stepwise alteration in the adhesive properties of the substratum can have a dramatic effect on the behaviour of cells. In this photograph (courtesy of Mike Lydon, Unilever Research), one half of the polystyrene dish has been treated with sulphuric acid, the effect of which is to make it more wettable and more adhesive for cells. The cells on the treated surface have flattened far more. The medium contained serum which will have formed a thin film, and which can be deformed (see Fig. 7.9b).

cellulose acetate) and that cells tended to remain on the surface of greatest adhesiveness. Similar experiments with nerve cells (Letourneau 1975) showed that the nerve growth cone would move preferentially on the more adhesive territory; in principle, therefore, it should be possible to guide nerve cells to particular sites either in morphogenesis or onto artificially introduced interfaces. Development of the ultimate man–machine hybrid may depend upon the exploitation of such guidance systems! Extracellular matrix components are known to affect the movement of neural crest cells, probably by altering the adhesive interaction between the cells and the surface over which they are moving (Erickson & Turley 1983), and such effects seem very likely to be important in directing morphogenetic movements.

By arranging adhesive and non-adhesive territories as parallel strips it is possible to produce a guidance field in which cells can move along the adhesive strips, but not across the non-adhesive areas. This effect only works if the adhesive (suitable) strips are fairly well separated; if a cell can span several strips then the effect is lost (Dunn 1982). Only if the adhesive strips are graded for adhesiveness along their length will there be a vectorial cue available to direct the cells along the strips in one direction rather than the other, but this is not impossible to envisage. The potential of this sort of experimental design has not been fully exploited. In a much simpler sort of experiment (Wilkinson *et al.* 1984) neutrophil leucocytes were studied at the boundary of more- and less-adhesive surfaces. The non-adhesive territory was glass coated with bovine serum albumin (BSA) (a protein coat on which the cells will adhere and move but on which they do not flatten) and the adhesive territory was similar but with the further addition of antibody (IgG) directed against the albumin. The antibody–antigen coated surface is one on which neutrophils attach and spread, probably through their Fc-receptors, which become redistributed to the lower surface of the cells. Adhesion through the Fc-receptor–Fc moiety interaction is considerably stronger than that of the cells to the BSA coat, as could be shown in an adhesion assay that eliminated the effect of spreading. On the more adhesive territory the speed of the moving cells is lower but the turning behaviour is unaffected. One consequence of the difference in speed on the two territories is that more cells cross the boundary onto the sticky surface than cross in the opposite direction, with the result that, with time, there are more cells in the 'slow' territory. Several other points emerge from considering this sort of simple boundary: not only does the speed of the moving cells decrease on the stickier surface but the proportion of time each cell spends moving decreases, and fewer cells are engaged in moving from place-to-place. A negative orthokinesis of this type has often been proposed but rarely demonstrated. Adhesion between the cell and the surface over which it is moving is affected not only by the properties of the inert substratum but also by the medium, which may affect the adhesiveness of the cell itself. The effects of gradients of some chemotactic factors in directing the movements of cells may partly be explained on the basis of the changes in adhesiveness which they bring about: this has been considered in detail for the behaviour of neutrophil leucocytes in the acute inflammatory response (Lackie 1982), but it remains speculative.

### 8.3.3 *Gradients of light*

Photokinetic or phototactic responses have only been described for a few cell types; photoreceptors are rather specialized structures. The best example is the response of the unicellular alga, *Chlamydomonas*, which has a prominent 'eyespot'. Despite the absence of specialized photosystems most cells are photosensitive (as anybody who has taken time-lapse film will confirm), but the phenomenon of the photosensitivity of tissue cells has received little attention.

Although the phototropic responses of plants are undoubtedly mediated at the cellular level, the responses fall more appropriately into the province of plant physiology and will not be considered here. Useful reviews of this area are given by Hader (1979) and Dennison (1984). The grex or slug of *Dictyostelium* shows a phototactic response (Poff & Whitaker 1979) without any obvious photoreceptor system, but we will omit this, too, on the grounds that it is a phenomenon of a multicellular system rather than of an individual cell.

## 8.4 Flow

Few cells are capable of swimming sufficiently strongly to resist a fluid flow, and generally we would not expect to find rheotactic responses amongst unicellular organisms. Some static unicellular organisms may align themselves in response to flow, and the endothelial cells of blood vessels tend to be aligned with their longest axes parallel to the direction of flow. I know of no other examples of responses at the cellular level, and it is difficult to see how a cell could perceive the direction of flow unless it was actually anchored. To recognize that you are being carried by a current it is necessary to have some external reference point (and is surprisingly difficult).

## 8.5 Magnetic and electric fields

Some flagellated bacteria respond to magnetic fields (reviewed by Blakemore 1982) and have been shown to contain permanent magnetic dipoles. Experimentally, it is possible to reverse the dipole and to show that the behaviour alters as a result. In the northern hemisphere the bacteria move northwards, which only makes sense as a behavioural response because there is a vertical (North downward) component of the Earth's magnetic field and their 'search' is not for the North Pole but for the anaerobic bottom ooze in which these Gram-negative organisms live. Species from the southern hemisphere have opposite polarity and a predilection for heading south: the prediction that equatorial species should be rare has not been tested so far as I know.

Galvanotaxes have been reported (see § 9.4) and there has been considerable interest of late in the effects of electrical currents on wound-healing in vertebrates. Cultured neural crest cells of *Ambystoma mexicanum* and *Xenopus laevis* have been shown to respond to DC fields by orienting their movement (Cooper & Keller 1984). In fields of 1–5 V cm$^{-1}$ cells retract both anode- and cathode-facing margins so that they lie with their long axes perpendicular to the field: at higher field strengths they move in a sort of 'crabwise' manner towards the cathode. The basis for this cellular response has not been discovered but other cell types are said to share the ability to respond to electrical fields.

## 8.6 Gravity

Plant cells undoubtedly respond to gravity and the mechanism of the geotropic response has received considerable attention (for review see Wilkins 1984). Again this topic is beyond the scope of the present text. Little is known about geotactic responses by free-living unicellular organisms, and although *Paramoecium* does seem to show a negative geotaxis the mechanism is obscure. For tissue cells the effect of gravity is probably negligible, and those few experiments which have been done in orbit around Earth do not suggest a crucial rôle for cellular geotaxis in multicellular animals.

## 8.7  Shape

It is far easier to observe and film cells on optically flat surfaces, but these are rare in all but laboratory environments. Within a multicellular organism surfaces are much more complex and natural substrata are much more likely to be rough, both microscopically and macroscopically (having both molehills and mountains). In addition, cells will often be moving within deformable matrices – but we will set this aside as a problem, temporarily at least. The curvature of a surface can affect cell behaviour very markedly and is the basis of the response known as **contact guidance**. In practical terms all deviations from planar can be considered as curvatures, even though the radius of curvature may be very small in some cases (at the edges of steps, for example). Only three geometrical forms need to be considered, cylinders (or segments of cylinders), tubes (or grooves) and spheres. The latter situation, the spherical surface, is the simplest example of curvature in two axes, the curvature being symmetrical. Of these forms the cylinder is probably the most important in biological systems since many connective tissue elements are fibrous.

If fibroblasts are plated onto cylindrical glass fibres then, if the radius of curvature is in the right range, they will align themselves axially. There is no preference for movement in one direction along the fibre rather than the other, but movement is restricted to the long axis with the cells being more or less accurately aligned to the axis of the cylinder. The accuracy of alignment depends upon the strength of the guidance, a function of the radius of curvature of the fibre. The fibre does not give vectorial information to the cell, it is just that the cell has a more restricted range of probable directions in which to move. The response of the cells to the curvature must presumably depend upon internal constraints in the operation of the locomotory machinery of the cell, which makes axially-disposed protrusion or contraction more probable or more likely to lead to an effective translocation of the cell body. The extent to which alignment occurs depends critically upon the radius of curvature, as shown in Figure 8.5, and, for chick heart fibroblasts, alignment is lost once the radius of curvature of the cylinder exceeds 100 μm. The dependence on radius of curvature, and the observation that fused silica fibres will serve equally well, makes it improbable that the cells are detecting the orientation of molecules within the fibre (which might have arisen in pulling the glass fibre and which is very commonly the case with natural fibres), and the explanation must be sought elsewhere.

Before trying to explain this behaviour another experiment should be described. An axially disposed extended fibroblast moving along the cylinder makes its new attachments in the same plane as the previous ones, whereas in attempting to move at right angles to the axis of the cylinder the new attachment would have to be below the plane of existing contacts: a similar situation arises when a cell attempts to cross the ridge of a prism (which resembles the roof of a house). Prisms with apical angles of just less than 180° were used by Dunn and Heath (1976) to test the abilities of cells to handle this sort of problem. Cells approaching the ridge at right angles crossed without deviation in their paths if the ridge angle was greater than 178°, but the cell swung to the left or right if the ridge was steeper. (By approaching a ridge

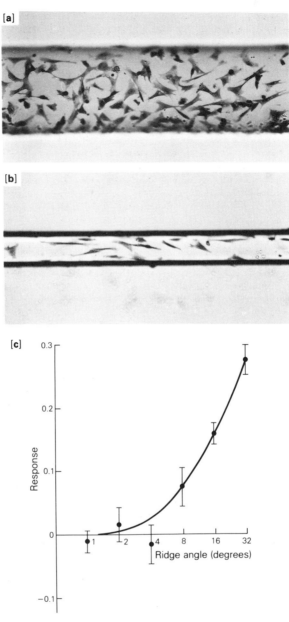

**Figure 8.5  Fibroblasts on fibres and prisms**  (a) and (b) Guidance on glass fibres; radius of (a) is 127 μm and of (b) 54 μm. On fibres of less than 100 μm radius the cells tend to be oriented parallel to the axis of the cylinder (from Dunn & Heath 1976). (c) The effect of an abrupt discontinuity in the surface: cells are tested for their ability to cross the ridge of a shallow-angle prism. With more than a 4° change (a prism with an internal angle of less than 176°) a chick heart fibroblast which approaches the ridge straight-on cannot cross without being 'refracted' (from Dunn & Heath 1976). The 'response' is derived from measurements on orientation of the nucleus and reference to the original paper should be made for the details.

obliquely the apparent angle is reduced.) Ridges with an angle of 164° or less proved an insurmountable barrier and cells were never observed to cross. The cells used in these experiments were chick heart fibroblasts, which have a remarkably constant morphology with a thin leading lamella less than 1 μm thick and some 10 μm long (see Fig. 8.5). If a microfilament bundle 15 μm long is to exert a contractile force then it must obviously remain within the cytoplasm, and there is a limit, set by the geometry of the leading edge, to the possible difference in level between pre-existing and new attachment points. This limitation will apply whether the cell is trying to cross a sharp ridge or attempting to crawl around a cylinder of low radius of curvature. Making a standard assumption about the geometry of the leading edge, it can be calculated that the apical ridge angle which causes a perturbation in the path corresponds closely to the effect of reducing the radius of curvature of a cylinder to the level at which axial alignment first becomes apparent. On this hypothesis, therefore, the axial alignment of the cell on a cylinder arises because the contractile elements of the motile machinery cannot operate to move the cell if the new attachments are too far below the plane of the leading lamella. If the new attachment is in the same plane (or very nearly so) then the motor will work and the cell will be able to move or spread in that direction. Inevitably, therefore, the cell will become aligned along the axis of the cylinder. Attractive though this explanation is, it does not seem to be entirely satisfactory in the light of the rôle currently assigned to microfilament bundles, as discussed in Chapter 6. Even if microfilament bundles are not involved in locomotion, however, the constraint on tension development which is set by the geometry of the leading lamella may well be significant for fibroblasts which are attempting to exert tension on elements of the surrounding matrix.

An alternative explanation for the phenomenon of alignment might be that the protrusion of new leading lamella is in the same plane as the existing lamella structure and that it will, if it deviates from this plane at all, tend to be pulled upwards and away from the substratum. The tendency to be pulled upwards might be because the contracting actin meshwork within the leading edge will meet less resistance on the dorsal surface; on the ventral surface, that nearest the substratum, some of the filaments become adhesively linked to the substratum and are therefore more restricted in their ability to move. This sort of explanation would also account for the formation of dorsal ruffles, but would only apply in the rather unusual circumstances of movement over a planar substratum and would not be the case for a cell moving through a three-dimensional matrix. If the protrusions formed at the leading edge are restricted to the same plane as the existing lamella, there will be a smaller chance of forming an adhesion with the surface if the ground is falling away in front of the cell, as will be the case in crawling around a cylinder. The opportunities for contact will be greater on a concave surface: fibroblasts, on this argument, should (like hares) be better at going uphill rather than downhill. If this argument *does* hold then a further prediction would be that in a tube the preferred alignment of the cells should be at right angles to the long axis. This prediction turns out to be correct, provided the internal radius of the tube is less than 100 μm, although a complication arises because some cells become stretched across the lumen of the tube and lose their orientation.

In favour of the former hypothesis is the observation that neutrophil leucocytes, which have a very much smaller and less flattened leading lamella, do not appear to show contact guidance on rigid cylinders although they do align along grooves (Wilkinson *et al.* 1982).

If the curvature of the substratum is in two dimensions, as with a spherical bead, then there should be a bead diameter at which spreading of a fibroblast is impossible because the machinery responsible for flattening the cell will not operate. This does seem to be the case, although there is a further possibility to be considered in this situation. Fibroblasts when they collide with one another show the phenomenon of contact inhibition of locomotion (see Ch. 10) and tend to move apart. A fibroblast spreading over a bead will meet itself coming the other way, and since self-inhibition of locomotion has been described it will retract following contact, and will never achieve a fully spread morphology. Clearly this sort of response is wholly inappropriate for a phagocytic cell, otherwise small particles could not be engulfed: phagocytic cells do not, on the whole, show the phenomenon of contact inhibition of locomotion. Phagocytosis by fibroblasts is sometimes described, but seems more likely to be pinocytosis involving **coated vesicles** or a misidentification of the cell type (fibroblasts are not very clearly defined except in morphological terms). If, however, cells which do show contact inhibition of locomotion are also capable of phagocytic engulfment of particles then some explanation must be sought of the mechanism by which the cell discriminates between self-contact, which should cause inhibition, and contact that should eventually be followed by membrane fusion. Whether an epithelial cell could handle this problem has not been tested, but an epithelial cell on a cylinder might tend to form a tube surrounding the cylinder because cell–cell contact is stabilized by the formation of specialized junctions, and retraction of the contacting areas of leading lamella does not occur. As part of a design-kit for morphogenesis the ability of epithelial cells to form tubules in this fashion might be very useful.

If we align cylinders, or segments of cylinders, in a plane then we generate a grooved surface with both convex and concave elements. The features of such a surface can be varied experimentally by altering the spacing of the grooves and by altering the dimensions of the pseudo-cylinders. Further variability can come from the geometry of the grooves and their roughness compared to the rest of the surface (the grooves are frequently scratched with an irregular point such as an industrial diamond). The alignment of fibroblasts on such surfaces has been described, although the mechanism remains obscure. One suggestion (Ohara & Buck 1979) is that the ridges constrain the direction of movement of fibroblasts because of the geometry of the focal adhesions which these cells form. Because focal adhesions are associated with the oblique insertion of an approximately cylindrical bundle of microfilaments into the plane of the plasma membrane, the adhesion has an elliptical shape, with the major axis of the ellipse lying parallel to the axis of the microfilament bundle. The argument would then be that a focal adhesion, which was aligned with the major axis of the ellipse parallel to a ridge on the substratum, would have a more secure attachment than an adhesion at right angles to the ridge, and that this difference would account for the tendency of cells to align and move along ridges. The determination of the direction of movement would arise from the competitive interaction of adhesions, much as with the haptotactic gradients discussed earlier in this chapter. The hypothesis does, however, predict (or require) that bundle formation should precede attachment, otherwise the adhesion is not necessarily elliptical, nor need all the microfilaments terminate in the same plane (there is no obvious reason for the adhesion zone on the plasma membrane to be flat rather than cup-shaped). The hypothesis also relies on bundle protrusion at the leading edge; in the light of current views on the rôle of microfilament bundles, the hypothesis seems less satisfactory than it did when it was first proposed. An alternative view would be that the groove spacing affects the probability of attachment, since the hollows between ridges are less likely to be contacted by the leading edge of a cell than are the crests of the ridges. If this were so then closely spaced grooves should not impose an axial orientation, provided the ridges are no more than a few microns apart: the width of the ridges should be irrelevant. Although these behavioural experiments provide interesting constraints on hypotheses which try to account for the organization and operation of the locomotory machinery, they are often rather difficult experiments to carry out, simply because surfaces of the right geometry can be difficult to manufacture.

## 8.8 Rigidity

So far we have considered only surfaces which are non-deformable (rigid) and that can resist any tension put upon them by the moving cell. Deformable surfaces may be of two kinds, those that deform in a plastic manner and those that have elastic properties and which spring back into shape when the stress is relieved. An intermediate kind of substratum would be one composed of a viscoelastic material that would respond elastically to low-level short-term stress but would deform irreversibly if the stress were maintained. Because these various kinds of surface have not really been distinguished in the experimental work done in this area, we will refer to deformable surfaces or materials as being simply 'non-rigid', but the possibility of plastic, elastic or viscoelastic properties should be borne in mind.

Guidance cues can be present on surfaces of anisotropic rigidity, and in principle tactic cues could be derived from gradients of rigidity, but such gradients have not yet been constructed experimentally. In complex extracellular matrices a gradient of rigidity might easily exist, especially if, for example, hydrolytic enzymes were being released and were causing degradation of the extracellular matrix components that are chiefly responsible for giving 'stiffness'. This might well occur around a site of infection or at a chronic inflammatory lesion. Although gradients of rigidity are difficult to construct, anisotropic rigidity is much easier to arrange and the effects of anisotropic rigidity can perhaps be visualized by considering the properties of guitar strings. On this analogy the strings resist tension (are more rigid) axially, whereas their resistance to lateral displacement is poor – enabling the string to be plucked. If a cell exerts tension across the strings then the net displacement achieved will be less than if tension is applied along the axis of the strings, because in the former case the attachment point displaces towards the cell.

Surfaces which are composed of oriented fibrils are by no means uncommon amongst biological materials, collagen or elastin fibres are often organized in this fashion. Experimentally, the effect of aligned collagen fibres on fibroblast orientation has been tested by Dunn and Ebendal (1978), who coated coverslips with reconstituted collagen and then drained them so that the gel 'dripped' or 'dribbled' down the coverslip and the collagen fibrils had an axial orientation (Fig. 8.6). If the collagen was subsequently air dried it collapsed to form a thin sheet with molecular orientation but regular microscopic topography; if carefully critical-point dried the collagen could be re-swollen to its original gel form. On the freshly drained gels and the reconstituted gels the cells showed a very clear orientation with their long axes parallel to the direction of drainage, parallel to the microfibrils of collagen. On the flat, air-dried sheets of collagen no orientation was shown

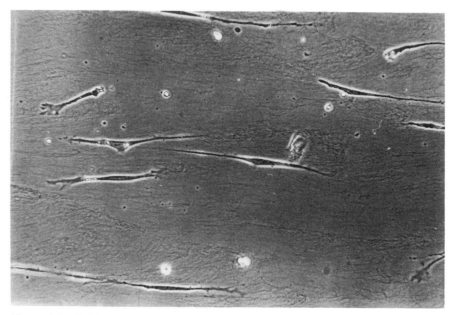

**Figure 8.6 Guidance on oriented collagen** Fibroblasts oriented on a coverslip coated with drained collagen; the guidance cue may be topographic or of anisotropic rigidity. The cells are aligned parallel to the collagen fibres which can just be seen (photograph courtesy of Graham Dunn, MRC Cell Biophysics Unit).

despite the molecular orientation of the collagen polymer. In this experiment it is difficult to separate the possible contributions of shape and rigidity, because the drained gel is both grooved (as can be shown by scanning electron micrograph examination of critical-point dried material) and oriented. To some extent this does not matter, the important phenomenon is the orientation of the cells, but it makes the experiment harder to interpret. One of the early test surfaces on which contact guidance could be demonstrated was the inner (cleaved) surface of a teleost epidermal scale (Fig. 8.7) where the collagen fibres are naturally aligned. Within the fish scale the collagen is embedded in extracellular matrix components, which may well contribute to the lateral rigidity of the fibrils, and the important environmental cue may be the shape of the surface: in the case of the drained gels rigidity may play a greater part.

Another aligned fibrillar substratum is the basis for the so-called two-centre effect, observed when chick heart explants are placed on a plasma clot, the standard tissue-culture milieu of early cell biology. The plasma clot is a fibrillar gel with fibrin forming the major structural element. Fibroblasts, which emigrate from the explants, exert considerable tension on the fibrin gel and produce stress-induced alignment of the fibrin, which can be detected with polarized light. Because the emigrating cells are then guided by the alignment of the fibrin, the cells move in much straighter paths than they would normally on an isotropic substratum, and their displacement is therefore greater than it would be otherwise (contact inhibition of locomotion also contributes to the tendency for outgrowth to be radial and should not be forgotten: we will return to it in Ch. 10). With two explants the extent of orientation in the fibrin between the explants, which is being tensioned by

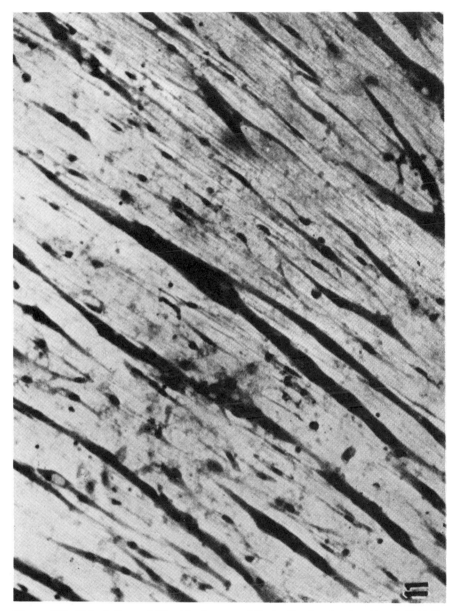

**Figure 8.7 Orientation of fibroblasts on a fish scale** A natural guidance field for fibroblasts is the surface of a fish scale which has been cleaved. The collagen fibrils in the scale are aligned and form a grooved (ridged) surface (from Weiss 1958).

the efforts of both, is proportionately greater, the fibre alignment more marked, the guidance more effective and the cell displacement greater. It therefore appears that cells move from the explant preferentially towards the other explant – the two-centre effect.

Early explanations for the guidance being shown by the fibroblasts invoked a recognition of molecular alignment by the cells as they were moving out, but this is clearly unnecessary; either the gross shape of the surface of the clot, or anisotropic rigidity, or a combination of the two could account for the guidance response. Having said earlier that guidance does not constitute a vectorial cue, it is perhaps worth commenting that cells from the explant all move *away* from the explant: the reasons for this have nothing to do with the guidance, they include a simple concentration-dependent diffusion (the fibroblast concentration is much higher in the explanted tissue than in the clot) and a contribution from contact inhibition of locomotion. The guidance cue contributes to the rate of outgrowth, however, because the guided cells achieve a much greater displacement than would cells moving in a classical random-walk fashion. The two-centre effect is more than a laboratory curiosity, the blood clot formed *in vivo* at a wound is composed of fibrin with platelets embedded within it, and the phenomenon of clot retraction brought about by the contractile activity of the platelet actomyosin motor will tend to align the fibrin in such a way that a radial guidance cue is available. Fibroblasts responding to the radial alignment of the fibrin will infiltrate the wound more rapidly and the wound healing response will therefore be facilitated.

*In vivo* the surface of the clot is by no means the only interesting part, the matrix of the clot is infiltrated by cells that also contribute to the processes of repair and reconstruction. Oriented fibrin in a blood clot and oriented collagen in connective tissue will influence the direction of movement of cells within the matrix as well as over the surface. In three-dimensional environments, such as these matrices, neutrophil leucocytes and lymphocytes will show guidance responses. Fibroblasts will also respond to orientation of the fibrillar matrix (Bard & Hay 1975) as will neural crest cells (Davis 1980). Neutrophil leucocytes do not seem to respond to convex curvatures in the same way as fibroblasts, and they have never been observed to align on isolated cylindrical fibres, possibly because the critical fibre diameter is too low to be experimentally testable, but they will align along the grooves of a haemocytometer (Fig. 8.8). In fibrin gels which have been aligned by tensioning them after they have formed, neutrophils show a clear preference for movement parallel to the fibre axes, and make relatively few excursions across the grain of the matrix (Wilkinson *et al.* 1982). If the cells are given a chemotactic (vectorial) cue at right angles to the axial guidance cue the cells will manage to displace across the grain, but they do seem to do so rather in the manner of a small boat 'tacking' against the wind, adopting a stepped path rather than a smooth diagonal (Wilkinson & Lackie 1983).

For cells moving in a three-dimensional matrix, anisotropic rigidity could clearly play a part in directing movement but another factor also comes into play. In a system of aligned fibrils a protrusion at the front of the cell will meet less resistance if it is axially directed and less deformation of the cytoplasm will be required. There is some evidence that the latter phenomenon may be important in the case of neutrophil leucocytes. Protrusions seem to be formed on the surface of the neutrophil with equal probability in any direction, with no appreciable bias along the fibril axis. Protrusions across the grain are, however, less pronounced than those in the axis and the larger protrusions are presumably more likely to contribute to

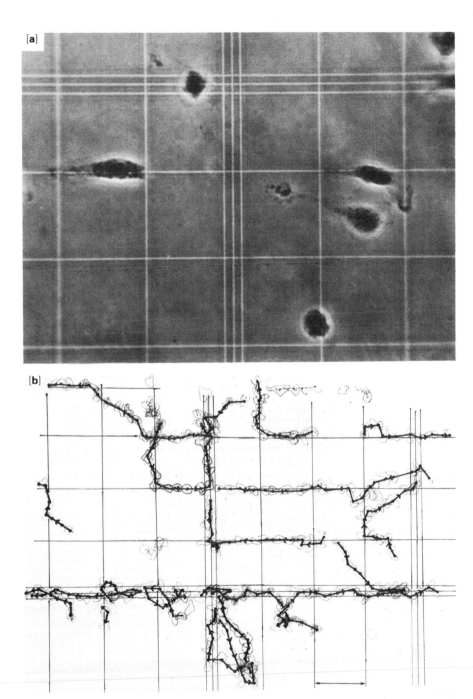

**Figure 8.8 The guidance of leucocytes by grooves** The behaviour of leucocytes on the grooves of a haemocytometer. The picture is of monocytes on a Neubauer chamber; the tracks are of human blood neutrophils moving on a Neubauer grid in the presence of 20 per cent plasma. The cell's position has been marked every 40 s (real time) (picture and tracks courtesy of Peter Wilkinson).

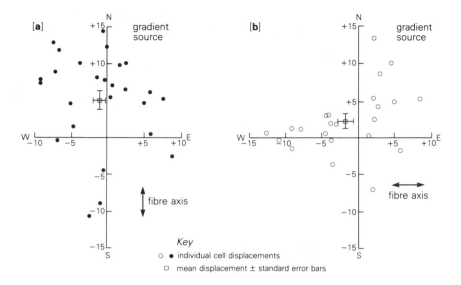

**Figure 8.9 Vector scatter diagrams** Vector scatter diagrams for neutrophils in an aligned fibrin gel, with a tactic cue (a gradient of fMet-Leu-Phe) parallel (a) or perpendicular (b) to the axis of alignment of the fibres. All the cells start at the intersection of the axes and their displacements are plotted on polar coordinates. Cells which move randomly will, as a population, displace only a small amount from the starting position and the expected random displacement can be calculated if the displacements are linked to form a sort of random track of their own. With a vectorial cue the population shifts up-gradient (if the gradient is of attractant). In the two scatter diagrams shown there is a net up-gradient displacement in both cases, but the overall displacement is greater if the guidance cue is parallel to the vectorial cue. Notice that some cells move down gradient, even though the population as a whole displaces up-gradient (from Wilkinson & Lackie 1983).

displacement. The speed of moving cells does not seem to be different whether they are moving perpendicularly to, or in parallel with, the axis of alignment. If a superimposed vectorial cue is offered the displacement is, however, greater if the vector is parallel to the axial cue (Fig. 8.9). The possible implications for cellular infiltration in an acute inflammatory response in differently organized connective tissue matrices can be appreciated, although they cannot be explicitly demonstrated.

More complex guidance cues are also possible in three-dimensional matrices. Cells may be restricted to moving in a particular plane rather than in an axis, depending upon the orientation and packing of linear elements. Such guidance cues seem very likely to play a part in morphogenetic movements, such as the ventral movement of the neural crest cells, but a detailed description of the factors influencing the distribution and fate of cells will require a much greater knowledge of the biochemical composition of the extracellular matrix, as well as an understanding of its organization and mechanical properties (Armstrong 1984; Erickson 1985). Thus, in order to predict the pattern of movement of cells in morphogenesis or in cellular invasion into adult tissues, we require a far greater knowledge both of the composition and organization of the environment and the way in which the properties of the matrix influence the behaviour of cells. Tumour infiltration along planes or lines of 'weakness' has frequently been described by histopathologists; the infiltration may partly be determined by the pattern of mechanical constraint and partly by guidance cues tending to direct cells along particular paths. The dispersion of cells from an area of high population density and the delivery of cells into areas in which they are needed are frequently observed: although the vectorial cue may derive simply from gradients of cell concentration or from gradients of diffusible substances, the effect of guidance cues will often be to improve the efficiency of the response by restricting the random wandering of the cells.

## 8.9 Summary

A wide range of environmental factors influences the motile behaviour of cells and it is no surprise to find that free-living unicellular organisms respond to light, to gravity, and so on. If such environmental factors are non-uniformly distributed then the distribution of cells will be affected, and the same will be true for the redistribution of cells which occurs during the process of morphogenesis in multicellular organisms. Such redistributions of cells may be population phenomena, the individual cell may move the wrong way, or the cells may respond to a gradient of some environmental property with a directed movement. For cells moving in the developing embryo or invading in the adult, the properties of the environment are just as important as for a free-living cell, although the factors involved are different. In recent years there has been an upsurge in interest in the properties of the extracellular matrix that surrounds cells in solid tissues and which may be influential in modifying many of their behavioural properties. Gradients of adhesion and of deformability of the matrix may well exist, and we are becoming more aware of the importance of supramolecular aspects of structure: the mechanical properties and organization. The frequent occurrence of aligned fibrillar structures in multicellular organisms means that guidance may often influence the way in which cells become distributed. In the next chapter we will consider in detail the responses of cells to diffusible substances in the environment, the chemokinetic and chemotactic behaviour of cells, which is so often invoked and has been so extensively studied.

## References

Dunn, G. A. 1982. Contact guidance of cultured tissue cells: a survey of potentially relevant properties of the substratum. In *Cell behaviour*, R. Bellairs, A. Curtis and G. Dunn (eds), 247–80. Cambridge: CUP. The only recent review I know.

Harris, A. K. 1982. Traction, and its relation to contraction in tissue cell locomotion. *Ibid*, 109–34. The rigidity problem put into biological perspective.

Keller, H.-U. 1981. The relationship between leucocyte adhesion to solid substrata, locomotion, chemokinesis and chemotaxis. In *Biology of the chemotactic response*, J. M. Lackie and P. C. Wilkinson (eds), 27–52. Cambridge: CUP.

# 9
# CHEMOTAXIS

# 9.1   General

Chemotaxis is the directed movement of a cell (or organism) in response to a chemical substance in the environment, usually a diffusible substance. The use of the term dates from 1884 when Pfeffer described the movement of the spermatozoa of bracken (*Pteridium*) towards the oogonia, and the problem of chemotaxis has fascinated many biologists since then. Unfortunately the term chemotaxis has often been used rather carelessly, and in many cases the only basis for the assumption that chemotaxis is occurring is that there is a local accumulation of cells. As we have seen in the previous chapter, accumulation need not be a result of directed movement and the criteria for defining a true chemotaxis have not always been fully appreciated. Purists will also point out that for bacteria the phenomenon described as chemotaxis is not a true chemotaxis at all but a klinokinesis with adaptation, but the usage has become so firmly established that it is unlikely to change.

A variety of cells exhibit chemotactic responses, and it will be impossible to give an exhaustive description of all the cases in this chapter. More interesting than a long listing of examples are the basic problems associated with perceiving a gradient and responding in the appropriate manner with a directed turn. Different systems have provided information about the various phases of the response and these systems are the ones we will consider in detail. The best known examples are bacterial chemotaxis, to which the methods of genetic analysis have been applied with conspicuous success, the tactic and avoidance responses of *Paramoecium*, in which the membrane transduction system has been investigated in greatest detail, the chemotactic behaviour of the cellular slime-moulds and the chemotaxis of mammalian leucocytes. The slime-moulds have many interesting features as models for differentiation, and their alternation between an unicellular phase and a multicellular 'organism', which has properties markedly different from those of the individual amoeba, offers an opportunity to investigate complex behavioural patterns. Mammalian leucocytes, because of their rôle in defensive reactions such as the acute inflammatory response, have received a great deal of attention, particularly in respect of the range of factors which are perceived, and in the mechanism of gradient perception.

In addition to these four systems, which will be considered in detail, the phenomenon is known to occur with various free-living unicellular organisms, the acellular slime-mould (*Physarum*), and the motile spermatozoa of many plants (for a review of some of these systems see Gooday 1981). It has also been described for a number of tissue cells, although it is not always clear that the possible contribution of other directional cues can be excluded. Few of these systems have been investigated in detail.

With the exception of bacteria, the majority of examples are of positive chemotaxis, which leads to the accumulation of cells in the area of highest concentration of factor; but negative chemotaxis may occur. Many things could be described as 'chemical' stimuli: a pH-dependent response, for example, or a response to the tonicity of the medium. In practice, the term is often taken to refer to situations that involve a specific substance for which the cell has a receptor system. Once the chemical is immobilized on a substratum, where it may affect the adhesiveness of the cells, it might perhaps more appropriately be discussed as haptotaxis, but we will mention surface-bound gradients of substances that normally would act in fluid-phase.

## 9.2  The problem – a theoretical analysis

Before discussing particular examples of chemotaxis let us analyze the problem faced by a cell in perceiving a gradient. The cell must have some means of measuring the concentration of the chemotactic factor, which means in most cases that there must be some form of cell-surface receptor. From what we know of other receptor-mediated phenomena, such as trans-membrane transport and the cellular response to peptide hormones and growth factors, the receptor is likely to be an integral membrane protein and is likely to have fairly high specificity for binding. This, however, restricts the ability of the cell to respond: it means that the exact composition of the chemotactic factor is known and familiar, which will not necessarily be the case if the response is directed to food, to pathogens or to damaged tissue (which may be abnormal in a multitude of ways). One solution would be for the receptor to be designed to recognize some common feature, perhaps some fundamental property of all chemicals in the category (Wilkinson 1982), but the gain in range will probably be at the expense of sensitivity. Because the cell must operate in a gradient of the factor the concentration-measuring system must work over a range of absolute concentrations, and this affects the sensitivity. It might seem easier to differentiate between zero and unity than between 99 and 100 but this neglects a fundamental aspect of receptor design. If the receptor system is to be used more than once then the interaction must be a reversible one (an equilibrium binding), or the receptor system must be continually renewed. Most specific binding interactions are reversible, and the status of the receptor, whether occupied or free, depends upon the concentration of free factor. Because this is a statistical phenomenon, at low concentrations the receptor is unoccupied most of the time (almost always zero and rarely one), and at high concentration the converse is true (nearly always 100 and rarely 99). Maximum sensitivity to change is at the mid-point when the absolute concentration is equivalent to the equilibrium binding constant for the receptor (Fig. 9.1) (Zigmond & Sullivan 1981). This argument will be familiar to those who have studied enzyme kinetics. An important implication is that the concentration-measuring device works most efficiently only at one position in the gradient, and both the top and bottom of the gradient are sub-optimal. It also means that the receptor system must be adapted to the concentration range which will normally be found in the environment, or else a number of receptor systems must be used (which would introduce other problems).

The task of the receptor system is not simply to measure concentration but to detect differences in concentration and to do so reliably. Because binding is an equilibrium reaction then receptor occupancy varies (in a predictable way: a normal or Gaussian distribution function) and the concentration measurement fluctuates; there is background noise over which the signal, a real change in concentration, must be detected. This sets limits on the ability of the system to respond although there are methods of improving signal recognition, the most common being averaging methods. In principle the cell could use either a time-averaging method, in which the statistical fluctuations ('noise') are smoothed by recording over a period of time or by averaging the signal from a large number of independent receptors (Segall *et al.* 1982, Lauffenburger 1982). An obvious restriction on time-averaging is that the cell is moving and, in a gradient, the external concentration might therefore change. A calculation of the time-constant that a neutrophil leucocyte would have to use, in order to detect signals of the size to which it has been shown to respond, suggests that its time-averaging system might have a time-base of

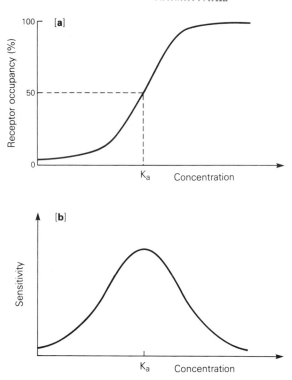

**Figure 9.1  Gradient perception at different concentrations of diffusible factor**  As the concentration of a chemotactic factor rises, the receptor occupancy changes as shown in (a). When the concentration is low, almost all the receptors are empty, and at high concentration almost all are full. Sensitivity is greatest where a small change in concentration causes the greatest change in receptor occupancy: around the level of the association constant, $K_a$. The ability of neutrophils to orient in a gradient (using a chamber such as that shown in 9.12) follows the pattern shown in (b) very closely, with an optimum around the $K_a$.

several minutes, comparable to the time-base for internal bias in the locomotory system, the persistence time mentioned in Chapter 7 (Lauffenburger 1982). A spatial averaging system will be limited by the distance between the receptors which, once large, introduces a time-factor for diffusion of the integrated signal.

Assuming the concentration can actually be measured, that the signal can be perceived, then the problem becomes one of detecting the gradient. Again

**Figure 9.2  Models for gradient perception**  Four possible methods by which cells might analyze the concentration of chemotactic factor in the environment are shown schematically. In pure spatial gradients there are three possibilities, (a–c). (a) Temporal sensing: this depends on the comparison of receptor occupancy at two different times, the cell having moved position between sampling. If the second concentration is greater than the first (which must be 'remembered'), then the cell is moving up-gradient. (b) Spatial sensing: depends on the cell comparing receptor occupancy on two parts of its surface at a single time: the side with the highest concentration is 'up'. (c) Pseudospatial sensing: relies on the cell making random protrusions up- or down-gradient. If the protrusion is up-▶

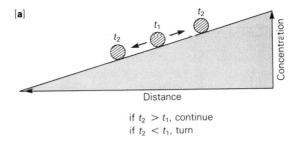

**[a]**

if $t_2 > t_1$, continue
if $t_2 < t_1$, turn

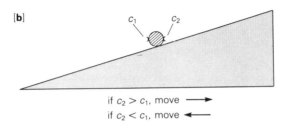

**[b]**

if $c_2 > c_1$, move $\longrightarrow$
if $c_2 < c_1$, move $\longleftarrow$

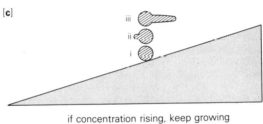

**[c]**

if concentration rising, keep growing
if concentration falling, retract

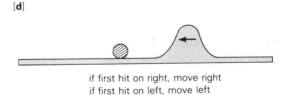

**[d]**

if first hit on right, move right
if first hit on left, move left

gradient then the concentration perceived by the tip will be increasing with respect to time; if increasing receptor occupancy promotes assembly of the protrusion then the cell will move in that direction (since movement involves protrusion). A down-gradient protrusion would not be reinforced. (d) True spatial gradients (in which the local concentration remains constant over a period of time) are probably rare, and it is worth considering the possibility illustrated, in which a 'wave' of high concentration is about to hit the cell. A simple rule would be to define the front as that portion of the surface which is first contacted by the wave.

this can be achieved in two ways, a spatial analysis or a temporal analysis (Figs 9.2 a & b) (Zigmond 1974). A spatial sensing mechanism would operate over the dimensions of the cell, comparing the concentration at two sites, and would require that there were at least two receptors or two sets of receptors. The introduction of more receptors would improve the spatial resolution but would reduce the difference between adjacent receptors, which are distributed over a finite surface. A temporal mechanism relies upon the movement of the cell to separate two measurements in space: the concentration measured at time $t_i$ is compared with the concentration at time $t_{(i-1)}$, where the unit time is such that the cell has had the opportunity to alter its position. Measurements must be made only at discrete time intervals, a continual updating of the 'remembered' concentration would remove the point of the system, and there must be some method for recording the concentration. Although it is somewhat difficult to visualize, this system could be used to affect the direction of turning: if the concentration decreased, a reversal of the previous turning direction would be the appropriate response; only by having more than one receptor–memory system could accurate steering be achieved.

A compromise solution is the mechanism described by Gerisch *et al.* (1975) for slime-moulds, by Zigmond (1982) and by Dunn (1981) for neutrophil leucocytes, which I have chosen to call 'pseudospatial' (Fig. 9.2c). By protruding a pseudopod or process the temporal change in concentration perceived by receptors on the protrusion as it moves through different concentrations of factor could be used to direct the cell using a simple rule: positive feedback to the protrusive system if the concentration increases with time, and inhibition of protrusion if the concentration remains unchanged or diminishes. In terms of the receptor system this rule is analogous to the adaptive behaviour of nerve cells which cease to respond quite soon to a constant signal but react to an increased stimulus with naive enthusiasm.

The latter mechanism also offers a solution to another problem, that of coupling gradient-perception to the motile machinery. If the receptor system is on the cell surface then the binding of factor to the receptor must be transformed into a cytoplasmic signal. The cytoplasmic signal must, in turn, act upon the motile machinery so as to alter the behaviour or distribution of the machinery, and so to affect the direction of movement. By analogy with other receptor systems we might expect that different chemotactic receptors interact with a common intracellular control system, which might also be the control system which is affected by chemokinetic stimuli.

There are, therefore, three major problems to be addressed in chemotactic responses: receptor design, gradient perception and receptor coupling to the locomotory machinery. These problems should be borne in mind when considering the systems described in the following sections.

## 9.3   Bacterial chemotaxis

Although strictly speaking this is a klinokinetic response with adaptation, the term 'bacterial chemotaxis' has become so firmly entrenched that it would be unreasonable to use any other. The description which follows relates particularly to the behaviour shown by *Escherichia coli*, that most popular of bacterial cells, but is probably applicable to other species. The basis of the response is an alteration in the direction of rotation of the bacterial flagellum (for details of the motor system see § 4.2 & § 5.2). Under normal circumstances the flagellum rotates anticlockwise for approximately 80–85 per cent of the time, causing the bacterium to swim in a smooth straight path with abrupt random changes in direction when, intermittently, clockwise rotation occurs. Since displacement in a random path depends upon turning frequency, as well as on speed, an alteration in the frequency with which an episode of clockwise rotation occurs will alter the pattern of movement (see Ch. 7, Fig. 7.1) (Berg & Brown 1972, Berg 1975, 1983). This in itself is sufficient to cause a redistribution of the population with accumulation at low concentration of the stimulant, where the displacement is lowest. To account for the directed response, however, an adaptive system is required (Hazelbauer 1981).

**Adaptation** may be defined as the resumption of the normal pattern of behaviour in the continued presence of the stimulant, and implies that the sensory system responds to changes in concentration rather than to the absolute concentration (though not to the rate of change) (Spudich & Koshland 1975). If the concentration of an attractant molecule is sharply increased then the probability of anticlockwise rotation of the flagellum is increased and the bacterium continues in a straight path because turning is inhibited. The bacterium will swim forward until adaptation occurs and the frequency of tumbling returns to normal. A further increase in the concentration of the attractant is necessary to maintain forward movement and inhibit the return to a normal frequency of turning. A decrease in attractant concentration will cause an increased frequency of turning which also adapts to the basal level with time. Thus a bacterium moving up a gradient of attractant and meeting higher concentrations will tend to persist in forward movement, a bacterium moving down-gradient will turn much sooner: only if the concentration begins to rise, as a consequence of the new path being followed, will the probability of a second turn be diminished. A cell moving at right-angles to the gradient will turn with normal frequency, but if it turns up-gradient further turns are less probable than if it turns down-gradient (Fig. 9.3). The mechanism of gradient perception is effectively temporal, it is the altered concentration in the new locality entered by forward movement that influences the probability of turning. The speed of movement during a period of forward smooth swimming does not seem to be altered. Turns are made in random directions, which is the reason for distinguishing this klinokinesis with adaptation from a 'true' chemotactic response in which the direction of the turn would be influenced by the vectorial cue offered by the gradient. Nevertheless, the overall path of the bacterium exhibits a bias in direction that leads to accumulation of cells in the area of greatest concentration of attractant. The converse is true for repellants.

The problems in bacterial chemotaxis are therefore to determine the basis upon which the motor is controlled, and to explain the phenomenon of adaptation. A variety of environmental stimuli lead to a directed response and these diverse stimuli must be integrated into a single 'gear-box' (forward/reverse) control: adaptation occurs at all levels of external concentration, therefore the response must be to change with respect to time rather than to the absolute values of attractant concentration.

There appear to be two distinct sets of environmental stimuli to which bacteria will respond: those such as light, temperature, oxygen tension and pH, which act directly on the cell to alter the cytoplasmic *milieu* and the motor, and those which are recognized by externally disposed receptor systems and which act directly upon the cytoplasm (Macnab 1982). The latter system may be regarded as a more sophisticated one by which the cell may maximize its opportunities for growth by moving into areas rich in nutrients

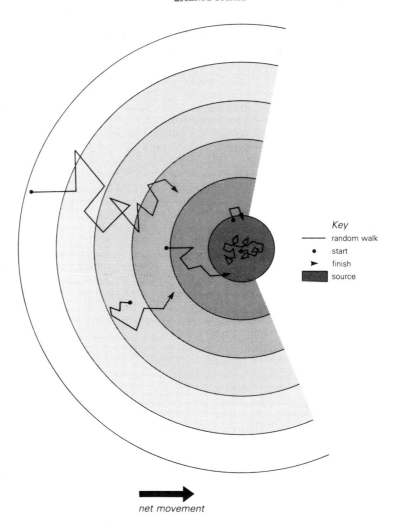

Key
⎯⎯⎯ random walk
● start
► finish
▨ source

net movement

**Figure 9.3 Bacterial chemotaxis** A diagrammatic illustration of biased movement of bacteria in a gradient of attractant. Straight runs are longer up-gradient than down-gradient, and if the cell is moving down-gradient the probability of turning is much higher. These simple rules lead to accumulation at the source provided there is adaptation (see text).

(amino acids and sugars being the attractants), the former as a means of avoiding life-threatening environments. Certainly the receptor-mediated system is more complex, and involves many more gene products, but the distinction between the two systems should not be taken too seriously – a photosynthetic bacterium undoubtedly improves its position by a positive phototactic response, although the photoreceptor acts by modifying the cytoplasm directly.

## 9.3.1 Responses which are not receptor-mediated

Not only does the motor system depend upon the proton motive force (PMF) for rotation (see Ch. 4), but the probability of reversal is sensitive to PMF-level (Macnab 1982). If the PMF decreases to very low levels, the probability of turning is reduced and the bacterium swims in a straight path, an escape reaction: moderate diminution of the PMF leads to a transient increase in the probability of turning, a tactic response which will maximize the chances that cells accumulate in regions where the PMF will be high. The status of the cytoplasm with respect to the PMF is, of course, a measure of the suitability of the environment, since the proton gradient is central to the function of the synthetic machinery. Because the operation of the motor depends upon the proton gradient, the effect of reducing PMF is to decrease the speed of rotation. It may be that only when the PMF is low enough to affect speed does it happen that the switching probability is sensitive to changes in PMF; this is not to imply that there is an orthokinetic component to the tactic behaviour, only that the PMF sensitivity becomes evident when PMF levels are limiting.

Various tactic responses are probably mediated through the sensitivity of the motor to the PMF-level, including phototaxis in photosynthetic bacteria such as *Rhodospirillum rubrum* and *Halobacterium halobium* where proton pumping depends upon light. The action spectra for the behavioural responses correspond closely with the absorption spectra of the pigments, and in *Halobacterium* decreasing the light intensity causes tumbling and is associated with a decrease in membrane potential. Taxis towards oxygen (aerotaxis) seems to depend upon the electron-transport-driven development of PMF and can be inhibited with cyanide and other inhibitors of respiration. Uncoupling agents, which interfere directly with the PMF, also induce tumbling and the more powerful **uncouplers** are more effective repellants. The PMF depends upon both membrane potential and proton concentration, and changes in extracellular pH may therefore affect behaviour by affecting the PMF. Changes in extracellular pH will, however, affect internal pH, which alters the motor through another mechanism and makes interpretation of the effects of extracellular pH-change more difficult, especially since changes in cytoplasmic pH are regulated through an homeostatic mechanism which takes a finite time to operate. Cytoplasmic pH changes induced by external pH change (by weak-acid repellants such as acetate and proprionate, and by weak-base attractants such as ethanolamine and $NH_4^+$) may act through the chemotaxis methylation system that processes the signal to act upon the motor (see next section). Temperature changes may also act through this system: a sudden decrease in temperature causes tumbling and sudden increase causes a smooth response.

## 9.3.2 Receptor-mediated chemotactic responses

Largely from the analysis of chemotactic mutants of *E. coli* (and a rather limited range of other bacterial species) it has been possible to identify four classes of receptor-mediated chemotactic responses, each operating through a different methyl-accepting chemotaxis protein (**MCP**) (see Fig. 9.4) (Springer *et al.* 1979, Hobson *et al.* 1982). In some cases the interaction of the chemotactic factor is directly with the MCP, in others (the sugars) the response is mediated by periplasmic binding proteins, which can also interact with the transport systems for sugars or amino acids. This illustrates very clearly one mechanism of acquiring (evolving) a chemotactic receptor system: the utilization of a receptor which was originally 'intended' for a

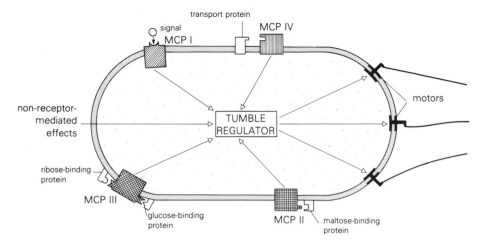

**Figure 9.4 Methyl-accepting chemotaxis proteins (MCPs)** The four MCPs respond to different factors, either directly or indirectly, and their response is integrated by the 'tumble regulator', which determines the probability of the flagellar motor going into reverse. Non-receptor-mediated responses are thought to act directly on the tumble regulator. Although only one MCP of each class is shown, there are many receptors of each type (see Table 9.1).

**Table 9.1** Methyl-accepting chemotaxis proteins

|  | Gene | Direct | Indirect | Sugar |
|---|---|---|---|---|
| MCPI | tsr | serine, acetate, threonine, (leucine, indole, temperature?) | | |
| MCPII | tar | aspartate (Co$^{2+}$, Ni$^{2+}$) | maltose-binding protein | maltose |
| MCPIII | trg | | ribose- and glucose-binding proteins | ribose glucose or galactose |
| MCPIV | tap | | enzyme II of phosphotransferase transport system | fructose |

In some cases the ligand interacts directly with the MCP ('direct'), in other cases the interaction is mediated by a binding protein which then interacts with the MCP ('indirect').

simpler process. Since sugars offer a valuable nutrient source, it is scarcely surprising that transport systems exist for these molecules; a bacterium which could also respond chemotactically to such nutrients by linking sugar binding to the motor system should then be at a distinct advantage. The four MCPs are listed, together with their ligands and the binding proteins with which they interact, in Table 9.1.

The MCPs are all integral membrane proteins (Wang & Koshland 1980) that have glutamyl residues exposed on the cytoplasmic face. They are the targets for methylation from S-adenosyl methionine by the *che X* gene product, a methyltransferase, and demethylation by the *che B* gene product, demethyl-esterase. At least three and possibly four sites may be methylated on each of these proteins, which offers the possibility of a complex and graded response (Hazelbauer 1981). Mutation in *che* genes leads to a loss of all receptor-mediated tactic responses, the response to temperature change and the response to weak acids. Mutation of one of the MCP genes leads to the selective loss of particular tactic responses, leaving others intact. MCPI may well be the primary system from which the other indirect systems were developed, and there is considerable homology in the nucleotide sequences of the genes for MCPI and MCPII.

Enhanced methylation of MCPs may result from a conformational change induced by temperature (MCPI), by direct binding of an amino acid (MCPI & II), or indirectly as a consequence of the conformational change in a sugar-binding protein which increases the affinity of a specific interaction between the binding protein and an MCP. Notice at this stage that we have linked a diverse set of receptor systems to a much simpler parameter, the methylation of certain glutamyl residues on the cytoplasmic face of the plasma membrane.

## 9.3.3 Adaptation

Adaptation is a phenomenon shown by many sensory systems in which the output of the sensory system returns to baseline levels despite the continued presence of the stimulus; further sensory output will only occur if there is a change in the level of the stimulus. Such an adaptive response in the sensory system enables us, for example, to ignore constant features of the environment and concentrate on alterations. Thus the bacterial flagellar motor adapts by returning to its normal frequency of reversal from the higher or lower turning frequency induced by increased or decreased concentration of a chemotactically-active substance (which would be negatively chemotactic if it raised the probability of turning when the concentration was increased). When an attractant is added the methylation of one of the MCPs increases to a new plateau level, whereas the addition of a repellant causes a decrease in methylation. An increase in the concentration of an attractant will, of course, decrease the frequency of turning at first, but if the concentration does not alter then the turning frequency will return to normal after a short time. The time required for the new plateau level of methylation to be attained is approximately the same as the adaptation time, and if methylation is blocked adaptation does not occur. Mutants which lack methylating activity are unable to adapt and are chemotactically incompetent, revealing incidentally the essential rôle of adaptation in the response. This system is, however, adaptive in its own right, the extent of methylation seems not to matter, only the process of change. One explanation would, of course, be that the transient conditions which favour a change in methylation are also the conditions that affect the behaviour of the motor and the methylation or demethylation is a side effect. However, this is not entirely consistent with the blockage of adaptation when the activity of the methyltransferase is inhibited.

Under normal, non-stimulated conditions the level of methylation reflects a balance between the activity of the methyltransferase and the demethyl-esterase. Attractants cause an inhibition of demethylation for several minutes, the period required for adaptation, and repellants enhance the demethylation reaction. These effects may be mediated by an attractant-induced rise or repellant-induced fall in the levels of cyclic GMP (cGMP). Exogenous manipulation of the cytoplasmic cGMP level alters the behaviour, and a behavioural mutant with high tumbling frequency has low cGMP levels. A rise in cGMP leads, however, to an increased methylation of all MCPs, as does the binding of an attractant; but the methylation of the MCP classes not involved in the binding of attractant falls to the basal level. Thus it may be that the change in cGMP is the excitatory stimulus for an alteration in motor activity, whereas the stabilization of the level of methylation constitutes the adaptatory response. Exactly how a steady state of methylation would cause the motor behaviour to return to the basal level is not clear and at present it seems that part of the puzzle is missing.

One possible explanation is that suggested by Boyd *et al.* (1982) who propose that the total population of MCPs produces a signal which acts on the motor system and that the signal strength is a function of the extent of methylation (Fig. 9.5). On first binding an attractant the signal is produced in supra-threshold amounts because the excitatory stimulus is acting on a population of MCPs adapted to a lower level. Methylation of the MCP particularly involved in the binding then reduces the output of that MCP, and the population response, in which the other MCPs are modified, brings the signal back to the normal level. The signal 'concentration' would be the product of the whole MCP population which has the effect of smoothing the response, a necessary feature when only three or four 'states' of each MCP exist. Because both increases and decreases of methylation can occur it is necessary to suppose that the normal state is around a threshold (rather than zero) and that random fluctuations in the signal strength, above and below the threshold level for motor reversal, would account for the random pattern of turning in a non-stimulated population of cells. The hypothetical signal could either be a molecule or a membrane potential; in view of the direct effect of PMF-change on the behaviour of the motor it is tempting to favour the latter.

In summary it seems that the mechanism of bacterial chemotaxis depends upon control of motor behaviour through a fairly complex biochemical system which incorporates feedback mechanisms, which integrates the response to different substances bound by different classes of receptor and which may graduate the response by averaging the output of a population of receptors – a means of improving the signal–noise discrimination. Even though bacterial motility seems fairly simple in behavioural terms, the products of some 40 genes are involved in the structure, rotation, and control of the flagellum: when we move to more complex behavioural patterns we can hardly expect to find fewer genes involved.

## 9.4   Chemotaxis in *Paramoecium*

All tactic responses in *Paramoecium* and probably in many other ciliates depend upon an avoidance reaction, first described by Jennings in 1906, which involves the transient reversal of ciliary beat (see Chs 3 & 5 for descriptions of the mechanism of ciliary beating and the control of reversal). Ciliary reversal leads to a brief rearward movement before restoration of the normal beat pattern; because the cilia are not wholly symmetrical in their beat pattern, nor in their contribution to forward propulsion, the shift from reverse to forward movement is associated with a turn (the organism pivots about its rear). The distribution of cilia, the duration of the reversal, and the precise timing of the resumption of a normal beat pattern in different areas of the cell surface may all contribute to the change in direction of movement, and since the beat frequency, both forward and reversed, can be varied in a

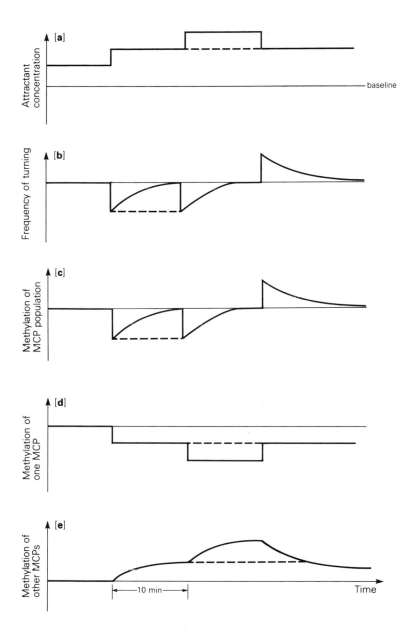

**Figure 9.5 Methylation of MCPs and adaptation** The adaptation system in bacterial chemotaxis may work in the manner shown here. Stepwise changes in concentration of a factor which binds to one MCP cause an abrupt alteration in the turning frequency, which then slowly reverts to the baseline level (adapts). Methylation of the MCPs increases as the cells adapt to the increased concentration of attractant and their tumbling frequency returns to normal. It may be that a rapid demethylation of the MCP to which the factor binds is followed by a slow increase in methylation of the other MCPs, so that the overall methylation level returns to normal once the adaptation is complete.

graded manner the possibilities for complex movements are obvious. This can be illustrated by an example. In response to a slight mechanical touch in the anterior region of the cell there is a weak ciliary reversal, and the greater contribution of the cilia of the oral groove leads to a turn in the aboral direction without rearward movement; a stronger mechanical stimulus elicits a more dramatic reversal and the cell moves backwards, slows down and resumes forward movement. During the shift from reverse to forward movement the oral groove cilia again become temporarily dominant and an aboral turn precedes forward movement (Fig. 9.6). Mechanical stimulation of the posterior of the cell increases beat frequency over the whole cell which accelerates its forward movement. Even this simple description is scarcely adequate since the cell actually moves forward in a helical fashion, with rotation of the whole cell body.

Spontaneous reversal of the ciliary beat does occur and the normal progression of the cell is one of 'straight' runs with intermittant turning. The response to environmental stimuli may be a klinokinetic response rather than a tactic one and there may be an orthokinetic component. The behavioural responses have received rather less attention than the transduction system, the excitation–response coupling.

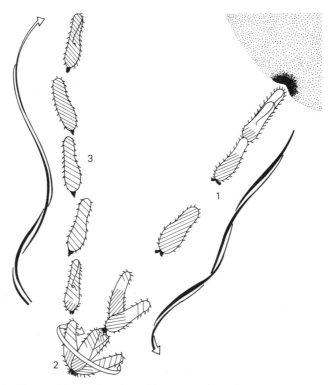

**Figure 9.6   The avoidance reaction of *Paramoecium***   The so-called avoidance reaction exhibited by *Paramoecium* is illustrated in this diagram which is taken from Grell (1973). The ciliate, having encountered a noxious stimulus (1), goes into reverse, rotates on its axis as normal beating is resumed (2), and sets off in a random new direction (3). The new direction is presumably determined partly by the relative rates of rotation and resumption of normal beating, but to a first approximation seems to be more-or-less random.

Ciliary reversal is probably controlled by changes in intracellular calcium concentration (Naitoh & Eckert 1974, see Ch. 5) and the common integrator of all tactic responses seems to be the plasma membrane. Intracellular calcium levels are normally below $10^{-6}$M but the membrane has voltage-sensitive calcium channels, and depolarization of the membrane from the resting potential of around $-50$ mV leads to an increase in calcium conductance. The influx of calcium ions contributes to the depolarization, which is shunted by the transient efflux of cations, chiefly potassium. The properties of the membrane are reminiscent of those of the membranes of electrically excitable cells such as nerves, although the depolarization is a graded response and not an 'action potential'. Restoration of intracellular calcium ion concentration to resting levels depends on some sort of pumping system, also located in the membrane, which has a longer time-constant than the transient influx of calcium which is moving down a steep gradient of concentration and of electrical potential. Adaptation is therefore a consequence of the slow return of the membrane potential to its resting level.

Various stimuli will induce a change in membrane potential, either a depolarization, which leads to ciliary reversal, or hyperpolarization, which causes an increase in beat frequency (Van Houten 1979). The range of responses of which membranes are capable, when modified by the incorporation of receptors or of light-sensitive pigments, can be gauged by considering the sense organs of multicellular animals, the responses of which are nearly all based on membrane depolarization. Direct modification of the resting potential of *Paramoecium* is probably the basis of galvanotaxis, and changes in membrane permeability as a result of mechanical distortion probably account for mechanosensitivity. It is, however, necessary to account for the differential responses over the membrane with the anterior region responding to mechanical contact with depolarization, the posterior with hyperpolarization.

Many chemoresponses of *Paramoecium* can be explained on the basis of directly induced membrane potential changes (cations produce an avoidance response more readily than anions, for example), but others probably require some receptor mechanism, an aspect which has attracted less attention. Obviously it would seem sensible to locate chemosensitivity in the anterior portion of the cell, and this does seem to be the case. The depolarization of one region of the membrane spreads electrotonically through the cytoplasm and all cilia will respond. From a consideration of the cable constants (the parameters which regulate the rate of transmission of an impulse along an insulated cable such as a nerve), treating *Paramoecium* as a cylinder, the decrement in depolarization is calculated to be less than 1 per cent over the length of the cell and should spread at 100 $\mu$m sec$^{-1}$, values which are in accord with experimental observations using microelectrodes inserted at front and rear. Although the depolarization is global, the ciliary response may differ locally and account for some of the behavioural reactions seen when the depolarization is small. Anterior cilia appear to have a lower threshold of depolarization for reversal, which offers complex possibilities for changes in direction of movement if the topographic distribution of sensitivity is asymmetric – as indeed it appears to be. A further possibility in control, more likely to be important in species where the ciliary beat pattern is not metachronous (see Ch. 5), is that the membrane of the cilium may be the site of receptors and each cilium operates as a unit, with little exchange of calcium between axonemal and cytoplasmic compartments.

An interesting mutant of *Paramoecium* is the so-called 'pawn' mutant, which fails to exhibit ciliary reversal and, like the chess piece, moves only forward. The lesion appears to be in the voltage-sensitive calcium channel, which in normal cells regenerates the depolarization (Oertel *et al.* 1977). Permeabilized pawn mutants will show reversal when the calcium level rises above $10^{-6}$M, indicating that the motor is not defective.

## 9.5 Chemotaxis in the cellular slime-moulds

The cellular slime-moulds provide a valuable model system in which to investigate a number of fundamental processes and, because large numbers can be grown easily in the laboratory, they are more amenable to biochemical study than many behavioural systems. Under favourable conditions the amoebae live as unicellular organisms feeding on bacteria; when food becomes scarce the unicellular amoebae aggregate into a multicellular slug, or grex, which then differentiates into a fruiting body. A proportion of the constituent amoebae form resistant spores and the remainder 'altruistically' contribute to the stalk and base of the fruiting body (notice the insidious anthropomorphism; a dangerous trap!). The simple life cycle is shown diagrammatically in Figure 9.7. Apart from remarking on the fact that the formation of a multicellular aggregate and its subsequent differentiation are models for the evolution of permanently differentiated multicellular organisms, we will largely restrict ourselves to a discussion of the mechanism by which aggregates are formed. The general aspects of slime-mould biology are reviewed extensively in Loomis' book (1982). If amoebae of different mating strains are present in the population a sexual phase of the life cycle may occur in which a diploid zygote is formed, which engulfs other amoebae before forming a resistant macrocyst. Subsequently, under favourable conditions, the macrocyst releases haploid amoebae, meiotic products of the zygote. Although the details of this sexual phase are irrelevant to the motile and chemotactic behaviour, it is an important aspect of slime-mould biology because it opens the possibility of genetic analysis of various behavioural activities.

Free living ('vegetative phase') amoebae show chemotactic responses both to their bacterial prey, and (negatively) to each other (Newell 1981). Positive taxis towards bacteria has been shown for several species, including *Dictyostelium discoideum* and *Polysphondylium violaceum*, and the attractant seems to be folic acid, an essential vitamin for the slime-mould, which it will normally obtain from bacteria. Both *D. discoideum* and *P. violaceum* release an enzyme which destroys folic acid, presumably to keep the background level low and to facilitate the detection of folic acid released by bacteria. This is analogous to the role of cholinesterase in the synaptic cleft of cholinergic neurons of vertebrates (such as the motor end-plate), and is a simple device to improve the signal : noise ratio. A negative chemotactic response to unidentified products of amoebae of the same species has also been described. Negative chemotaxis in the vegetative phase will aid dispersion of the amoebae and therefore permit better utilization of the environment. (An exception may be the carnivorous slime-mould *D. caveatum* (Waddell 1982).)

The chemotactic response which has received by far the most attention is, however, that observed in the aggregation of amoebae to form the multicellular body, which will occur on agar plates under laboratory conditions and can therefore be studied very conveniently (Devreotes 1982). Several interesting points emerge from considering this chemotactic system. Some 4–6 h after starvation the chemotactic system is switched on and

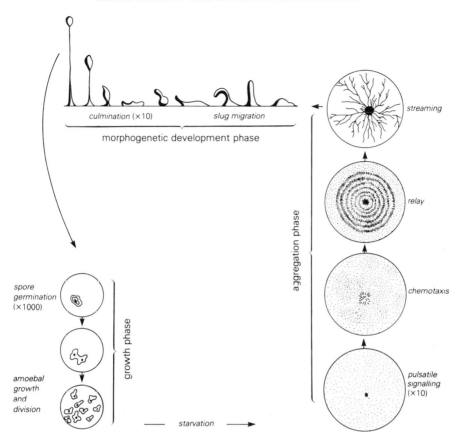

**Figure 9.7 Life cycle of the cellular slime-mould *Dictyostelium discoideum*** Spores germinate to produce amoebae, which are free-living phagocytic cells, about 10 μm in diameter. Starvation induces the aggregation phase in which a few cells begin producing pulses of cAMP. In response to these pulses, aggregation-competent amoebae begin moving periodically towards the aggregation centre, relaying the signal and producing a pattern of concentric bands of moving and stationary amoebae (see Fig. 9.8). Eventually the simple concentric pattern breaks down and amoebae move towards the aggregation centre in streams of cells (which are 'contact-following'). Once the cells have aggregated a slime sheath is produced and the grex, or slug, migrates rather like an inch-worm or looper caterpillar. The grex may move for many days before culmination, a process in which the differentiated cells sort out into a stalk and a mass of spore-forming cells, which climbs to the top of the stalk. The spores may be dispersed and are a resting phase; their germination completes the cycle (from Newell 1978, but reproduced in Newell 1981).

individual amoebae begin to move towards collecting centres. The collecting centres are, presumably, those cells which first reach the 'active' stage. Movement of amoebae towards the collecting centre is discontinuous, with rhythmic forward steps separated by quiescent periods: the microscopic appearance of the cells under dark-ground illumination is different in the motile and quiescent stages, so the distribution of movement over a territory can be visualized directly. The direction in which the cells move is determined by a chemotactic signal which initiates motile activity in

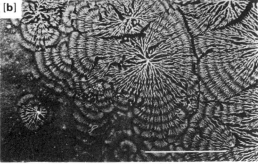

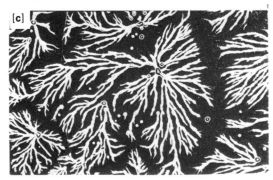

**Figure 9.8 Phases in the aggregation of *D. discoideum*** In (a) concentric waves of amoebae can be seen moving towards aggregation centres in response to a pulse of cAMP (to which they will respond by releasing cAMP, thereby relaying the signal). Between pulses the cells move more slowly and randomly and appear dark in the dark-field optics used here. Once streams of cells begin to move towards the aggregation centre (b) the amoebae show contact following. The stream provides a strong signal and individual cells may move towards a stream and away from the centre, although in the end they will join the aggregate. Eventually all the cells become incorporated into streams, (c). The photographs shown here are of the same field at one hour intervals; scale bar represents 10 mm (from Newell 1981).

quiescent cells and which diffuses outwards as a centrifugal wave from the collecting centre. In the simplest case, shown diagrammatically in Fig. 9.8, pulses of chemotactic factor, cyclic AMP (cAMP) for *D. discoideum*, are released from the centre and the chemical simply diffuses out. Quiescent amoebae receiving a pulse of cAMP move up-gradient for some 1–2 min and then stop. Not only do the cells receiving this pulse move toward the centre, but they are also stimulated to release a pulse of attractant themselves, thereby acting as relay stations.

This relay signalling is efficient at propagating the signal over considerable distances: the concentration in a simple diffusion gradient from a point source falls exponentially with distance from the source, whereas a relayed signal will give rise to a linear gradient. Thus, if there is a minimum concentration of factor required to initiate a response, this minimum concentration will be located much further away from the primary source. The rhythmic movement of amoebae towards the centre, and the outward propagation of the pulse, depend upon several factors: an initial pulsatile release from the aggregation centre, the duration of the refractory period during which amoebae are unable to respond to a second signal (quite a short period, half-time around 1 min), and the relay-refractory period (a rather longer period, half-time approximately 5 min) in which they are unable to relay a second signal. The relay response, the release from the cell of cAMP, depends upon an increase in cAMP binding, and the release brought about by the rise in receptor occupancy continues for some time provided that the extracellular cAMP concentration remains at the new high level. The response to increased cAMP concentration is, however, adaptive and even if the stimulus persists the relay signal diminishes with time (Fig. 9.9). A second stimulus of comparable intensity to the first produces a much smaller release in response to the alteration in cAMP level.

For a relay system to operate there must clearly be a refractory period in the motile response, otherwise the movement of the cells would reverse as cells further away from the centre release their pulse of attractant. As with many receptor-based signalling systems the 'noise' level will rapidly smother the signal unless some mechanism exists for destroying the attractant. In *D. discoideum* the cAMP signal is broken down by **phosphodiesterases**, one of which is bound to the external surface of the plasma membrane, and one released to the medium. The importance of the phosphodiesterase in

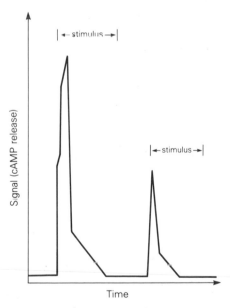

Time

**Figure 9.9 Adaptation in *Dictyostelium*** The release of cAMP by amoebae, which have themselves been stimulated with cAMP, decreases with time even if the signal remains constant; a second stimulus elicits a smaller response if it is of the same intensity as the first signal. In a sense two kinds of adaptation occur: adaptation to a constant stimulus (the decline of the first release to zero over a 30 min period), and the adaptation to intermittent signals of constant intensity (based on Newell 1981, Fig. 8).

maintaining signal detectability is well illustrated by the occurrence of mutants with lesions in the phosphodiesterase system, which will only aggregate successfully if exogenous phosphodiesterase is added.

Gradient perception will involve, as in systems discussed earlier, the binding of the attractant to receptors, thereby triggering both a polarized locomotory response and, because of the relay system, triggering a burst of release of attractant. The cAMP receptors of *Dictyostelium* have received a considerable amount of attention and are interesting in that they exhibit rather complex binding characteristics (Newell & Mullens 1978). A **Scatchard plot** of binding data is non-linear, indicating either that there are multiple classes of receptor, or that receptor affinity is variable because of cooperativity. Variability in the binding may be biologically important in enabling the amoebae to respond over a wide range of attractant concentration and thus to make use of 'longer' gradients. This sort of variability in receptor affinity may also exist in other chemotactic receptor systems, such as the formyl-peptide receptor of mammalian leucocytes. The intracellular sequelae of receptor occupancy which occur in the right temporal range (a few seconds), and which could account for the observed signalling and response characteristics of the amoebae, include increases in the intracellular concentration of cGMP, a rapid influx of calcium ions and a rapid proton release. Any of these might link receptor binding to locomotor response, and changes in calcium concentration are known to affect actomyosin-based locomotory machinery (see Ch. 2).

It seems likely that for *D. discoideum* the outward wave of attractant may be a prerequisite for a directed response, and there is some evidence that a static gradient of cAMP is not interpretable (Vicker *et al.* 1984). The front of the cell is defined, thereby directing the locomotion, as the region in which increased receptor occupancy first occurs, rather than the region of highest receptor occupancy. Thus the temporal form of the gradient is important as well as the concentration gradient itself.

The simple pattern of concentric rings of moving and quiescent amoebae shown in Figure 9.8a (which shows the centrifugal pattern of the waves), often breaks down and a spiral pattern may develop. The reasons for this are quite complex and those interested should consult more specialized reviews such as that of Cohen and Robertson (1971). After a time both concentric and spiral patterns are modified to give radial streams of amoebae moving towards the centre (see Fig. 9.8c): this is the result of inhomogeneities in the distribution of amoebae which makes the relayed signal non-uniform. This is more obvious once cells have formed a stream; the signal simultaneously released from the cells in the stream is a much stronger stimulus than that from the centre and the stream will recruit more cells, further enhancing its signalling capacity. In consequence, individual cells may move towards a stream even if this involves movement tangentially with respect to the original centre or even centrifugally. Because the stream continues to move toward the prime centre the end result is still an accumulation of cells at a single centre. Amoebae in the radial streams move in a head–tail fashion which is partly stabilized by specialized contact sites (contact sites A; Gerisch *et al.* 1975, 1980) on the front and tail of aggregation-competent cells, adhesion sites that are not expressed on vegetative phase amoebae. The nose–tail movement can be seen in the radial streams, the phenomenon known as **contact following** (Fig. 9.10), depends upon the non-random distribution of contact sites A, over the cell surface, and provides a strong extrinsic polarity. Once a cell is committed to a stream then movement towards the aggregation centre is stabilized.

Competition between adjacent aggregation centres will occur, and it is fairly clear that the concentration of chemotactic factor will vary in a complex fashion over a crowded territory. The size of the multicellular aggregate (which in *D. discoideum* is approximately $1-2 \times 10^5$ cells) will be affected by competition between centres, and mutants with disturbed patterns of aggregation may form much larger fruiting bodies. Competition between centres is probably not solely through gradient interactions, there is some evidence for 'dominance' by older established centres in a territory, although the mechanism is unknown. In the natural habitat the aggregation process must elicit a species-specific response, and in experimentally mixed cultures the different species aggregate independently as might be expected. Different chemotactic factors would be one obvious method for generating specificity, although the evolution of a new receptor system, linked in an appropriate fashion to the locomotory machinery, would seem to render speciation rather an improbable event.

The directed movement of the multicellular aggregate, the slug or grex, poses other problems, most obviously that of coordination: 100 000 dissenting voices are scarcely a recipe for united action. The grex does in fact exhibit so-called tactic responses to light, temperature, humidity, oxygen and ammonium ions, and there must therefore be some fairly sophisticated coordination. Biologically it is clear that tactic responses will be of adaptive value for the grex: the vegetative amoebae, living in damp forest litter, are poorly placed for dispersal and the grex must move up to the surface layers, into the light, to produce the stalk and cap. (Dispersal depends on the sticky spores adhering to a passing animal.) Exactly how movement of the grex and the cessation of movement are coordinated is not clear, although there is some evidence for pulsatile movements in the body of the grex with a periodicity comparable to that of the aggregation-phase signalling, and in *D. discoideum* gradients of cAMP within the grex are known to be important. Amoebae within the grex are arranged in rows with cells joined head–tail, as in the aggregation streams, and there are also some lateral adhesions: some degree of mechanical coordination and coherence is therefore possible. At least three differentiated zones can be recognized within the grex, and cells from the different zones will give rise to the various parts of the fruiting body.

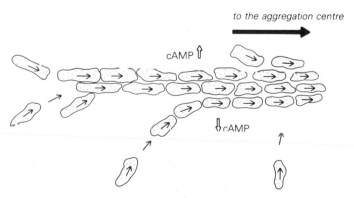

**Figure 9.10 Contact following**  A diagram to illustrate the orientation of individual cells within and approaching an aggregation stream. Cells within the stream adhere end-to-end through 'contact sites A', a new class of adhesion molecules which are expressed as aggregation competence is acquired.

The tip of the grex, which is of pre-stalk cells, may signal to the posterior zones by means of relayed signals, and is also presumably the area in which sensory analysis of the environment takes place. Fascinating though the problems of coordination of movement in the multicellular aggregate are, they are beyond the scope of this book and we will leave them here.

## 9.6 Chemotaxis in Myxobacteria

A very similar pattern of behaviour, the aggregation of individuals to form a fruiting body, is also seen in the Myxobacteria, Gram-negative rod-like bacteria which exhibit a gliding motility (Kaiser 1979). The movement of these bacteria is associated with the production of slime-trails and has been analyzed genetically; at least 30 gene loci are involved in movement and the control of movement. The motor mechanism remains obscure although the system may be basically the same as in other bacteria (see § 4.3). Like the cellular slime-moulds, the Myxobacteria aggregate under starvation conditions and produce resistant myxospores; the behaviour is remarkably similar in many respects although the scale is much smaller. Whether chemotactic gradients play any part in the aggregation has not been proven: those interested will find the topic reviewed by Clarke (1981).

## 9.7 Chemotaxis of neutrophil leucocytes

The neutrophil leucocyte, the primary phagocytic cell of the blood, has probably received more attention in regard to chemotaxis than any other cell type, partly because of its medical importance, and partly because it is an easily accessible cell with well-developed locomotory ability. Its rôle *in vivo* is most dramatically obvious in acute inflammatory reactions, although it probably plays a rôle in general tissue maintenance (Wilkinson & Lackie 1979). Following tissue damage or local infection, neutrophil leucocytes emigrate from the post-capillary venules adjacent to the lesion and accumulate at the focus of the inflammation where they will phagocytose bacteria and digest, intra- or extra-cellularly, damaged tissue components. Their emigration from blood vessels is rapid and they constitute the first line of defence against insult; neutrophil malfunction whether in locomotion (lazy leucocyte syndrome), or in bacterial killing (chronic granulomatous disease of childhood, for example) has serious consequences and recurrent bacterial infections commonly prove fatal within the first decade of life. Even though the neutrophil leucocyte has clearly been demonstrated to be capable of a chemotactic response *in vitro*, the evidence that movement towards the lesion *in vivo* is directed by chemotaxis remains largely circumstantial. Although it would facilitate the accumulation of cells, chemotaxis is not the only possible mechanism, as has been shown experimentally *in vitro*, and those who invoke chemotactic cues whenever cells accumulate do well to realize how difficult it is to prove that a directional response by the cells is actually involved. Nevertheless, it is generally held that chemotaxis is important for neutrophils and we will proceed on this basis.

### 9.7.1 Factors

The range of insults which may require a defensive reaction by the body is extremely wide and an obvious problem is in designing a sensory system which will handle all the cues (Wilkinson 1981 & 1982). The substances which, *in vitro*, elicit a chemotactic response from neutrophils fall into four major categories:

(a) complex lipid derivatives, products of arachidonic acid metabolism, particularly the leukotrienes;
(b) small N-terminal-blocked peptides, of which the best known is the synthetic tripeptide formylmethionyl-leucyl-phenylalanine (fMet-Leu-Phe);
(c) a complement-derived peptide C5a of *ca* 9 kd generated by proteolytic cleavage of C5 in both the classical and alternate pathways of complement activation;
(d) denatured proteins.

## LIPID FACTORS

The lipid factors are products of the complex enzyme system which acts upon arachidonic acid to generate prostacyclin, prostaglandins and thromboxanes, all of which have important effects on the tissues and which are produced in varying amounts by many of the cells of the reticulo-endothelial system. Only for 5,12-dihydroxytetraenoic acid (5,12-diHETE or LTB-4) is there evidence for a chemotactic response by neutrophils, but the lipid metabolites are receiving so much attention from pharmacologists that this may prove to be the tip of an iceberg. One possibility, since neutrophils themselves produce LTB-4 (Ford-Hutchinson *et al.* 1980), is that this is an internal amplification system rather than a major environmental cue, particularly since the half-life of such lipid metabolites is short.

## FORMYLATED PEPTIDES

The formylated peptides are extremely potent chemotactic and chemokinetic factors and although a formylmethionyl (fMet) start is common it is not essential. One speculation is that these synthetic peptides are being recognized by a system designed to respond to prokaryotic proteins, which differ from eukaryotic proteins in having a fMet start. This would be a simple criterion by which to recognize microbial proteins and is therefore a seductive hypothesis.

## COMPLEMENT-FRAGMENT C5a

The complement-fragment C5a has, in addition to its effect on neutrophils, an effect on smooth muscle, causing bronchoconstriction and vasoconstriction. Here too, as with the lipid metabolites, the signal substance is involved in triggering a range of defensive responses. The production of C5a from C5 occurs whether complement activation is by the classical pathway, which involves antibody–antigen complexes, or by the alternate pathway, which is initiated by various surfaces such as those of yeast particles. The alternate pathway is an immediate humoral response to a range of foreign materials and enlarges the repertoire of response by the neutrophil: the classical pathway, although requiring antibody directed against the foreign object and therefore of value to the primary phagocytic response only in reinfection by a pathogen, also enlarges the range of the sensing system. By responding to C5a the neutrophil borrows, as it were, the recognition system of the humoral immune response.

DENATURED PROTEINS

Denatured proteins occur in tissues through wear-and-tear as well as specific insult. An almost infinite range of denatured proteins must exist, since any protein can be damaged in any number of different ways. This poses an obvious problem for a clearance system, and the recognition of denaturation must rely upon some feature common to most or all denatured proteins, but absent from the native molecule. The tertiary structure of most proteins depends in part upon exposing hydrophilic residues to the aqueous environment. Denaturation, which affects the tertiary structure, leads hydrophobic residues to be exposed and might therefore constitute a marker (Wilkinson 1973). By covalently linking residues of variable hydrophobicity onto native human serum albumin Wilkinson (Wilkinson & McKay 1972, reviewed by Wilkinson 1981) was able to demonstrate that as the hydrophobicity of the protein increased, judged by its physical properties, so did the effectiveness of the protein as an attractant. Denaturation of haemoglobin by removal of the haem moiety to form globin produces a potent chemotactic factor, but as the haem is added back the restoration of the normal quaternary structure of the globin correlates with the loss of activity as an attractant.

## 9.7.2   Receptor design

The four groups of factors discussed above pose a range of problems in receptor design. For C5a the problem is relatively straightforward and conventional, a single fairly large molecule is being bound and the complement system is fairly well conserved in evolution (Chenoweth & Hugli 1978). The lipid metabolites, too, are easily recognizable and the location of hydroxyl-groups and other determinants will permit specific interactions with a receptor molecule. For the formylated peptides the situation is rather better understood: the critical parameters for a ligand to bind to this receptor seem to be that the N-terminus is blocked (acetylation will serve, though not as effectively as formylation); methionine need not be the N-terminal amino acid, f-tri-Tyr is active, and the peptide is more active if the C-terminal is a bulky hydrophobic amino acid, such as phenylalanine or tyrosine (Showell et al. 1976). That a single receptor is involved can be shown by competitive deactivation studies, which also show that this receptor is different from that which binds C5a (Williams et al. 1977).

For denatured proteins the problem is far greater. Even if the hydrophobicity of the denatured protein provides a simple recognizable change, the shape and size of the molecule will be very variable. The binding of these denatured proteins may not in fact be by a true receptor–ligand interaction: the term receptor is being used too loosely (Wilkinson & Allan 1978a).

The hallmark of a receptor-mediated binding is that it should be saturable, that it should be possible to define both a receptor affinity and a receptor number. The binding of denatured proteins does not fully satisfy these conditions, and the interaction with the cell is apparently a direct binding to the surface, perhaps through an hydrophobic interaction. The absence of an identified specific receptor system should not, however, be allowed to obscure the fact that denatured proteins are effective as attractants, although not as potent as the small peptides or C5a.

Receptors for formyl-peptides have been identified in the membrane and are integral proteins with a molecular weight of around 60 kd (Neidel & Dolmatch 1983). The formyl-peptide receptor seems to be restricted to a few species, notably man and rabbit, and it does not seem to be present on the leucocytes of many domestic animals (pigs, cattle, chickens and probably horses). Estimates of the receptor number have been made by various workers and a generally accepted figure is in the range of around $10^5$ cell$^{-1}$. The reason for some of the diversity in these estimates is that the cell can modify the expression of the receptors on the surface by a process known as down-regulation, and because receptors are internalized and recycled (Zigmond & Sullivan 1981). Further problems also arise because there may be a reservoir of receptors which are only expressed following fusion of the 'specific' (secondary) granules with the plasma membrane, and because receptors may be switched from low- to high-affinity (Fletcher *et al.* 1982, Snyderman 1983). The dynamics of the receptor population are complex, and may improve the sensitivity of the system, but the topic is rather esoteric and interested readers should consult the primary literature. The 'C5a-receptor' has proved more elusive but from binding studies it appears that there are approximately $10^5$ sites cell$^{-1}$ (Chenoweth & Hugli 1978).

## 9.7.3 Responses of the cell to binding of attractant

The neutrophil has a number of functions in an inflammatory response and different activities must occur at various stages. If the cell is directed chemotactically by a factor diffusing from the focus of damage, then the other activities of the cell might well be regulated in part by the concentration of attractant. At low levels a directed movement is required; at high levels, near the site, the neutrophil must become immobilized and begin to destroy bacteria and digest damaged tissue. We should not, therefore, be too surprised if the responses to chemotactic factors are rather different at high concentrations, although we might hope to account for the differences on the basis of quantitative rather than qualitative alterations in some cellular control mechanism.

In response to the binding of a formyl-peptide there seems to be an alteration in the permeability of the plasma membrane, which can be detected, for example, as an efflux of $^{45}Ca^{2+}$ from cells pre-loaded with the radioactive isotope (Naccache *et al.* 1979). This also indicates that the level of freely diffusible calcium within the cytoplasm has increased, a particularly interesting change in the light of the rôle calcium plays in regulating the locomotory machinery and the secretory response. Other ions such as $K^+$ may affect the response, suggesting that membrane depolarization may be important. Some of the changes are transient, which accords with similar membrane perturbations in other sensory events.

At a behavioural level, chemotactic factors in non-gradient conditions probably influence movement by increasing the speed of the moving cell, rather than by altering the turning behaviour, and there are few exceptions, if any, to the general rule that chemotactic factors are also chemokinetic (the converse is not true, some native proteins, for example, are chemokinetic but do not elicit a directional response). Adhesion changes induced by chemotactic factors are time- and dose-dependent (Smith *et al.* 1979, Lackie & Smith 1980). Initial exposure to low levels of fMet-Leu-Phe (fMLP) causes a transient increase in adhesion lasting a few minutes, after which the cell is refractory (adapted) unless the concentration is increased. After the transient increase the adhesiveness seems to be reduced for a time before increasing again (Fig. 9.11). High levels of factor induce a rapid increase in adhesiveness and the period of reduced adhesiveness may be lost between the first and second phases of enhanced adhesion. High levels of fMLP, besides increasing

243

adhesion, also stimulate secretory activity (release of primary and secondary granule contents) and metabolic activity associated with the production of toxic oxygen species (superoxide, peroxide and hydroxyl-radicals). Not only does the neutrophil release bacteriostatic and bacteriocidal compounds and degradative enzymes which will digest damaged tissue, but there may also be the release of cell-derived chemotactic factors which recruit other cells into the area (Zigmond & Hirsch 1973).

The adhesion changes are of some interest because of the varying requirements for adhesion as the neutrophil adheres to the wall of the blood vessel, moves through a three-dimensional matrix and then becomes trapped in the lesion. Decreased adhesion at low chemotactic factor concentration might facilitate movement in a matrix, and the increased adhesion later (or at high concentration) might contribute to accumulation of cells by a haptotactic cue (see Ch. 7). Other features of an inflammatory lesion may also contribute to adhesive trapping: the deposition of immune complexes, for example, and phagocytosis, which itself increases the adhesiveness of the cells.

### 9.7.4 Chemotactic responses by neutrophils

The ability of neutrophils to respond in a directional fashion to gradients of **chemoattractants** has been shown in various ways. Stable gradients of diffusible substances are rather difficult to maintain and various methods have been adopted to evade these problems. One assay system in particular, the micropore filter or Boyden chamber assay, will be discussed in detail (next section) because it illustrates the difference in behaviour elicited by chemokinetic and chemotactic cues. A direct visual demonstration of chemotactic responsiveness can be achieved by a system such as that used by Zigmond and Hirsch (1973), in which a strip of aggregated immunoglobulin is used as a gradient source. Neutrophils adjacent to this source direct their movement towards the highest concentration of diffusible factor and the first displacement of a previously stationary cell is almost invariably in the right direction, implying that a spatial rather than temporal sensing mechanism is involved. As the cells move up the gradient their turning behaviour is non-random and turns are made preferentially towards the gradient source.

Because the direction of movement of a cell can be judged on the basis of the morphology of the cell, it is possible to assess the orientation of cells in a gradient, the basis of another visual assay (Zigmond 1977). In the 'orientation chamber' a fairly stable gradient is set up across a narrow bridge between two large wells (Fig. 9.12). The dimensions of the bridge are such that diffusion is the rate-limiting process in equilibration of the two wells, one of which contains a higher concentration of attractant. Using this system the ability of cells to orient in known gradients at different absolute concentrations can be measured, an important test because the cell must be able to perceive the directional cue both at the bottom and the top of the gradient up which it is moving. Using this assay system, Zigmond was able to show that the ability of the cell to respond to a gradient of fNle-Leu-Phe was maximal when the average concentration of attractant was such that half the binding sites would be occupied (the $K_a$ or $K_d$ depending on whether binding or dissociation are being measured). This is, of course, exactly what would be predicted since the ability to discriminate between small changes in concentration will be greatest at this concentration (see § 9.2). Knowing the dimensions of the bridge, the concentrations of factor in the two wells and the number of binding sites, it is possible to calculate that at best the cell is making a response when the concentration difference between the front and

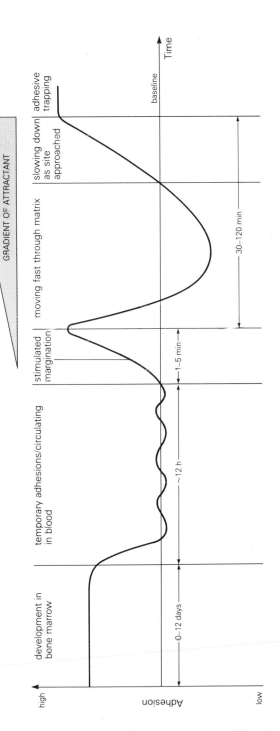

**Figure 9.11  Adhesion changes in the life of a neutrophil**  During the development of the neutrophil in the bone marrow it remains in place, and is presumably adherent. The mature neutrophil enters the circulation where it forms transient adhesions to the walls of post-capillary venules until stimulated by a factor such as fMLP. There is presumably a transient increase in adhesiveness, which allows the cell to stop, and a factor such as fMLP will induce a marked stimulation of the motor. If the motor is activated during this brief period of adhesion the neutrophil may move through the endothelium, and will not therefore re-enter the flow. Once the endothelium is penetrated the movement of the cell will be facilitated by low adhesiveness – in the three-dimensional matrix adhesion anchorages are unimportant, and may slow the cell down. As the focus of damage, the source of factor, is approached the adhesiveness increases, and the cells may be trapped. It is at the centre of damage, the top of the gradient, that the cell may have to adhere to particles in order to phagocytose. (Note: 'adhesion' in the above is used rather loosely; it is often impossible to estimate the strength of the adhesion.)

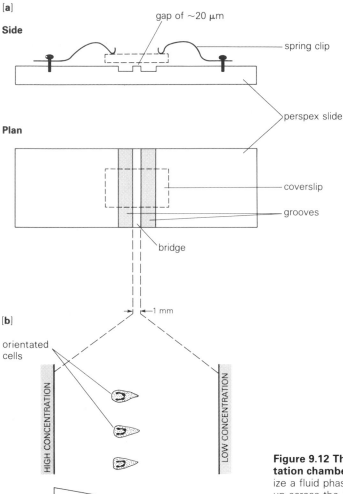

[a]

Side

gap of ~20 μm

spring clip

perspex slide

Plan

coverslip

grooves

bridge

1 mm

[b]

orientated cells

HIGH CONCENTRATION

LOW CONCENTRATION

GRADIENT OF ATTRACTANT

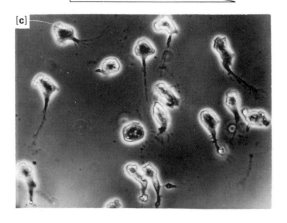

[c]

**Figure 9.12 The Zigmond orientation chamber** In order to stabilize a fluid phase gradient it is set up across the 'bridge', which is a channel 1 mm wide and only 15–20 μm deep, in which convection is minimized. Cells are allowed to adhere to a coverslip which is then placed over the bridge and held in place with spring clips. One groove is filled with buffer, and the other with medium containing the test factor. Because cells will orient without moving and their orientation can be recognized on morphological grounds, (c), the response can be scored as the proportion of cells which are aligned up-gradient. The photograph, (c), was kindly supplied by Peter Wilkinson. (Orientation of neutrophils can also be seen in Figure 6.19.)

rear of the cell differs by only 1 per cent, and the difference in the number of receptors occupied may be very small, only a few tens of receptors. Knowing the association and dissociation rate constants it is possible to estimate the signal : noise ratio under such conditions, and a calculation, based on a number of assumptions, suggests that time-averaging over a period of 2–5 min would be required (Lauffenburger 1982).

Although this figure can only be a rough 'order-of-magnitude' estimate, it is nevertheless a very interesting figure: the implication must be that the cell has some form of signal-integrating device which operates on a time-scale of minutes. Neurophysiological integration at the membrane level usually operates at a much faster rate, of the order of seconds, and it is therefore unlikely that membrane permeability or conductance provides the analytical device. Two cellular activities do, however, operate on the correct time-scale for time-averaging, the adhesion changes induced by initial contact with chemotactic factor (see § 9.7.3) and the reorganization of the locomotory machinery involved in pseudopod formation. The latter seems a particularly attractive possibility and is supported by the observation that stepwise increases in chemotactic-factor concentration induce a transient activation of pseudopod formation which relaxes to the basal level over a few minutes. The pseudospatial gradient-sensing mechanism alluded to previously could both integrate the signal and direct the motile activity. A pseudopod which is protruded up-gradient will receive additional stimulation because increased concentrations of factor are encountered as it extends, this will lead to further protrusion and the dominant pseudopod will be the leading edge of the cell. Pseudopods moving down-gradient will not receive the positive feedback of increased receptor occupancy and are therefore unlikely to become dominant.

This simple explanation does not account for all of the behaviour. If the gradient of factor concentration is reversed, the cell does not commonly respond simply by an exchange of dominance between front and rear but rather by making a tight turn, maintaining the intrinsic polarity of the cell (Zigmond et al. 1981). One explanation of this behaviour would be that pseudopod formation is more favoured in the anterior portion and that this is the region most likely to direct the movement. Following gradient reversal a pseudopod on one side of the 'front' will become more active and swing the cell around to face the new source. In considering gradient perception the assumption is generally made that receptor distribution is uniform and that the full width or length of the cell can separate sampling sites. If only the anterior portion is active in gradient perception, then the problem of discrimination becomes greater, although a number of things might improve the situation. Receptor asymmetry, with most receptors at the leading edge, or the incorporation of new receptors only at the front (as has been suggested by Gallin et al. 1983), would improve the sensitivity especially if receptors are inactive once they have bound the factor and have to be moved to the rear and be recycled before being used again. The problem of sensing a diffusible gradient is clearly quite a difficult one, but it is encouraging to see the emergence of hypotheses by which levels of receptor occupancy might affect the locomotory machinery and control movement.

## 9.7.5 Taxis versus kinesis

As stressed in previous sections the accumulation of cells in a particular site does not constitute evidence for a tactic response, and separating the effects of a gradient of chemokinetic factor from a chemotactic response is important (Wilkinson et al. 1982). An assay for neutrophil chemotaxis which has been widely used is the Boyden chamber, in which two compartments are separated by a micropore filter (Fig. 9.13a). Micropore filters have a complex meshwork structure with the pore-size being based upon the density of the mesh rather than upon discrete holes (in contrast to nucleopore filters which do have holes of fixed dimensions). The filters are quite thick (150 μm) and neutrophil leucocytes can penetrate into filters with pore-sizes of 1.1 μm or more. Movement into such a rigid matrix depends upon deformation of the cell and is an active process. Provided there is no hydrostatic gradient, the filter serves also to stabilize a diffusion gradient between the two compartments. Cells placed in the upper chamber move into the filter and their distribution can be measured at the end of the incubation period having stained and cleared the filter.

Under non-gradient conditions the cells will move randomly and the population distribution (Fig. 9.13b(i)) can be accounted for by simple diffusion, the incubation period is sufficiently long to ensure that the internal bias of the cells does not affect the dispersion. With a gradient of a purely chemokinetic substance (which increases the speed of locomotion) the distribution will be altered because cells accelerate as they move into the higher concentrations deeper in the filter (Fig. 9.13b(ii)). Negative gradients, in which the cells decelerate as they move further, will also produce a different distribution. In contrast to the distributions found with kinetic factors, a positive chemotactic gradient will shift the whole population (Fig. 9.13b(iii)).

Notice that if in such an assay one were to score the number of cells arriving at a particular level, for convenience the bottom of the filter perhaps, then the positive chemokinetic gradient increases delivery of cells, although not as effectively as a chemotactic response. A method for distinguishing the chemokinetic and chemotactic effects of a substance, the so-called checkerboard assay, has been devised and is widely used (Zigmond & Hirsch 1973). The Boyden chamber assay does not, of course, tell us about the equilibrium distribution of cells, but perhaps in a defensive reaction it is early arrival that matters.

## 9.7.6 Surface-bound gradients

Neutrophils have been shown to respond to surface-bound gradients of chemotactic factors, which illustrates rather clearly the interaction between various tactic cues and the arbitrary nature of their classification (Dierich et al. 1977, Wilkinson & Allan 1978b). A planar, surface-bound gradient of factor might act in exactly the same way as a diffusible fluid-phase gradient. Receptors on the ventral surface of the cell could bind factor and signal in the normal way. Alternatively, the receptor–ligand interaction might contribute

---

**Figure 9.13 Distribution of neutrophils in a micropore filter** Penetration of cells into ▶ the micropore filter which separates the two compartments of a Boyden chamber (shown diagrammatically in (a)) is used as a method for scoring their responses to different concentrations of substances (without a gradient present), and to gradients. (b) The distribution of cells through the filter can be determined by scoring the number of cells at different levels. If the substance is chemokinetic and increases the rate of movement then the population will diffuse faster and the distribution will resemble (ii) rather than (i), but there will be no population displacement. If, however, there is a chemotactic gradient then cells will move up-gradient and there will be a net population displacement, as shown in line (iii).

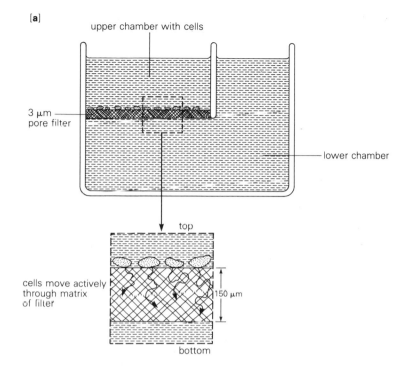

[a]

upper chamber with cells

3 μm pore filter

lower chamber

top

cells move actively through matrix of filter

150 μm

bottom

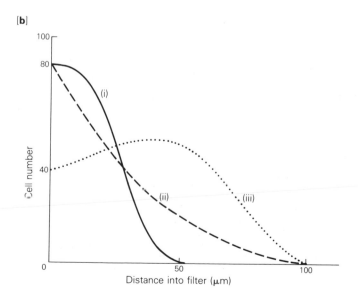

[b]

(i)

(ii)

(iii)

Cell number

Distance into filter (μm)

to the cell–substratum adhesion and the surface-bound factor would then resemble a gradient of adhesiveness, a haptotactic gradient. Secondary amplification of the adhesion change due to the receptor–ligand interaction might come from the adhesion changes which chemotactic-factor binding induces in the cell. This argument can then be extended to the fluid-phase gradient: the substratum might remain isotropic but the changes in cell adhesion generated by the interaction with soluble factor might produce a gradient in the cell–substratum adhesion, which would be the operational cue for directing movement. At present these problems remain unresolved.

Although we are beginning to understand the details of some parts of the response of neutrophils to chemotactic factors, the overall biology should not be forgotten. The neutrophil has a range of problems to overcome in order to arrive at the inflammatory lesion and directing the locomotion is only one aspect. At low levels of chemotactic factor gradient perception will be difficult, if not impossible, but if low levels increase the speed of movement then the sooner it will enter a region where the concentration is higher. This may seem to conflict with the normal outcome of increased diffusion, but the environment is non-isotropic and cells which enter the high-concentration region randomly escape non-randomly, up-gradient rather than down-gradient, having encountered a directional stimulus. Near the source of the gradient the direction of movement becomes less critical, the important thing is that the cells, having entered this region should remain in place and carry out their defensive rôle. Superimposed on the kinetic and tactic cues there may be guidance cues from tissue architecture or from stress-induced orientation of connective tissue fibrils, following clot-retraction when haemorrhage and haemostasis are associated with the inflammatory insult. It is often worthwhile to stand back from the details of the cell behaviour and ask about the outcome of the behavioural response in the context of the biological system.

## 9.8 Summary

In this chapter we have seen how a number of cell types respond to spatial and temporal gradients of diffusible substances in the environment, and have tried to describe some of the problems involved in gradient perception. It is remarkably difficult to set up unambiguous tests to distinguish between the various behavioural responses that may be involved in cell accumulation at a focus, and it is important to realize that some terms, such as chemotaxis, have specific implications and should not be used unless there is some evidence as to the mechanism which is actually being used by the cells. On the whole, few cells seem to be able to respond chemotactically and this may reflect the difficulties of the tasks involved in designing and operating a gradient-sensing system.

## References

Lackie, J. M. and P. C. Wilkinson (eds) 1981. *The biology of the chemotactic response.* Cambridge: CUP. (Soc. Exp. Biol. Seminar Series *12*.) Reviews of most systems.
Koshland, D. E. 1980. *Bacterial chemotaxis as a model behavioral system.* (Distinguished Lecture Series of the Society of General Physiologists.) New

York: Raven. An extensive but approachable coverage of bacterial chemotaxis. (See also Berg 1983.)

Wilkinson, P. C. 1982. *Chemotaxis and inflammation*, 2nd edn. Edinburgh: Churchill Livingstone.

Gallin, J. I. and P. G. Quie (eds) 1978. *Leucocyte chemotaxis: methods, physiology, and clinical implications*. New York: Raven.

# 10
# CELL–CELL INTERACTIONS

## 10.1   Introduction

Many of the most fascinating problems of cell behaviour are those which involve cells interacting with the other cells in their immediate environment; these are the problems of, among other things, morphogenesis, tissue-reconstruction, and invasiveness by tumour cells. Essentially the question we wish to answer is simple – what happens when a cell meets another cell? Sometimes cells behave as though indifferent to collisions, in other cases they may recoil or again may fasten irreversibly to their partner. We know little about the molecular bases for the differences in behaviour, and relatively little is being done to remedy this deficiency, yet a multitude of problems in the body arise through alterations of cellular behaviour.

Previous chapters have dealt with relatively simple and well-defined situations, and the information gleaned from experiments on cell behaviour in artificial environments may well be the way to approach the more complex problems encountered in dealing with cell populations. We know less about movement *in vivo* than *in vitro* because of the technical problems of observation and because we cannot define even the non-cellular components of tissues in sufficient detail. Some cell–cell interactions can, however, be studied *in vitro* and these studies have provided many insights.

## 10.2   Contact inhibition of locomotion

Before embarking on a discussion of the phenomenon of contact inhibition of locomotion it is necessary to stress that we are talking about alterations in the motile behaviour of the cells and not about the growth (or more properly division) of cells. Confusion arises because cells, when grown *in vitro*, tend to stop proliferating when the surface of the container is covered with cells, and when most cells are in contact with other cells. The cessation of proliferation occurs for reasons that are now generally accepted not to depend simply upon contact: it happens that growth stops when cells are in contact. By manipulating the culture conditions it is quite possible to stop division at a stage when the cells are well separated and not in contact. The confusion is worse confounded by careless scientists who write about 'contact inhibition' meaning '. . . of division' while referencing, presumably unread, papers which deal with inhibition of movement. Be certain: we are dealing in the following section with alterations in the motile activities, our analogical vehicle will not divide nor will it be inhibited from reproducing by colliding with another vehicle.

Unlike the products of Detroit or Dagenham, collisions between cells do not lead to extensive, expensive, bodywork damage, but there are quite clear rules about what should happen after a collision and these were first described and are probably best understood for fibroblasts.

---

**Figure 10.1   Contact inhibition of locomotion**   A time sequence of phase contrast ▶ micrographs to illustrate contact inhibition of locomotion following the collision of two chick heart fibroblasts;   (a) 5 min post-collision,   (b) 10 min,   (c) 15 min,   (d) 25 min, (e) 30 min; scale bar represents 25 μm. Following the collision further movement in that direction ceases and other regions of the perimeter gain dominance so that the cells move apart (from Abercrombie 1980).

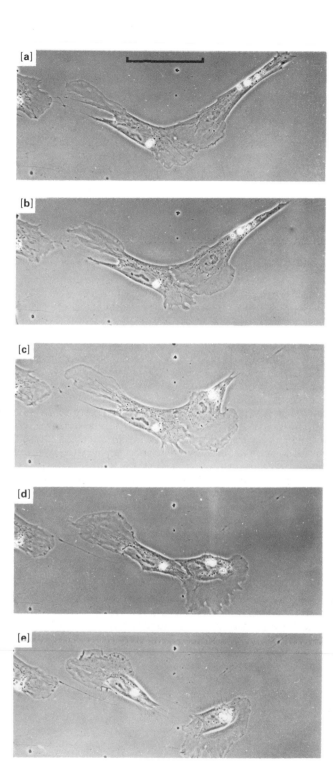

## 10.2.1 Fibroblasts: contact inhibition Type 1

When a fibroblast collides with another fibroblast an adhesion forms between the two cells, there is a cessation of protrusive ruffling activity in the region of contact, and then the two cells separate from one another (Fig. 10.1). The separation of the two cells is often abrupt and the impression is very much of the sudden failure of adhesion between the two cells. The most striking feature of the phenomenon is that there is an apparent paralysis of the locomotory machinery in the contact area. The overall effect is to alter the direction in which the cell is moving, although the cell may make repeated attempts to persist in the direction in which it was moving just before the collision occurred. Eventually, however, another region in the cell becomes dominant in protrusive activity and the cell sets off in another direction. The change in direction may be a turn to left or right, or the cell may literally turn tail with the former tail region becoming the new leading edge. Frustrated attempts to persist in the direction which led to the collision do not, of course, achieve anything in terms of movement and in consequence the overall speed of the cell is reduced. Head-on collisions tend to be more dramatic, with both protagonists exhibiting paralysis and retraction, whereas in head–side collisions there may be little change in the cell which is contacted laterally (in an area which is relatively quiescent in protrusive activity and which may be less adhesive than the leading edge).

Although descriptions of the phenomenon, such as that given above, sound straight-forward, the reality is often less easily interpreted. Students are often disappointed when they see time-lapse film of collisions because the events, even when speeded up, are far from instantaneous (some 10 min of real time may be involved), and the collision is much more complex than that of billiard balls. In many cases a considerable degree of **underlapping** occurs and, particularly in head–side collisions, the leading edge of one cell may protrude some distance under the other. The retraction is, however, an unmistakeable response and is extremely convincing. Because the collision can be scored only in a qualitative fashion and because only two cells are involved, transiently at that, it is a much harder phenomenon to study than, for example, the mass migration of a population of cells. Even so, by careful experiment and much film making it is possible to learn rather more about contact behaviour.

Contact inhibition will occur when the cells are of different phyletic origin: mouse muscle fibroblasts are inhibited by, and inhibit, chick heart fibroblasts even though the tissues of origin, too, are different (Abercrombie *et al.* 1968). A cell can also inhibit itself when, as happens occasionally, the leading lamella splits and the two portions converge further forward (Ebendal & Heath 1977). An obvious question to ask concerns the way in which the collision event is 'signalled', and the way in which the locomotory system is altered in response to the collision. One hypothesis would be that there are signals which pass between cells; signals which might be of diffusible chemicals or of electrical depolarization, for example. Such a signalling hypothesis could accommodate the observations of interphyletic *and* autologous inhibition, but a further observation decisively rules out such a signal hypothesis. Collisions between live fibroblasts and dead fibroblasts, provided the latter have been fixed with zinc chloride, leads to contact inhibition of the live cells (Heaysman & Turin 1976). Earlier attempts to demonstrate inhibition between live and fixed cells had been unsuccessful; the key seems to have been the use of a fixation procedure which is analogous to the methods used for stabilizing membranes for isolation and subsequent biochemical analysis. This means that an explanation for the contact response must be sought elsewhere.

Another approach to studying the problem of contact inhibition of locomotion has been to examine the ultrastructure of the contact area at different times after the collision (Heaysman & Pegrum 1973a). Technically of course this is far from simple: it requires patient direct observation of the cells, their rapid fixation and then the successful relocation of the colliding pair once they have been embedded in resin for sectioning. Some uncertainty must inevitably exist as to the precise time of contact, because of the limited resolution of the light microscope and the very low contrast of the thin leading edges of cells. A reasonably good time sequence has been established. Very shortly (20 s) after the collision occurs there is specialization in the contact region (Fig. 10.2) and the cytoplasmic faces of both plasma membranes become distinctly more electron dense. Associated with these zones of increased electron density is the appearance, within 30 s, of aligned microfilaments which, after two minutes, are organized as quite conspicuous bundles. It is as though two focal adhesions, of the type formed between a fibroblast and a rigid substratum (see Ch. 6), have been apposed. It seems reasonable to suppose that these microfilament bundles exert a force on the adhesion zone, and that the eventual failure of the adhesion leads to an abrupt elastic recoil or retraction.

Why, however, should the apposition of two focal adhesions lead to paralysis of the locomotory machinery and the later separation of the two cells? It could be argued that further protrusion is inhibited because the assembly of microfilament bundles has a higher 'priority' than the assembly of actin gel, and that depletion of the sub-unit pool (pre-formed micro-filaments rather than G-actin) prevents further protrusion. Several possible reasons for the failure of the adhesion between the two cells can be proposed. It could be that two bundles of microfilaments exert more stress on the adhesion plaque than one, and that failure is essentially mechanical. The stress would, of course, derive from the activity of other stress fibres, which are anchored to the substratum over which the cells are moving, and would be transmitted (and possibly increased) by the activities of the apposed bundles. Alternatively, it could be that the focal adhesion formed by each of the cells is smaller, and has a higher local density of microfilament insertions, because the substratum for each cell is of low lateral rigidity so that redistribution of dispersed adhesion sites can occur more easily. On this latter argument we would have to suppose that the life-time of the focal adhesion was sensitive to the packing of microfilament insertions, or that the local stress resulting from contraction exceeds that which can be resisted. It could perhaps be that the half-life of the focal adhesion in the cell–cell contact area is much shorter because the 'adhesion sites' are attacked from both sides. None of these explanations can be ruled out.

If we consider that the contact area is one in which microfilament bundles from the two cells are apposed end-to-end, then the similarity with a *zonula adhaerens*-type junctional specialization should be obvious. In epithelia these are more-or-less permanent structures and, indeed, some fibroblasts may form semi-permanent contacts of this type as in the bovine lung (Sims & Westfall 1982) (Fig. 10.3). In the latter case we might suppose that contact inhibition of locomotion has merely stopped at the adhesion and paralysis stage, and considerations of this type lend some support to the idea that it is reduced half-life of the focal adhesion (which is just clustered diffuse adhesions) which is important for breakdown of the contact and then the retraction. It would then be necessary to suppose that the half-life was altered

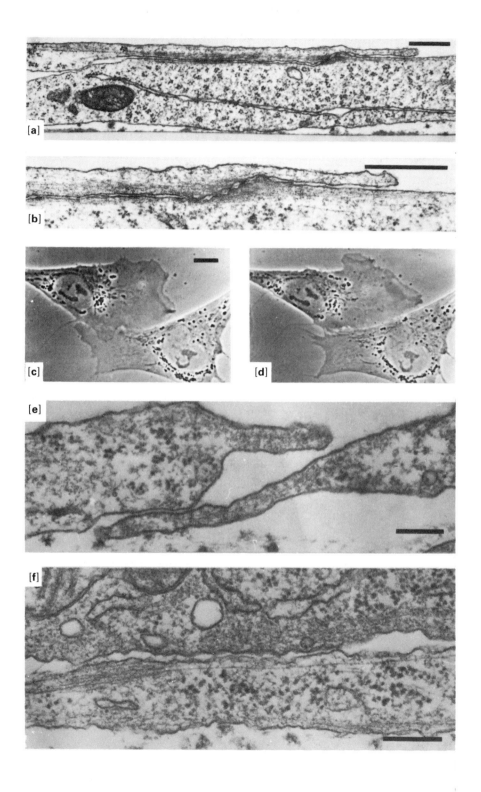

[a]

[b]

[c]　　[d]

[e]

[f]

by some method that did not rely upon identity of origin of the cells or, indeed, upon both cells being alive. Conversely, it would be necessary to propose that the decay of the adhesion can, under certain circumstances, be inhibited.

At present the mechanism of contact inhibition remains in the realm of speculation: almost the only biochemical work which indicates that membrane-associated proteins are involved is that of Weston and co-workers (see below).

## 10.2.2   Fibroblasts: contact inhibition Type 2

Although the behaviour of colliding fibroblasts certainly does involve paralysis of the locomotory machinery and retraction of the cells, there is an alternative mechanism which could affect the direction of locomotion following collisions. If the leading edge of a fibroblast is protruded over an area where adhesions cannot be formed, the edge of an agar overlay, for example, then the cell cannot obtain an anchorage and will be unable to move forward. Overlapping a non-adhesive area is of course an extreme case; if the leading edge formed contacts of different strength then the cell might well move preferentially towards the stronger adhesions, which slip less and provide more efficient traction. Thus, if the surface of another cell were less adhesive or of lower rigidity than the solid substratum over which the cells were moving, then the tendency of the cells to crawl over one another would be much reduced. The mechanism would then be of adhesive 'preference', or of the requirement for a rigid substratum, and if the surface of a fixed cell were cross-linked and therefore more rigid or more adhesive the inhibition might be released. Whether zinc-fixed fibroblasts retain their membrane fluidity and lateral mobility of their integral membrane proteins has not, so far as I know, been tested.

The extent to which chick heart fibroblasts **overlap** one another in a crowded culture (the **overlap index:** Abercrombie & Heaysman 1954, see § 10.3.1) can be modified in two ways: by altering the adhesive properties of the substratum on which the cells are growing and by inducing an alteration in the cell surface properties. If fibroblasts are grown on less adhesive surfaces then the extent of overlapping increases, partly because the cells are less flattened and underlapping is much easier, and partly because their mobility is greater. The latter is rather a subjective impression but the speed of movement is certainly affected by reducing the adhesiveness (within

---

◀ **Figure 10.2 Ultrastructure at the contact point** (a) Vertical longitudinal section through the collision zone between chick heart fibroblasts (CHFs) more than two min after contact. The upper cell which is overlapping a thicker region of the lower cell, has already begun to show some specializations in the contact. A well developed contact is particularly conspicuous near the front; bar represents 0.5 µm. (b) A higher power view of a specialized contact zone between two chick heart fibroblasts. The microfilament bundle inserting into the contact area is particularly conspicuous in the cell on the left; bar represents 0.5 µm. (c) Phase contrast micrographs of two contacting cells showing some underlapping; bar represents 10 µm. (d) The same pair of cells as in (c), four min later, showing separation in the region where the underlapping was seen in the previous picture. (e) Early contacts (60 s after collision) between a chick heart fibroblast (above) and a sarcoma (S180) cell. The absence of specialization in the contact area is obvious; bar represents 0.25 µm. (f) Section through a CHF–S180 contact more than ten min after collision. The S180 cell is the upper one; bar represents 0.25 µm. Photographs courtesy of Joan Heaysman (from Heaysman & Pegrum 1973a & b).

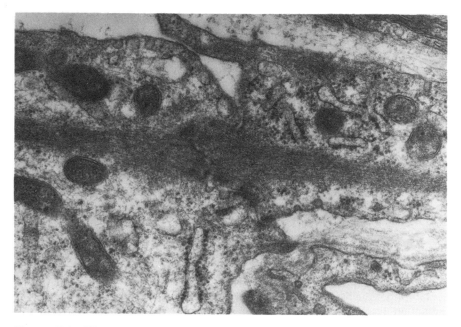

**Figure 10.3 Fibroblast contacts in bovine lung** Well developed and possibly semi-permanent adhaerens-type junctions between fibroblasts in the bovine lung. Notice the resemblance between these junctions and the specialized contacts formed rapidly between fibroblasts *in vitro*, as seen in Figure 10.2 (from Sims & Westfall 1982).

limits): the contribution of increased movement to the increased overlapping has not been explicitly defined.

Modification of the cell surface is a more interesting manipulation because, potentially at least, it may give some indication of the molecular interactions which are important. The surface properties of cells are certainly altered following viral or spontaneous transformation, although the extensive research effort directed at defining the nature of the change has contributed disappointingly little to our understanding of the altered behaviour of neoplastic cells. The social behaviour of fibroblasts can, however, be modified, in a reversible fashion, by removing surface components with urea. The addition of 0.2 M urea to the growth medium will increase the overlap index, and restoration of the normal overlap index following removal of urea depends upon protein synthesis. Alternatively, normal behaviour can be restored by exogenous addition of proteins from the urea-containing medium; the essential component seems to be a single protein of molecular weight around 60 kd (Weston *et al.* 1979). This protein appears to be shed into the medium even without treatment, but the mechanism by which this protein brings about the change in behaviour is not yet clear. A more recent approach, and one which offers some exciting possibilities, involves masking the cell surface with monoclonal antibodies. An antibody which modifies the **invasion index** (as defined by Abercrombie & Heaysman 1976) has recently been described (Steinemann *et al.* 1984).

Underlapping, in which the moving cell retains its contact with the solid substratum, will not be affected if the mechanism of inhibition is purely one of adhesive preference, since there has been no change in the substratum and there should be no reason to stop. On this argument, therefore, the underlapping partner in a collision should not show retraction. For fibroblasts this does not seem to be the case; although quite extensive underlapping does occur, even in some head–head collisions, the underlapping cell does generally retract. For other cells this may not be the case (see § 10.4). Similarly, removing the choice, by offering a substratum composed solely of the other cells, should not inhibit locomotion although the speed might well be affected. Fibroblasts plated onto confluent layers of fibroblasts seem, however, to be unable to spread, which is generally interpreted as being due to contact inhibition of the locomotory machinery. This latter observation also rules out the possibility that clambering aboard another cell is simply a problem of mechanical hindrance or geometrical constraint by analogy with the problems faced by cells on prisms (see Ch. 8).

## 10.2.3   Epithelia

Since one of the characteristics of epithelial sheets is that the cells are in close contact with their neighbours it should be obvious that the collision behaviour is likely to be different from that of fibroblasts which live more solitary lives. Collisions between epithelial cells do lead to the formation of an adhesion and local paralysis of the locomotory machinery but there is no retraction phase (Middleton 1982). Following contact the leading lamellae of the epithelial cells are gradually lost, a process which can clearly be visualized when the collision is between pigmented cells of the embryonic chick retina (PRE). The leading edge of these cells excludes pigment granules, in much the same way as the leading edge of a fibroblast is kept free of granular cytoplasm, but shortly after the collision the pigment granules begin to enter the marginal zone until the boundary between the two cells becomes indistinguishable from boundaries deep within the sheet. The inability of pigment granules to enter this region in a moving cell is presumably a consequence of the presence of an actin gel meshwork which excludes granules, although this has not been explicitly demonstrated. In a normal epithelial sheet, though not in pigmented retinal epithelium, the contact between the cells will be stabilized by the formation of desmosomes which give mechanical strength to the sheet as a whole.

A curious feature of contact between two isolated epithelial cells is the phenomenon of **contact-induced spreading** described by Middleton (1977). An isolated epithelial cell is in an abnormal situation and, unlike fibroblasts, epithelial cells do not seem to have a well-defined polarity (see § 7.2.2); no one area of the margin becomes dominant in terms of locomotory activity. The cell attempts, as it were, to move simultaneously in all directions and, apart from very small excursions, does not achieve a significant displacement. If two such cells do manage to contact one another the contact region is 'stabilized' and the two cells spread out so that the area covered by the two cells together is greater (2.2 ×) than the sum of the areas covered when they are separated (Fig. 10.4). This may well be because a polarity has now been defined for each cell, each of which can then concentrate protrusive and spreading activity in a more limited region, and can therefore exert more stretching tension on the doublet.

The differences in contact behaviour between fibroblasts and epithelia are very much as might be predicted by considering the rôles of these cells *in vivo*. Fibroblasts are solitary cells which move through the connective tissue matrix, whereas epithelia, by definition, provide boundaries. For a boundary

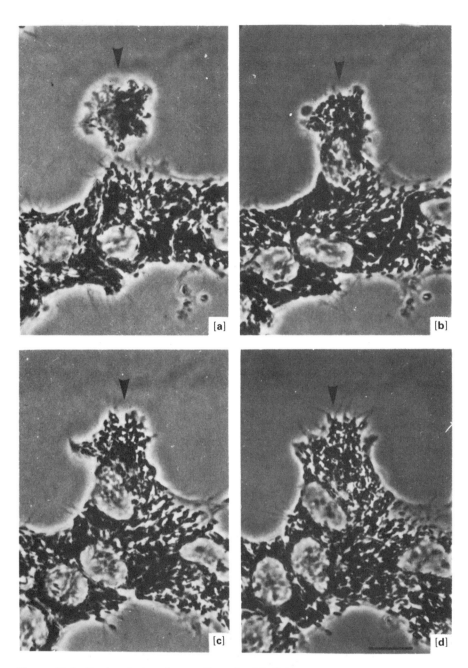

**Figure 10.4  Contact-induced spreading of epithelial cells**  A series of phase-contrast photographs of contact between a single pigmented retinal epithelial cell (arrowed) and an island of the same cells. The cell shows a marked increase in its spread area once it has contacted the sheet. The pictures were taken at 0 min, 30 min, 85 min and 125 min; the scale bar represents 10 μm (photograph courtesy of Adam Middleton; from Middleton 1982).

layer of cells to be effective cell–cell contact must be stable, and the sheet should have a 'built-in' tendency to cover any gap and restore the integrity of the barrier. Thus, in a wound-healing response the epithelial edges must come together and join, whereas the fibroblast must move into the wound area and restore the connective tissue matrix which, being only sparsely populated by cells, does not require permanent cell–cell contacts. Of course this view of wound healing is very oversimplified, and a range of other cellular activities is involved, but in general terms the contact behaviour of the two cell types is such as to generate the appropriate restoration of the tissue architecture.

## 10.2.4   Fibroblast–epithelial cell collisions

Thus far we have restricted discussion to homologous collisions, those between more-or-less identical cells, but it is likely that heterologous collisions will occur *in vivo*. One interesting sort of interaction is that between fibroblasts and epithelia in which the effects seem, in those few situations which have been investigated, to be non-reciprocal. One of the better known examples involves cells from the eye of the chick embryo. The retina is composed of three layers of cells: on the outside a layer of choroid fibroblasts, then a layer of pigmented retinal epithelium and, adjacent to the vitreous humour, the neural retina (Fig. 10.5). Because the chick eye develops early and can be taken apart into its constituent cell types relatively easily, it has been a favourite tissue in which to look at the behavioural rules governing the maintenance of cellular organization. As we have already seen in the previous section, the contact behaviour of pigmented retinal cells has been studied in detail and choroid fibroblasts show the normal behaviour expected of fibroblasts. Collisions between choroid fibroblasts and PRE, which have been filmed on a number of occasions, lead to an inhibition and retraction of the fibroblast, while the PRE apparently remains unaffected (Parkinson & Edwards 1978).

On this basis we would predict that an epithelial sheet would spread successfully over a territory occupied by choroid fibroblasts but that

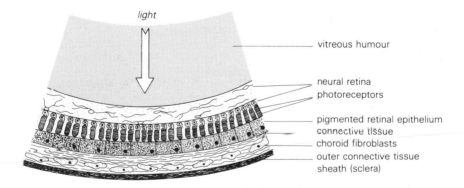

**Figure 10.5  Cell layers in the eye**  A schematic diagram of the different cell layers in the eye. The eye develops rapidly in the chick embryo and has been a favourite source of cells which must, *in vivo*, organize themselves into separate layers. The pigmented retinal epithelium phagocytoses outer rod segments from the photoreceptors of the neural retina.

fibroblasts would not move onto the apical surface of the retinal epithelium. (Since the eye must not turn inside-out if it is to develop normally this 'rule' seems sensible.) We can also extrapolate the rule, because most animals can be considered as fibroblast-containing matrices covered with sheets of epithelia (arranged in rather a complex fashion) and this simple rule will contribute to maintenance of the normal organization. Further support for this interpretation comes from experiments in which epithelial cells are plated onto confluent layers of fibroblasts and *vice versa*. The epithelial cells flatten successfully on a substratum of fibroblasts, whereas fibroblasts on the apical surface of an epithelial cell sheet remain rounded.

## 10.2.5  Leucocytes

Another heterologous interaction between normal cells which has been investigated is the behaviour of various leucocytes, particularly neutrophil granulocytes (the major class of polymorphonuclear leucocytes of the blood) in collisions with fibroblasts and with **endothelium**, the epithelial lining of blood vessels. Here we are dealing with a cell type, the neutrophil, which is capable of moving through other tissues as part of its normal function in the clearance and resolution of acute inflammation.

When a neutrophil collides with a chick heart fibroblast there appears to be no contact inhibition of either cell, although, because neutrophils are very fast moving cells compared with fibroblasts, the collision may be too transient for the fibroblast to respond (Armstrong & Lackie 1975). Despite the absence of contact inhibition of locomotion, a collision rarely leads to the neutrophil moving onto the dorsal surface of the fibroblast even though it is capable of moving over such a substratum. The reason seems to be related to the relative adhesiveness of the dorsal surface of the fibroblast and the serum-coated substratum on which the fibroblasts are growing. The adhesion of neutrophils to protein-coated glass is much greater than to the upper surface of the fibroblast. This adhesive difference probably accounts for the non-random distribution of neutrophils on a field partly covered by fibroblasts: there are fewer neutrophils overlapping fibroblasts than would be predicted on the basis of chance, taking into account the relative areas available and predicting a Poisson distribution. (This inhibition of overlapping has been referred to as contact inhibition Type 2 and differs from Type 1 in that it does not involve contact inhibition of the locomotory machinery, see § 10.2.2.)

Collisions between neutrophils and endothelial cells differ from those with fibroblasts in that the neutrophil is quite likely to move onto the dorsal surface of the endothelial cell (Lackie & DeBono 1977). In a field that offers equal areas of protein-coated glass and endothelial cell surface, the neutrophil will spend an approximately equal time on each of the two substrata. Predictably, if the argument is accepted that adhesive preference plays a part, the adhesion of neutrophils to endothelial cells is comparable to their adhesion to the protein-coated glass. The difference between the adhesion of leucocytes to fibroblasts and to endothelial cells is probably due to the fact that fibroblasts have fibronectin on their dorsal surfaces, a protein to which neutrophils do not adhere at all well.

Neutrophils placed onto layers of fibroblasts or endothelium will move around but will also squeeze between the cells so that they come to lie between the ventral surfaces of the cells and the solid substratum. Here they continue to move, restricted only by the areas of close apposition of fibroblasts to the substratum; indeed, they can be used as 'probes' for the distribution of focal adhesions.

The behaviour of other leucocytes (lymphocytes, blood monocytes) is substantially the same as that of neutrophils, although lymphocytes move poorly on simple planar substrata and move much better when constrained between the ventral surface of a cell and the solid substratum.

One reason for being particularly interested in the behaviour of leucocytes in respect of their collision responses is that these are 'professionally' invasive cells, cells whose normal function requires that they move through territories occupied by other cells: were they inhibited by collisions then their normal function would be jeopardized. There are, of course, other cells whose normal behaviour requires that they should continue moving even after collision with another cell type. Among these normally invasive cells are capillary endothelial cells, whose migration is important in neovascularization, and the motile tip of a growing neuron (nerve growth cone), which innervates muscle long after the tissue has been established as a 'territory' occupied by muscle cells or which forms connections between, for example, the optic cup and the optic tectum in the developing brain. Slightly less 'normal' perhaps is the spreading of embryonic trophoblast in the formation of the placenta. In morphogenesis many cell movements and migrations occur in which some contact between the migrating cell and other cells is inevitable, and the pattern of movement may well be defined partially by differential collision behaviour.

## 10.3   Consequences of contact inhibition of locomotion

Introducing a major traffic rule which forbids cells to crawl over one another has a considerable effect on the behaviour of a population. Three main consequences are often discussed: the formation of monolayers, the stabilization of position, and enhanced outgrowth from areas of high population density.

### 10.3.1   Monolayering

Because cells are unable or unwilling to persist in moving when this would involve crawling over another cell, the tendency is for cells in a crowded dish to form a monolayer in which every cell is in contact with non-cellular substratum. If fibroblasts are plated in a Petri dish they will divide, provided the medium contains appropriate nutrients and growth factors, and they will become evenly distributed even if the initial plating is non-uniform. Often the formation of a monolayer coincides with the cessation of cell division, but this is only a coincidence in so far as locomotory activity is concerned. Once the population density becomes such that there is a limited area available for each cell then contact inhibition of locomotion prevents the cells from flattening out as much as they would if they had unlimited room. The effect of limiting the area available for the cell to flatten, and in consequence on its ability to absorb growth factors, may well influence the probability of division. If contact inhibition of locomotion did not occur, then the cells would spread to their normal 'uncrowded' area but would then overlap one another. The even distribution of cells in a monolayer is not, of course, random: there is far less overlapping than would be expected. A measure of this is the overlap index. The number of nuclear overlaps that should be found can be predicted knowing the projected area of the cell nucleus, the area of surface available for the cells, and the number of nuclei in this area. In making the prediction it is necessary to assume that there is a Poisson distribution. The observed number of overlaps can be scored, and this can be done at leisure on fixed, stained preparations. The nucleus is usually chosen rather than the whole cell because its margins can be distinguished and the area of the nucleus can be calculated easily because it is a simple shape. Corrections to the predicted number of overlaps may need to be made to allow for 'edge effects', but the index is a relatively straightforward measurement to make. If there is no constraint on the distribution of cells, then the index should be unity; if the constraint is absolute, so that no overlapping occurs at all, then the value of the index will be zero. The latter situation should be the case with simple epithelia. In practice, for chick heart fibroblasts the index is between 0.05–0.2, varying with population density and the nature of the substratum on which the cells are grown. The overlap index has often been used as a measure of contact inhibition of locomotion but is not entirely satisfactory as such. Inhibition of overlapping may arise either from contact inhibition involving paralysis of the locomotory machinery (contact inhibition of locomotion Type 1) or from differences in cell–cell and cell–substratum adhesion (contact inhibition Type 2, or substratum-dependent inhibition of locomotion (Heaysman 1978)), or from a combination of the two types of inhibition. Thus, neutrophil leucocytes show an overlap index significantly below 1 on chick heart fibroblasts probably because of adhesive preference and not because of effects on the motile machinery. Despite the difficulty of interpreting the overlap index it is clear that it does vary between cells and that it is a crude measure of some behavioural characteristics. The piling up of virally-transformed cells, which gives a characteristic morphology to clones, is reflected in their higher overlap index, which may be in the 0.5–0.9 range. An index which does not enable us to predict the mechanism giving rise to the spatial distribution is, however, an unsatisfactory one.

## 10.3.2    Stabilization of position

It is obviously important that tissues remain discrete in the organism (that heart cells do not wander off into the liver) and that the microanatomy of the tissue is stable. Failure of cells to keep to their place is often a malign symptom. Once the tissue or organ has reached its appointed size (and the factors controlling this are now beginning to be understood) then mobility of the cells is unnecessary, except for repair. An obvious consequence of contact inhibition of locomotion is that a cell which is surrounded on all sides by other cells will be inhibited from moving, and will therefore be fixed in position. For an epithelial cell the problem does not really arise because contacts with other cells are stabilized by specialized junctions, particularly desmosomes, which are more-or-less permanent and which do not permit an individual cell to leave. The more autonomous fibroblasts do need some restraint and this may be through an inhibition of locomotion. As with the phenomenon of contact inhibition itself, the effect is not absolute and cells in some tissues may retain the capacity to exchange position. This is rather difficult to estimate in undisturbed tissue, although the blurring of boundaries between differently pigmented clonal populations of melanocytes derived from neural crest might serve as an experimental system. Do the black and white bits of a Friesian (Holstein) cow have a bigger grey boundary as the cow gets older? If fragments of embryonic chick heart are apposed in hanging drop culture then there is some intermingling of cells at the boundary (Abercrombie & Weston 1966) (Fig. 10.6), and labelled single cells collected onto a pre-formed aggregate of homotypic chick heart fibroblasts

[a]

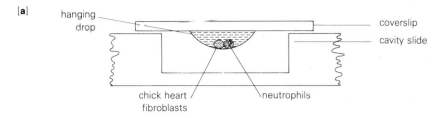

[b]

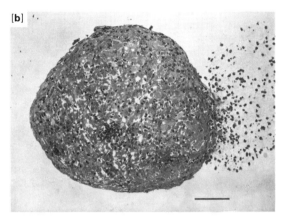

**Figure 10.6  Invasion of a chick heart fibroblast aggregate by neutrophil leucocytes *in vitro*** (a) A schematic of the hanging-drop method used to bring two cell aggregates into contact. (b) A section through a chick heart fibroblast aggregate which has been invaded by neutrophil leuco-cytes from the loose aggregate on the right. The process is rapid, the aggregates were fixed after only two hours in contact. Neutrophils show up as dark-staining bodies in the fibroblast clump. Notice that the fibroblast aggregate, which was originally irregular, has rounded off, and the peripheral cells are organized in a circumferential manner; bar represents 100 µm (from Armstrong 1977).

may later be found deep within the unlabelled territory (Wiseman & Steinberg 1973, Gershmann & Drumm 1975).

## 10.3.3  Radially directed outgrowth

A common method for obtaining primary embryonic chick heart fibroblasts is to explant small pieces of heart onto coverslips and incubate them for a time in an appropriate growth medium with serum present. From the tissue come fibroblasts, as though the edge of the explant were the edge of a wound, but not myoblasts, despite the ability of myoblasts to move. If there were no behavioural constraints and fibroblasts moved in a random fashion then the outgrowth of cells could be modelled as a diffusion with predictable kinetics. *In vivo* there will be an advantage in healing wounds as rapidly as possible and if fibroblasts can be delivered more efficiently then tissue repair will be speeded. One consequence of the phenomenon of contact inhibition is that the cells are directed to a radial outgrowth pattern and, because their movement is directed, the movement of cells from the outgrowth is faster than if a simple diffusive spread occurred. The radial movement is a consequence of the higher probability of collision if the cell moves tangentially or centripetally: by moving centrifugally collisions are much less likely to occur (Fig. 10.7). Thus, contact inhibition of movement will actually enhance invasion of the empty wound area. The spreading of tumour cells will also be enhanced if they show homologous contact inhibition (see § 10.4.1). A comparable effect is seen in the outgrowth of neurons from

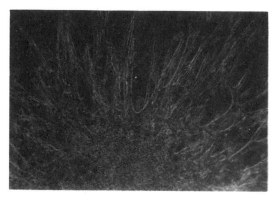

**Figure 10.7 Radial outgrowth from an explant** A phase contrast photograph showing the outgrowth of cells from a circular explant of bovine retina placed on a collagen gel. The pattern is essentially the same as that shown by fibroblasts emigrating from a fragment of chick heart, although from this adult tissue the response is much slower. Not only do cells move out over the surface of the gel, but they also invade the gel matrix (photograph courtesy of Caroline Kerr, Cell Biology Dept, University of Glasgow).

explanted dorsal root ganglia and, since the pattern of innervation *in vivo* differs considerably from this simple pattern, we would expect to find other behavioural constraints playing an important rôle in directing the movement of the nerve growth cone.

### 10.3.4 Formation of parallel and orthogonal arrays

A characteristic feature of confluent cultures of fibroblasts is their 'watered-silk' appearance to the naked eye. This pattern arises because the cells are arranged in territories which do not blend smoothly into one another. Within each territory the cells lie in parallel, but at the boundary the cells are arranged approximately at right angles in an orthogonal fashion. It seems that the gross morphology is a consequence of the behaviour of individual cells when they collide (Erickson 1978). In a broadside collision between the leading edge of one ('active') cell and the lateral margin of another there are two possible outcomes, which depend on the angle at which the cells collide. With BHK cells collisions at less than 55° result in the 'active' cell turning so that it lies parallel to the 'passive' one; if the collision is more nearly at right angles then the result is retraction of the leading edge and reversal. Similar rules seem to operate for the territories, which will only merge if the contact is at less than 55°. The critical angle for collision does vary between cell types and for BHK cells, which have a very broad leading lamella, the angle is high. Foetal lung fibroblasts, which have a much narrower front, will only move into a parallel position at angles of less than 20°. The arrangement of fibroblasts in the developing cornea may partly be specified in this way, and the collagen fibrils which are laid down by the cells are therefore in orthogonal arrays (Bard & Hay 1975).

## 10.4 Escape from normal contact inhibition

Although invasiveness, which we might define as the movement of cells into a territory normally occupied by another cell type, is not restricted to pathological situations much attention has been directed to invasion by neoplastic cells. One of the hallmarks of malignancy is the occurrence of secondary tumours at sites which are remote from the primary lesion, the phenomenon of metastasis. In some cases, though by no means all, metastatic spread is preceded by local invasion so that the tumour cells become distributed through adjacent normal tissue. Often the local invasion may contribute to the tumour cells becoming detached into blood or lymph vessels. Once into such a vessel the cells can be passively transported to lodge, and possibly become established, elsewhere. Access to the bloodstream will also be facilitated by the vascularization of the growing primary tumour, which is invaded by endothelium in response to the production of tumour **angiogenesis** factor.

One of the difficulties faced by the cell biologist studying the contact behaviour of tumour cells is that of quantifying the invasive potential of particular tumour cells. Although the pathologist can score the malignancy of a particular tumour on a rough scale, this is hardly sufficient if we are to attempt to relate contact behaviour to the invasive behaviour *in vivo*. Further difficulties in establishing experimental model systems arise because the time-scale of invasion *in vivo* is probably much greater than of feasible experiments *in vitro*. Obviously a cell which has lost normal contact inhibition of locomotion should be capable of invasion, but it might be sufficient that a small proportion of collisions does not lead to inhibition, it need not be that the response is totally lost. If this is so, then scoring individual collisions in time-lapse film may not reveal a partial deviation from normal behaviour. A possible solution is to confront populations of normal and abnormal cells and look at the end result of multiple collision events, the distribution of the two populations after some time has elapsed (Fig. 10.8). Under these circumstances, however, it is impossible to distinguish whether the distribution arises from alterations in locomotory behaviour (contact inhibition Type 1) or changes in adhesiveness (contact inhibition Type 2).

Despite the problems associated with the study of contact behaviour *in vitro*, some valuable insights have come from looking at the behaviour of **sarcoma** cells, malignant cells derived from normal connective tissue cells such as fibroblasts. Comparable studies on the far more prevalent tumours of epithelial origin, the **carcinomata**, have been sadly neglected.

## 10.4.1 Sarcoma cells

In considering the collision behaviour of normal and abnormal cells we must be careful to distinguish between homologous and heterologous collisions and, in the latter, we need to determine whether the response is reciprocal or non-reciprocal. Thus in the heterologous collisions between fibroblasts and epithelial cells discussed in Section 10.2.4 we had a non-reciprocal heterologous response, although homologous collisions lead to inhibition in both cases. The mouse sarcoma, sarcoma 180 (S180), has probably been the subject of more attention in regard to collision behaviour than any other. The tumour is maintained as a sub-dermal solid tumour by passage through mice, although it can adopt an ascitic form, in which the cells grow dispersed in the peritoneal cavity. There is some reason to suppose that the modern S180 differs in some respects from the original, an almost inevitable consequence of maintained passage, but this need not invalidate the more general messages from studies on these cells.

Collisions between S180 and normal chick heart fibroblasts lead to a reciprocal absence of contact inhibition Type 1, neither cell is affected; whereas both S180 cells and chick heart fibroblasts in homologous collisions show inhibition. This apparently paradoxical situation is not entirely unexpected since homologous inhibition will contribute to directed outgrowth from the primary tumour, and will be adaptively advantageous for the

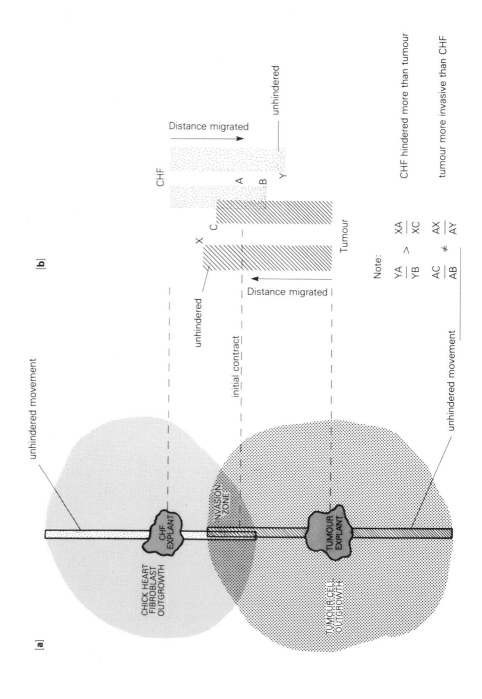

[a]

unhindered movement

CHICK HEART
FIBROBLAST
OUTGROWTH

CHF
EXPLANT

INVASION
ZONE

TUMOUR
EXPLANT

TUMOUR CELL
OUTGROWTH

unhindered movement

[b]

Distance migrated →

unhindered

CHF

A
B
Y

C

X

Tumour

unhindered

initial contract

← Distance migrated

CHF hindered more than tumour

tumour more invasive than CHF

Note:

$$\frac{YA}{YB} > \frac{XA}{XC}$$

$$\frac{AC}{AB} \neq \frac{AX}{AY}$$

tumour. An examination of the ultrastructure of contact areas reveals that the S180 cell does not respond to contact with the chick heart fibroblast by developing the cytoplasmic density and microfilament bundling characteristic of the fibroblast contact region, although the fibroblast with which the collision has occurred does show signs of microfilament organization at the contact point (Fig. 10.2). The absence of a contact inhibition in the fibroblast, despite the development of the adhesion site specialization, lends some support to the view that apposition of focal adhesions may be important for the normal response, although it can be interpreted in other ways as the reader will realize.

The altered distribution of microfilaments in transformed cells has often been commented upon (Royer-Pokora *et al.* 1978), and is particularly interesting since the absence of bundles will alter the local stress put upon the substratum and may well enable these cells to move over substrata of lower rigidity. The diffuse distribution seen in transformed cells resembles that of leucocytes, another invasive cell, and of fibroblasts immediately after they begin to leave an explant, the stage at which their speed may be greatest (Couchman & Rees 1979). In transformed cells the altered distribution of microfilaments and adhesion sites may be a consequence of the increased phosphorylation of vinculin, which in the normal non-phosphorylated state contributes to the stability of bundles (see § 6.4.4). A further interesting aspect of this hypothesis is that epidermal growth factor also affects phosphorylation of vinculin and in principle one can see a possible link between altered growth properties in tumours, which may produce their own growth factors, and the altered behaviour which would affect invasive potential.

Collisions between cells of the methylcholanthrene-induced mouse sarcoma (MCIM) and chick heart fibroblasts are non-reciprocal, with the MCIM cell remaining unaffected while the fibroblast shows paralysis and retraction. *In vivo* the MCIM tumour has marked invasive capacity, possibly more so than S180. When, however, explants of MCIM and chick heart are confronted (Fig. 10.8), the extent of heterologous overlap is only marginally greater than for the homologous confrontation. This may be because the MCIM cells move more slowly than the normal fibroblasts and the decreased speed offsets the reduced inhibition following contact, or that the chick cells retreat in front of the invading horde. It may be that a much longer incubation period would be required to reveal the enhanced heterologous overlapping that would be expected. The experimental confrontation is also rather different from the situation that would be found *in vivo* because the normal cells are able to proliferate and move.

There seems to be considerable variability in the contact behaviour exhibited by tumour cells and in some cases the tumour cells appear normal in their response. Thus Guelstein *et al.* (1973) found no differences in the contact behaviour of normal embryonic mouse fibroblasts and the same cells

◀**Figure 10.8 Confronted explants and the invasion index** Diagram (a) shows two explants, the cell outgrowths from which have confronted one another. In the contact zone the extent of intermingling is (at least in part) a consequence of the extent to which reciprocal contact inhibition of locomotion occurs. Once fixed and stained the explant can be examined, and measurements made of the distance of cell outgrowth at the sides, unhindered by collision, and towards the other outgrowth. The measurements can be illustrated in histogram form, as in (b), where it will be seen that the movement of the chick heart cells has been hindered more than that of the tumour cells, and that there is some mutual invasion – there is intermingling in the contact zone.

transformed by Moloney mouse sarcoma virus, whereas Vesely and Weiss (1973) found neoplastic rat fibroblasts exhibited normal homotypic inhibition and non-reciprocal heterotypic inhibition of the normal cell. The general picture is made more complicated by the use of overlap indices, which may reflect the combination of both types of contact inhibition operating, and the failure to distinguish carefully between head–head and head–side collisions in the analysis of some time-lapse studies. A further problem is that the tumour-forming potential of some transformed cells has not actually been tested. It would perhaps be overly optimistic to suppose that a complicated phenomenon such as invasion by a population of cells can be interpreted simply in terms of easily defined parameters of behaviour that can be scored quantitatively *in vitro*. Nevertheless it is probable that contact inhibition of locomotion is important in restricting the ability of tumour cells to invade locally. Whether the response depends upon inhibition of locomotion *per se*, or is mediated by adhesive preference, or by a combination of the two cannot so readily be determined and almost certainly varies, depending on the particular cell type involved.

## 10.5 Invasiveness as a general phenomenon

Because many studies of the locomotion of tissue cells have as a strategic goal the understanding of the mechanism of tumour invasiveness, we will digress briefly to consider the general problems of invasion. It is important to realize that the environment *in vivo* is considerably more complex than that usually used *in vitro*, and that the pathologist, in speaking of tumour invasiveness, is referring to the behaviour of a population of cells which may well be heterogeneous in properties. As a starting point we will follow Abercrombie (1979) in defining invasion as the movement of a population of cells into an area occupied by another population. A number of factors have been proposed as contributory or predisposing characteristics for invasion to occur:

(a)  growth pressure,
(b)  heterotypic adhesive preference,
(c)  protease release,
(d)  failure of contact inhibition of locomotion,
(e)  negative chemotaxis.

We can easily exclude growth pressure since benign tumours of high growth rate may remain firmly restricted in position, and some malignant tumours (e.g. disseminated carcinoma of the intestine) invade faster than they proliferate so that the primary tumour is hard to localize. The release of proteases may well contribute to the ease of penetration: degradation of extracellular matrix components may remove a mechanical constraint on infiltration by the tumour cells, and modification of the surface properties of the tumour cell by released proteases may contribute to altered behavioural responses. A counter argument is that some invasive cells, such as lymphocytes, produce almost no proteases, whereas tissue macrophages which produce ample supplies of hydrolytic enzymes tend to be fixed in position. Negative chemotaxis might serve to direct the movement but does not constitute a causative mechanism. The idea that heterotypic adhesion preference might contribute to invasion is essentially the converse of the 'sorting out' of tissues. The segregation of differentiated cells into homogeneous territories of similar cells during morphogenesis (often referred to as 'sorting out') seems very plausibly explained as being the outcome of homotypic adhesion preference as suggested by Steinberg (1962a, b & c, 1970). The sorting out of an experimentally mixed population of cells in an aggregate or randomly disposed over a surface, into territories of homotypic cells depends, according to the Steinberg hypothesis, on the homotypic adhesion of one cell type being thermodynamically preferable, i.e. more probable, than the heterotypic adhesion between dissimilar cells (Fig. 10.9). If

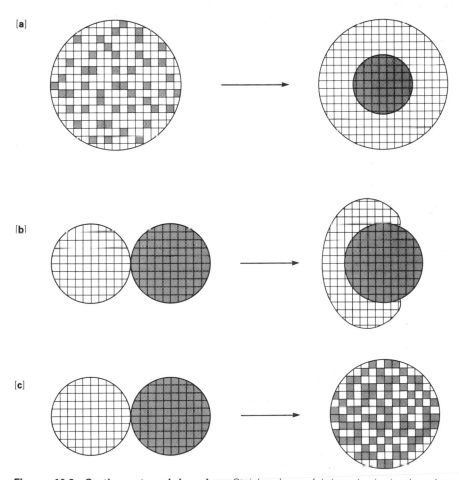

**Figure 10.9 Sorting-out and invasion** Steinberg's useful hypothesis is that the 'sorting out' behaviour is a consequence of cells maximizing their preferred contacts. Thus an aggregate made of two cell types as in (a), will sort out into two isotypic territories, with one cell type surrounding the other. If two aggregates are put in contact then one will spread over the other, as in (b), and in general the cells that sort out into the internal territory from a mixed aggregate will be the ones that are enveloped. The assumption is that these processes are aimed at maximizing homotypic contacts which are more stable or stronger. So, if the heterotypic adhesion is preferred, then two aggregates put together might engage in 'reverse sorting' and form a mixed aggregate: one cell would invade the homotypic territory of the other (see Fig. 10.6).

heterotypic adhesion were stronger than homotypic adhesion then the equilibrium position, in which heterotypic contacts were maximized, would be of intermingled populations; the tumour would be invaded by normal cells as it concurrently invaded adjacent tissue. Although this is an attractive hypothesis there is little experimental evidence to favour it and at least one piece of evidence against. Neutrophil leucocytes will become distributed through an aggregate of chick heart fibroblasts, yet the adhesion of neutrophil to fibroblast seems to be weaker than the homotypic adhesion of neutrophil to neutrophil and fibroblast to fibroblast (Lackie & Armstrong 1975). That we can dismiss some possibilities in some systems does not mean that they may not operate or contribute to invasion in other situations.

This leaves as a major contender for explaining invasive potential the loss, by the invasive cell, of the normal constraint of contact inhibition of locomotion.

## 10.5.1 Directional cues for invasive movement

As pointed out in an earlier section, the invasive spread of neoplastic cells may be enhanced by homotypic contact inhibition of locomotion, which would direct the movement centrifugally. Although invasive neoplastic cells have altered social behaviour they will still respond to many of the directional cues in the environment, as will the 'professionally' invasive cells such as leucocytes. Pathologists have long recognized the tendency of tumours to invade along 'planes of weakness' of connective tissue. An important cue may be contact guidance, a phenomenon which is certainly shown by leucocytes (Wilkinson *et al.* 1982) and probably by neoplastic cells. In confronted explants of mouse sarcoma (MCIM) and chick heart fibroblasts the outgrowth of the sarcoma cells is greater in the contact region than at the unobstructed margin, and it seems that they are responding to the guidance cue offered by the upper surface of the chick heart cells which are, of course, radially disposed. Although a guidance cue is, by definition, only an axial one, the vectorial cue can be provided by the population distribution – the concentration of cells is high at the focus (explant) and diffusive spread is enhanced by the directional constraint of the guidance field.

More recently the implications of guidance cues in wound healing responses have been considered. Clot retraction brought about by platelets in the fibrin meshwork of a blood clot will tend to align the fibrillar matrix which can then serve to guide both leucocytes and fibroblasts. Subsequent contraction in the wound, brought about by fibroblasts in the second, tension-generating, phase of their activity (see § 6.4.1) may cause further alignment of the adjacent tissues and affect the movement of the epithelial edge – although this latter aspect has not been investigated.

## 10.5.2 Invasion in the chick limb-bud

Although almost beyond the scope of this book, it would seem inappropriate to leave the discussion of cell–cell interactions without mentioning an assay for invasiveness which has given some important insights, and which models far more closely the natural situation (Tickle 1982). The limbs of avian embryos develop as 'limb-buds', loose mesenchymal tissue covered by a simple epithelial sheet of **ectoderm**. Much is known about the specification of positional information in these limb-buds, the information which is required to specify the disposition of skeletal elements, and the grafting techniques developed for studies on positional information can be modified to permit the implantation of various cells into the mesenchymal tissue. The distribution of the introduced cells can be determined by histological examination of the limb-bud after one or two days. The wing mesenchyme is invaded by cells of many types, particularly those which are invasive as part of their normal behaviour (trophoblast, leucocytes, endothelium) (Tickle & Crawley 1979), many 'transformed' cells (PyBHK, S180), and some 'normal' cell lines (BHK, Nil 8, MRC5, for example). Normal embryonic tissue cells do not invade but, surprisingly, cells of three carcinomas did not invade the mesenchyme either. The carcinoma cells (which are epithelially-derived) became associated with the limb ectoderm and infiltrated the ectodermal epithelium so that the basement membrane was shared by normal and tumour cells (Tickle *et al.* 1978a & b). The extent to which the behaviour of the different cells is influenced by cell–cell contact and the contribution of the extracellular matrix to determining behaviour remain unknown but it is clear that differences in cell behaviour play an important part in regulating the distribution of cells.

One thing that emerges clearly from the studies on the chick limb-bud assay is that no single property or alteration in behaviour characterizes those cells which are invasive and those which remain in position. It is important to realize that apparently complex interactions can be specified by relatively simple rules, but equally important to recognize that we do not understand all the rules nor have the interactions been fully described even on the phenomenological level.

## 10.6  Summary

It should be clear from the description of contact inhibition of locomotion that the rules governing cell behaviour following contact are both complex and fascinating. Complex because the behaviour is influenced in a variety of ways depending upon the cell types involved, and fascinating because of the implications for the behaviour of cells in multicellular organisms. The idea that cells possess or gain invasive capacity because they fail to obey the normal rules is extremely attractive, but it would be wrong to suppose that other factors are not likely to be important. We know remarkably little about cell–cell interactions but, as information about behaviour of cells as individuals accumulates, it will become easier to interpret the behaviour of populations.

## References

Heaysman, J. E. M. 1978. Contact inhibition of locomotion – a reappraisal. *Int. Rev. Cytol.* 55, 49–66.

Abercrombie, M. 1979. Contact inhibition and malignancy. *Nature (Lond.)* 281, 259–62.

Bellairs, R., A. Curtis and G. Dunn (eds) 1982. *Cell behaviour.* Cambridge: CUP. Particularly the chapters by Heaysman and Pegrum, Stephenson, Middleton, and Tickle.

Trinkaus, J. P. 1984. *Cells into organs,* 2nd edn. Englewood Cliffs, NJ: Prentice-Hall. Trinkaus devotes two chapters to the topics covered in this chapter, and writes authoritatively on the developmental aspects.

Armstrong, P. B. 1984. Invasiveness of non-malignant cells. In *Invasion, experimental and clinical implications*, M. M. Mareel and K. C. Calman (eds), 126–67. Oxford: OUP.

Armstrong, P. B. The control of cell motility during embryogenesis. *Cancer Metastasis Reviews* (in press). An introduction to the problems of controlling cell migration in morphogenesis – problems that have not really been discussed in this book.

# GLOSSARY

Words in bold type may also be found as separate entries.

**A23187**   The most commonly used calcium ionophore.

**A-band**   That portion of the sarcomere in which the **thick (myosin) filament** is located. Anisotropic in polarized light.

**acrasin**   Name originally given to the chemotactic factor produced by cellular slime-moulds: now known to be cAMP for some species.

**acrosomal process**   A long process actively protruded from the acrosomal region of the spermatozoon following contact with the egg: assists penetration of the gelatinous capsule.

**acrosome**   The lysosomally-derived vesicle at the extreme anterior end of the spermatozoon.

**actin**   A protein of 42 kd, very abundant in eukaryotic cells and one of the major components of the **actomyosin** motor.

　**F-actin**   A filamentous polymeric form of actin; 5–7 nm diameter. The actin is sometimes associated with **tropomyosin** (**thin filaments** of striated muscle, **microfilaments** in **stress fibres**).

　**G-actin**   The globular monomeric form of actin; 6.7 × 4.0 nm.

**α-actinin**   A protein of 100 kd normally found as a dimer: may link **actin** filaments end–end with opposite polarity. Originally described in the **Z-disc**, now known to occur in **stress fibres** and at **focal adhesions**.

**β-actinin**   A protein (35 kd, normally dimeric) which is thought to bind to the end of the **thin filament** furthest from the **Z-disc**, serving to block disassembly. Might be homologous to **acumentin**.

**actinogelin**   Protein (115 kd) from Ehrlich ascites cells which gelates and bundles **microfilaments**.

**actomyosin**   Generally: an adjective applied to a motor system which is thought to be based on **actin** and **myosin**. More specifically: a viscous solution formed when actin and myosin solutions are mixed at high salt concentrations. The viscosity diminishes if ATP is supplied and rises as the ATP is hydrolysed. Extruded threads of actomyosin will contract in response to ATP.

**acumentin**   Protein (65 kd) of analogous function to β-**actinin**, isolated from vertebrate **macrophages**.

**adaptation**   Used specifically in the context of sensory behaviour to mean the loss of responsiveness to a persistent stimulus.

**adhaerens junctions**   Specialized cell–cell junctions into which are inserted **microfilaments** (known to classicists as *zonulae adhaerentes*) or **intermediate filaments** (*maculae adhaerentes* or 'spot desmosomes').

**aequorin**   Protein (30 kd) extracted from jellyfish (*Aequorea aequorea*) which has the valuable property of emitting light in proportion to the concentration of calcium: it can therefore be used to measure calcium concentrations, but has to be microinjected into cells.

**aerotaxis**  Tactic response to oxygen (air), nothing to do with airborne taximeter cabs!

**angiogenesis**  The process of vascularization of a tissue involving the development of new capillary blood vessels.

**anisotropic**  Not the same in all directions, non-uniform.

**ankyrin**  Globular protein (200 kd) which links **spectrin** and an integral membrane protein (band 3) in the erythrocyte plasma membrane. **Isoforms** probably exist in other cell types.

**antiplectic**  Pattern of **metachronal** coordination of ciliary beat in which the waves pass in the opposite direction to that of the active stroke.

**apocrine**  Secretory process which involves the production of membrane-bounded vesicles from the apical region of the cell.

**arrowheads**  Fanciful description given to the pattern of **myosin** molecules attached to a filament of **F-actin**. Hard to see, but define the polarity of the filament.

**axoneme**  The central **microtubule** complex of eukaryotic **cilia** and **flagella** with the characteristic '9 + 2' arrangement of tubules when seen in cross-section.

**axopod (pl. axopodia)**  Thin processes (a few microns in diameter but up to 500 μm long), supported by complex arrays of **microtubules**, which radiate from the bodies of *Heliozoa*.

**axostyles**  Ribbon-like bundles of **microtubules** found in certain parasitic protozoa: may generate bending waves by **dynein**-mediated sliding of microtubules.

**basal body**  Structure at the base of a **cilium**; anchored to the cytoskeleton in some way. Has an appearance reminiscent of a centriole.

**BHK cells**  A cell line (putatively normal) derived from baby (Syrian) hamster kidney. Although fibroblastic in appearance, they may be smooth muscle derived. Most common stock is 'BHK-21 clone 13'.

**birefringence**  Some substances have optical properties such that the refractive index is different for light polarized in one axis compared with light polarized in the orthogonal axis. The effect is as though there had been rotation of the plane of polarization (although this is not the case unless the material is also dichroic): it may arise from molecular organization (form birefringence) or as a result of flow or stress causing **anisotropy** in a normally random material.

**brush border**  The apical surface of, for example, intestinal **epithelial** cells which have many **microvilli**.

**caldesmon**  Protein (150 kd, normally dimeric) from gizzard smooth muscle: a calcium-sensitive **F-actin** cross-linker.

**calmodulin**  Ubiquitous calcium-binding protein (17 kd). 'Ancestor' of **troponin C, leiotonin C,** and **parvalbumin**.

**calsequestrin**  Protein (44 kd) found in the **sarcoplasmic reticulum** of muscle: sequesters calcium.

**capping**  (a)  The movement of cross-linked cell-surface material to one pole of the cell (or to the perinuclear region);
(b)  the intracellular accumulation of **intermediate filament** protein in the pericentriolar region following **microtubule** disruption;
(c)  blocking further addition (or removal) of sub-units to linear polymers by binding to the free ends.

**carcinoma (pl. carcinomata)**  Malignant tumour of **epithelial** origin, much more common than **sarcoma**.

**cell line** Cells which have been maintained in culture and have passed the crisis phase, become established and effectively immortal. They may be one step along the road to malignancy, and cannot be considered normal.

**cell strain** Cells adapted to culture, but retaining finite doubling potential. They may in time become a line, especially if derived as a primary culture from a rodent.

**centriole** Organelle in animal cells which is located near to the nucleus and divides prior to mitosis: daughter centrioles separate to opposite poles of the spindle. Structure: a pair of orthogonal cylinders with 9 triplets of microtubules forming the walls, very similar to the **basal body** of a **cilium**.

**chemoattraction** Non-committal description of cellular response to diffusible chemicals; preferable to terms which imply a known mechanism.

**chemokinesis** A response to a soluble chemical substance which involves a change in the rate or frequency of movement or a change in the frequency or magnitude of turning behaviour.

**chemotaxis** A response of motile cells or organisms to a concentration gradient of a soluble chemical in the environment which affects the direction in which a turn is made and thus the direction of movement.

**cilium (pl. cilia)** Motile appendages of eukaryotic cells: contain an **axoneme** and differ in length and beat pattern from most eukaryotic **flagella**.

**cirri** Large motor organelles of hypotrich ciliates: formed from fused cilia and contain many **axonemes**.

**clathrin** Protein (180 kd plus 34 kd & 36 kd light-chains) which forms the 'basket' of triskelions around a **coated vesicle**.

**coated vesicle** Vesicle, formed from the plasma membrane (by closure of a coated pit), which is surrounded by a **clathrin** basket. Associated with the vectorial transport of membranes between organelles.

**colcemid** Methylated derivative of **colchicine**.

**colchicine** Alkaloid (400 d) isolated from the Autumn crocus, binds to the heterodimer (but not the monomer) of **tubulin** and thereby inhibits its incorporation into a **microtubule**.

**comb-plates** Large flat plates formed by the fusion of **cilia** which are arranged in vertical rows: the motile appendages of ctenophores.

**concanavalin A (Con A)** A protein (tetramer: $4 \times 27$ kd) which binds carbohydrates (a lectin). Isolated from the Jack bean: highest affinity for $\alpha$-mannoside residues.

**connectin** Cell-surface protein (70 kd) from mouse fibrosarcoma cells: binds **laminin** and **actin**.

**contact following** The behaviour shown by individual slime-moulds when they join a stream of cells moving towards the aggregation centre. Cell–cell contacts at head and tail stabilize position ('contact sites A' involved).

**contact guidance** Directed locomotory response of cells to an axial cue – the shape of the substratum: see Chapter 8.

**contact-induced spreading** Phenomenon exhibited by epithelial cells 'in which the whole is greater than the sum of the parts' (see Ch. 10). Two cells in contact spread more than they would individually.

**contact inhibition** Requires to be qualified '. . . of movement': reaction in which the direction of movement of one cell is altered following collision with another. May involve contact paralysis of the locomotory machinery.

**contact inhibition of growth** A misleading term: see **density-dependent inhibition**.

**cortical meshwork** The sub-plasmalemmal array of **microfilaments** which gives support and encouragement to the surface of the cell.

**cyclosis** Cyclical streaming of the cytoplasm of plant cells, best seen in the giant internodal cells of characean algae.

**cytochalasins** A family of fungal metabolites which have inhibitory effects on the addition of **G-actin** to the barbed end of **microfilaments** and on 'temporary' **actin**-based motors. At higher doses some affect glucose transport.

**cytokeratins.** A generic term for the proteins found in the **intermediate filaments** of epithelial cells.

**cytokinesis** The process of division of the cytoplasm, as distinct from the (mitotic) division of the nucleus.

**Density-dependent inhibition of growth** The phenomenon exhibited by most cells in culture which stop dividing once a certain critical cell density is reached. The cessation of population growth may depend upon limitation of growth factor availability, or on ill-defined factors, such as substratum availability.

**Desmin** Intermediate-filament-protein isoform (55 kd) of striated muscle cells. Also known as skeletin.

**desmosome** Specialized cell–cell junction (*macula adhaerens*): associated with **tonofilaments**.

**displacement** Generally used to mean the direct distance ('as the crow flies') between the start and finish of a track.

**distribution** Various specific and well-defined kinds of distributions exist (Gaussian, Poisson, binomial, for example) and the term should be used with caution.

**dynein** Large multimeric protein (600–800 kd) with ATPase activity: found in the **axoneme**, responsible for **microtubule** sliding in the **cilium**.

**ectoderm** The outer of the 'germ layers' of the embryo which will give rise to outer epithelia, nervous tissue, and many other tissues.

**ectoplasm** That portion of the cytoplasm of amoebae located nearest the plasma membrane: normally granule-free and probably mostly composed of **actin** meshwork. Also used in spiritualism.

**EDTA** Ethylenediaminetetraacetic acid (or the disodium salt). Commonly used chelator of divalent cations.

**EGTA** Ethyleneglycol-bis ($\beta$-aminoethyl ether) N,N,N′,N′-tetraacetic acid. Like **EDTA** this compound chelates divalent cations – but its affinity for calcium (log $K_{app}$ 6.68 at pH 7) is much higher than its affinity for magnesium (log $K_{app}$ 1.61 at pH 7). (However, it does bind magnesium and many other divalent cations and is less specific than is often assumed.)

**endoderm** Inner of the embryonic germ-layers: **epithelium** of the intestine is the major tissue derived from this layer.

**endoplasm** The inner cytoplasm of amoebae.

**endothelium** The epithelial lining of blood vessels, lymphatics and other fluid-filled cavities, such as the front chamber of the eye. Unusual in being a **mesodermally**-derived **epithelium**.

**epithelium** Cell layer in which the cells are associated with a basement membrane and are linked (usually) with **desmosomes**. Useful histological classification which seems to have predictive value for the behaviourist.

**fascin** **Actin**-binding (bundling) protein (58 kd) from sea-urchin eggs.

**fibroblast** Cell of connective tissue (**mesodermal**). Secretes collagen and collagenase. Usually applied rather unsatisfactorily to a cell in culture 'which looks like a fibroblast'.

**fibronectin** Large (220 kd) protein found in plasma and in extracellular matrix. Promotes **fibroblast** spreading. Also known as LETS-protein, cold insoluble globulin, CSP, CAP.

**filament** Any long thin structure; intracellularly may be 'thin' (**actin**), 'thick' (**myosin**), or '**intermediate**' (various).

**filamin** **Actin**-binding protein (250–270 kd, dimeric) involved in actin gel and bundle formation: forms flexible cross-links.

**filopodia** Thin protrusions from **fibroblasts** or other tissue cells, possibly linear equivalents of leading lamella.

**fimbriae** Filamentous appendages of the cell wall of (generally) Gram-negative bacteria. 7–10 nm thick, several microns long. Made of pilin; also known as **pili**.

**fimbrin** Protein (68 kd) from **microvillus** core.

**flagellin** Protein (40 kd) sub-unit of bacterial **flagellum**.

**flagellum (pl. flagella)** Motor appendages of bacteria and of many eukaryotes, though structures totally different. In eukaryotes differ from cilia only in length and number per cell.

**fMet-Leu-Phe** The original and most popular synthetic chemotactic peptide (formyl-methionyl-leucyl-phenylalanine). Binding of the peptide to the neutrophil receptor initiates many intracellular events, not just directed locomotion. It is nice to think that it is 'modelling' bacterial signal sequence.

**focal adhesion** Small area ($0.2 \times 1$ μm) of very close apposition of a cell to its **substratum**; assumed to be an adhesion site.

**fodrin** Tetrameric protein ($\alpha$ 240 kd, $\beta$ 235 kd): isoform (probably) of **spectrin**, from brain.

**fragmin** **Actin**-binding protein (42 kd) from *Physarum*: calcium-sensitive severing/**capping**.

**galvanotaxis** Directed cellular response to an electrical field.

**gap junction** Small specialized cell–cell junction which permits passage of ions and small molecules between the cells.

**gel** Deformable semi-solid matrix made of cross-linked polymers. Gelatin gel of lurid colour usually popular with small children at parties.

**gelsolin** **Actin**-binding protein (90 kd) from **platelets** and **leucocytes**: calcium-sensitive severing/**capping**/nucleation.

**geotaxis** Directed response to gravity.

**GFAP** Glial fibrillary acidic protein: sensibly non-committal and informative name for intermediate filament protomer (50 kd) isolated from glial cells.

**granulocyte** White blood cell of the myeloid series. Polymorphonuclear granulocytes may be **neutrophils**, eosinophils, or basophils, depending upon their affinity for the components of Romanovsky stains such as Giemsa.

**guidance** Response of cells to axial cue, often referred to as **contact guidance**.

**H-zone** Central 'light' region of the **A-band** of the sarcomere. May represent region of **thick filament** without cross-bridges.

**haptotaxis** Directed response of cells to a gradient of adhesiveness.

**HeLa cells** Perhaps the most famous line of human cells. Derived originally from a cervical carcinoma (the patient being Helen Lane or Lange – the history is obscure) and therefore epithelial in origin.

**heterodimer** Dimer made up of two dissimilar monomers. **Tubulin** is the classic example.

**hispid**  Hairy.

**histogenesis**  In embryonic development the process of formation of tissues.

**HMM**  Heavy meromyosin: soluble tryptic fragment of **myosin** which has ATPase activity and binds to **F-actin**. Can be split by papain into S1 and S2 subfragments.

**homology**  Relationship which implies common ancestry or origin: frequently misused.

**hyaline**  Clear, transparent, used to describe the anterior cytoplasm in amoebae.

**I-band**  The **isotropic** part of the sarcomere which contains only **thin filaments** and which appears to shorten during contraction. Isotropic (not **birefringent**) in polarized light.

**intercalated disc**  The specialized intercellular junctional region in cardiac muscle: contains **adhaerens** and **gap junctions**.

**intermediate filaments**  Tension resisting filaments of 10 nm thickness found in most cells. Protomer varies according to the origin of the cell. See **desmin**, **GFAP**, **cytokeratins** and **vimentin**.

**invasion**  Movement of one cell type into a territory normally occupied by another.

**invasion index**  An index defined by Abercrombie and Heaysman (1976) to estimate the invasiveness of cells in confronted explants. The index is the ratio of the estimated untrammelled movement and the actual invasive movement in the contact zone.

**isoform**  Shorthand for 'protein having the same function, similar or identical sequence – but product of a different gene': rather less committing than **homologous**, but having that implication.

**isometric**  Of constant length: applied to tension generated in a muscle without any shortening (sliding) occurring – cross-bridges are being made and broken with the same **actin** molecule each time. Should not be used as an adjective applying to contraction!

**isotonic**  Of constant force: applied to muscle contractions. (Also of equal tonicity when referring to solutions.)

**isotropic**  The same everywhere, uniform.

**junctional complex**  The set of junctions found in **epithelial sheets**. Moving from apical to basal: a tight junction (*zonula occludens*), an **adhaerens junction** (into which are inserted the microfilaments of the **terminal web**), **desmosomes** (*maculae adhaerentes*), and **gap** (communicating) **junctions**.

**kinase**  An enzyme which phosphorylates something.

**kinesis**  Locomotory response in which the characteristics of the path are altered but there is no directional component.

**kinetochore**  Region of the chromosome to which **microtubules** attach: no longer thought to be an **MTOC**.

**klinokinesis**  Locomotory response in which the frequency of turning or the magnitude of turns is altered.

**lamella**  Literally, thin sheets or plates: applied to cells, the thin region of cytoplasm protruded at the front of a moving cell.

**lamellipodium**  Flattened projection from the surface of a cell, in contrast to **filopodia** which are finger-like.

**laminin**  Multi-subunit protein (400 kd, plus 3 × 200 kd) associated with collagen Type IV in the basal lamina of an **epithelium**. Perhaps an analogue of **fibronectin**.

**lectin** Generic term for any carbohydrate-binding protein: originally lectins were isolated from plants but the term is now used in a more general sense. The best known examples are proteins such as **concanavalin A** and wheat germ agglutinin.

**leiotonin** The analogue (perhaps) of **troponin** in smooth muscle. Complex of two sub-units, leiotonin A and leiotonin C (18 kd; homologous with **calmodulin** and **troponin C**).

**leucocyte** White blood cell: generic term, needs qualification since the functions of these cells are so diverse. See **lymphocyte**, **neutrophil**, **monocyte**, and **granulocyte**.

**ligand** That which is bound; often, but not necessarily, implying the involvement of a specific receptor: conventionally the smaller partner.

**light-chain** Two kinds of proteins are referred to in this way: parts of the immunoglobulin complex and part of the **myosin** complex. The latter set can be subdivided into 'regulatory' and 'essential' light chains (see Ch. 2).

**LMM** Light meromyosin: insoluble tryptic fragment of **myosin**; forms the backbone of the **thick filament**.

**lumicolchicine** Photo-rearranged form of colchicine which will bind to membranes but not to **tubulin**. Useful (essential) as a control.

**lymphocyte** White blood cell (**leucocyte**) of the lymphoid series which, according to set or subset, may be involved in immunoglobulin production (B-cells) or cell-mediated immunity (T-cells). T-cells may help, suppress or remember the appropriate immune response or be involved directly in cytotoxicity ('helper', 'suppressor', 'memory', & 'cytoxic' cells).

**MAP** Microtubule-associated protein: MAP-1 (350 kd) and MAP-2 (300 kd) co-purify with **tubulin**: may be part of the electron-lucent zone around the tubule and may interact with other cytoskeletal components.

**MCP** Methyl-accepting chemotaxis protein: protein on the cytoplasmic face of the bacterial plasma membrane which becomes methylated in response to binding of a chemotactic factor. At least four classes have been described. (A confusing term to use in feminist company?)

**M-band** Central region of the **A-band** of the sarcomere.

**M-line** Central part of the **A-band** of the sarcomere of striated muscle. Contains M-line protein (**myomesin**: 165 kd). Thought to help in mainten-ance of spacing of **thick filaments** since myomesin binds **myosin**. Contains creatine kinase (40 kd) and glycogen phosphorylase b (90 kd).

**MLCK** Myosin light chain kinase: **calmodulin**-sensitive kinase (variable molecular weight depending on source: 130 kd in non-muscle cells) possibly involved in control of **myosin** activation.

**MTOC** Microtubule organizing centre: necessary but mysterious zone or structure involved in specifying the number and position of **microtubules**. In mammalian cells the major MTOC is the pericentriolar region.

**macrophage** Phagocytic cells of the myeloid series found in tissues. Different types are found in the peritoneal cavity, the alveolar spaces of the lung, in connective tissue (histiocytes), and in the liver (Kupffer cells): now generally believed to be differentiated from blood monocytes. Sometimes referred to as mononuclear phagocytes to distinguish them from granulo-cytes. Become 'activated' when stimulated.

**magnetotaxis** Response directed by a magnetic field.

**mastigonemes** Projections from eukaryotic flagella which may be stiff, and modify the hydrodynamics of propulsion, or flexible (flimmer filaments), and increase the effective diameter of the flagellum.

**matrix** Material in which cells are embedded or which separates cells in a tissue. Used in this book to mean a three-dimensional acellular medium which is not fluid, although it may perhaps be viscoelastic. A **gel** is a matrix in this sense.

**Meromyosin** Tryptic subfragment of **myosin**. May be heavy (**HMM**, containing the ATPase activity) or light (**LMM**, the insoluble portion important for **thick filament** assembly).

**mesoderm** Inner or middle of the germ layers, lies between **ectoderm** and **endoderm**: gives rise to skeleto-muscular system and to various connective tissues.

**metachronal** Happening at a later time; applied to the beat pattern of **cilia** in which successive cilia perform the active part of their beat-cycle just behind (after) the adjacent one. Different patterns of metachronal rhythm are found, and are named according to the relationship of the wave of coordination in the field and the direction of the active stroke of each cilium (see Ch. 5).

**microfilament** Cytoplasmic filament 5–7 nm thick composed of **actin**: can be decorated with **HMM** and may be associated with other proteins.

**microspikes** Filopodia.

**microtrabecular network** A lattice-work of proteins visualized using high-voltage electron microscopy following rapid fixation or freezing of cells. It is suggested that the whole cytoplasm is composed of this matrix, the gaps being filled with watery solution of small molecules. (Not everybody wants this idea to be correct.)

**microtubule** Cytoplasmic tubule 25 nm in diameter, composed of **tubulin**. Microtubules are found in eukaryotic **cilia** (& **flagella**), in the mitotic spindle, and in the cytoplasm.

**microvillus (pl. microvilli)** Finger-like projection, usually one of many, from the apical surface of an **epithelial cell**. Supported by a cytoplasmic 'rod' of bundled **microfilaments**.

**monocyte** Circulating form of the mononuclear phagocyte (a **leucocyte** of the myeloid series) which will differentiate into a **macrophage** once it enters the tissues.

**monopodial** One-legged: used of amoebae with only one **pseudopod** (or false leg).

**morphogenesis** Process of shape-formation: the formation of a complex body from the division products of a single cell during embryonic development.

**myoblast** The embryonic cell which will differentiate into striated muscle, a process which may involve cell fusion in vertebrate skeletal muscle.

**myofibril** Intracellular array of sarcomeres, arranged in series. There are many myofibrils in a single muscle fibre.

**myogenesis** Process of muscle formation (during embryogenesis, not exercise).

**myomesin** See **M-line**.

**myosin** Multimeric protein which has an **actin**-activated ATPase activity, and which probably changes shape as part of its rôle in the mechano-chemical cycle.

**myotube** Syncytium formed by the fusion of several **myoblasts** in a longitudinal array.

**N-line** Structure seen in the sarcomere.

**nebulin** Very large protein found in the **N-line**(s) of the sarcomere.

**neutrophil** Commonest of the leucocytes, a short-lived phagocytic cell of the myeloid series which is one of the first cells to respond in the acute inflammatory response. Strictly speaking should be 'a neutrophil granulocyte' or a polymorphonuclear leucocyte which stains with both acid and basic dyes. Often referred to as a PMN or PMNL by those who write in jargon.

**Newtonian** Obeying the rules of Newtonian physics; or rather, behaving in a straightforward way according to simple mechanics and not changing properties unexpectedly. A Newtonian fluid has the same viscosity at all flow rates.

**nexin** Protein which links the adjacent doublets of the ciliary **axoneme** together.

**occludens junction** *Zonula occludens* or **tight junction**.

**orthokinesis** Response in which the speed or frequency of movement is altered. Can be positive or negative.

**ouabain** (G-Strophanthin.) Cardiac glycoside (585 d) which inhibits the sodium–potassium pump of the plasma membrane.

**overlap index** A measure of the extent of multilayering of cells based on fixed and stained preparations. The predicted extent of **overlapping** is calculated knowing the projected area of the nucleus, the area available and the number of nuclei in that area, assuming a **Poisson distribution**. The actual extent of overlapping is then determined and expressed as a ratio. A value of 1 implies no inhibition of multilayering; lower values imply some constraint but not its source.

**overlapping** When applied to cells implies that one cell is moving over the surface of another – in contrast to **underlapping**.

**parenchyma** Plant tissue composed of large relatively undifferentiated cells.

**parvalbumin** Calcium-binding proteins (12 kd) found in teleost and amphibian muscle: have sequence **homology** with **calmodulin** but only two calcium-binding sites.

**patching** Passive process in which integral membrane components become clustered following the addition of a cross-linking polyvalent **ligand** such as **Con A**.

**persistence** Tendency to continue moving in the same direction, shown particularly by **leucocytes** and to a lesser extent by **fibroblasts**.

**phalloidin** Active principle of *Amanita phalloides* (Death Cap). A cyclic peptide (789 d) which rapidly promotes the polymerization of **G-actin** and stabilizes **F-actin**. Too large to penetrate the plasma membrane easily but fluorescent derivatives can be used on permeabilized cells.

**phosphatase** Enzyme which dephosphorylates, thereby reversing the action of a **kinase**.

**phosphodiesterase** Generally used to mean the enzyme which breaks down cAMP, although this is only one of a class of these enzymes.

**phototaxis** Directional response to light.

**pilus (pl. pili)** Hair-like projection from the surface of a bacterium. Involved in attachment to surfaces and to other bacteria (sex-pili) in conjugation. Also known as **fimbriae**.

**plasmalemma** Old name for the plasma membrane (and often including the **cortical** layer of cytoplasm).

**platelet** Small anucleate cell (a fragment of a megakaryocyte) found in mammalian blood: involved in the formation (and retraction) of a blood clot.

**PMF**  Proton motive force: the proton (pH) gradient across a prokaryote membrane which provides the coupling between oxidation and ATP synthesis, for example.

**podophyllotoxin**  Toxin (414 d) which binds to **tubulin** and prevents assembly.

**polarity**  Orientation, literally 'having poles'.

**polypodial**  'Many-legged' – in contrast to **monopodial** or 'one-legged'. Used of amoebae.

**profilin**  Actin-binding protein (15 kd, pI 9.3) which forms the complex 'profilactin' with **G-actin**: prevents the activation (nucleation) step of F-**actin** formation rather than the elongation process.

**protomer**  Sub-unit from which a polymer is formed, not necessarily monomeric (thus the **tubulin heterodimer** is the protomer for **microtubule** assembly).

**pseudopod**  Blunt projection from a cell. Used originally for the protrusions from amoeba.

**PtK1 cells**  An established cell line derived from female potoroo kidney. The cells have an epithelial morphology; they are popular because the nucleus has only a few large chromosomes. (Potoroos are marsupials, also known as kangaroo rats, in case you think this is a spoof.)

**receptor**  Site to which things bind. Implies selectivity in the binding, that there is a consequential outcome of binding, and that there are a finite number of sites. The binding data is often analyzed using a **Scatchard** analysis. The term 'receptor' is often used even when there is no evidence for the existence of a discrete site.

**retraction fibres**  Thin cytoplasmic processes, attached to the substratum at one end, which are under tension. Found at the tail of a moving cell, also anchoring a cell which has rounded up at mitosis or because of experimental treatment. Often have **microfilament** bundles inside.

**Reynolds' number**  A dimensionless constant, the ratio of inertial and viscous drag on a body moving through a fluid medium.

**rheotaxis**  Directional response to a fluid flow.

**ruffle**  Apt description of the protrusion at the front of a moving **fibroblast** which, in time-lapse film, seems to be 'ruffling'.

**S1**  Soluble fragment (102 kd) of **HMM** produced by papain cleavage: has the ATPase and **actin**-binding activity.

**S2**  The other fragment of **HMM**. Fibrous but soluble.

**saltatory**  Jumping from place-to-place abruptly: mode of locomotion of the kangaroo and flea.

**sarcoma**  Malignant tumour derived from cells of connective tissue. Sometimes prefixed to indicate more precisely the cell-type of supposed origin (e.g. osteosarcoma, derived from bone cells).

**sarcoplasm**  Cytoplasm of a striated muscle cell.

**sarcoplasmic reticulum**  Smooth membrane system of a muscle cell responsible for sequestration of calcium using an integral membrane calcium pump.

**Scatchard plot**  Way of handling binding data where the binding is saturable. x-axis, free concentration; y-axis, ratio of bound : free; intercept on x-axis is number of sites, slope is inversely (and negatively) proportional to the binding affinity.

**scruin**  Actin-binding protein found in the undischarged acrosomal process of *Limulus*: blocks the **HMM**-binding site.

**spasmin** Constituent protein (20 kd) of the **spasmoneme**. Thought to change shape when it binds calcium.

**spasmoneme** Contractile organelle found in certain vorticellids. See Chapter 4.

**spectrin** Membrane-associated dimeric protein (240 & 220 kd) of the erythrocyte: forms a complex with **ankyrin** (band 4.1) and **actin**, and is responsible for the restriction of lateral mobility of some integral membrane components. Isoforms in other tissues: **fodrin, TW 240–260 k protein**.

**spokein** Constituent protein of the radial spokes of the eukaryotic **cilium**.

**stress fibre** Bundle of **microfilaments** and other proteins found in **fibroblasts**. Known to be contractile; has regular **sarcomere**-like organization.

**substratum** Solid surface on which a cell moves: used in order to avoid confusion by biochemists who use the word substrate in a different sense.

**symplectic** Pattern of **metachronal** coordination in which the wave of movement over the ciliary field is in the same direction as the active stroke.

**T-tubule** Invagination of the plasma membrane in striated muscle: responsible for transmission of the electrical depolarization deep into the fibre and triggering the release of calcium from the **sarcoplasmic reticulum**.

**tactic** A tactic response is one in which the cell responds by turning in the appropriate direction. See **chemotaxis**.

**talin** Protein which binds **vinculin** (but not **actin**) and which needs a function.

**tau** Protein of 60–70 kd which co-purifies with **tubulin** and which is thought to be important for assembly.

**taxol** Drug which stabilizes **microtubules**.

**terminal web** Misleading term for the region of cytoplasm at the base of the **microvilli** in intestinal **epithelial** cells. Contains **actin, myosin** and other components of an **actomyosin** contractile system.

**thick filament** The **myosin**-containing filament of striated muscle, approximately 15 nm in diameter.

**thin filament** The thinner filament (6 nm diameter) of striated muscle: contains **actin, tropomyosin** and, periodically, **troponin**.

**tight junction** Close apposition of plasma membranes of two cells (separation of 1–3 nm) which prevents leakage of fluid through the intercellular space in an **epithelium**, for example (syn. *zonula occludens*): a gasket.

**tonofilaments** **Intermediate filaments** inserted into a **desmosome**. Constituent protein usually keratin.

**topoinhibition** Another term for **density-dependent inhibition of growth**.

**treadmilling** Process of continual assembly at one end of a polymer and continual disassembly at the other, so that protomers move along the length of the polymer.

**triad** The triple array of vesicles (tubules) seen in striated muscle fibres, adjacent to the **Z-discs** of the **myofibrils**; composed of **T-tubules** and **sarcoplasmic reticulum**.

**tropomyosin** Group of proteins which bind **F-actin**. The skeletal muscle form is dimeric with $\alpha$ and $\beta$ sub-units of 33 kd and binds to seven **actin** molecules. Non-muscle forms may bind only six actins.

**troponin** Complex of three sub-units found on the **thin filament** of striated muscle: Tn-C (18 kd) binds calcium and has sequence homology with **calmodulin**, Tn-T (30 kd) binds to **tropomyosin**, and Tn-I (22 kd) inhibits the actin-activated myosin ATPase.

**tubulin**  Constituent protein of **microtubules**: protomer for tubule assembly a heterodimer of α- and β-**tubulin** (both 55 kd).

**TW 240–260 k protein**  Component of the **terminal web** of intestinal **epithelial** cells, probably a **spectrin** isoform. (By now it will have been christened, by a scientist hoping for citations, with some pseudoclassical name.)

**twitch-muscle**  Striated muscle adapted for rapid contraction rather than sustained tension.

**uncouplers**  Inhibitors of energy metabolism which work by 'disconnecting' the oxidative breakdown from the synthesis of 'high-energy' intermediates such as ATP.

**underlapping**  Contact between two cells in which one crawls underneath the other, maintaining its contact with the acellular **substratum**.

**villin**  Actin-severing and -capping protein (95 kd) from the **microvilli** of the intestinal **epithelium**. Calcium sensitive.

**vimentin**  **Intermediate filament** protein (58 kd) from mesodermally-derived cells including muscle.

**Vinca alkaloids**  Tubulin-binding compounds isolated from the periwinkle (*Vinca sp.*). Used as anti-tumour drugs. Commonest is vinblastine (811 d) but vincristine also used.

**vinculin**  Actin-binding protein (130 kd) from muscle and HeLa cells: found associated with cytoplasmic face of plasma membrane at focal adhesions. Actin-binding capacity may not be physiologically relevant and is disputed.

**Z-disc**  That part of the sarcomere into which the **thin filaments** are inserted. α-**actinin** is a major component. Often referred to as the Z-line.

**zonula adhaerens**  Cell–cell adhesion involving **actin** filaments. Similar in composition to **focal adhesions** to the **substratum** (involves **vinculin** and α-**actinin**). Also referred to as **intermediate** or **adhaerens junctions** (although the latter includes **desmosomes**).

# BIBLIOGRAPHY

Abercrombie, M. 1979. Contact inhibition and malignancy. *Nature (Lond.)* **281**, 259–62.

Abercrombie, M. 1980. The crawling movement of metazoan cells. *Proc. R. Soc. Lond.* B **207**, 129–47. (Reprinted in Bellairs *et al.* 1982.)

Abercrombie, M. and G. A. Dunn 1975. Adhesions of fibroblasts to substratum during contact inhibition observed by interference reflection microscopy. *Exp. Cell Res.* **92**, 57–62.

Abercrombie, M. and J. E. M. Heaysman 1954. Observations on the social behaviour of cells in tissue culture. II. Monolayering of fibroblasts. *Exp. Cell Res.* **6**, 293–306.

Abercrombie, M. and J. E. M. Heaysman 1976. Invasive behaviour between sarcoma and fibroblast populations in cell culture. *J. Nat. Cancer Inst.* **56**, 561–70.

Abercrombie, M. and J. A. Weston 1966. Cell mobility in fused homo- and heteronomic tissue fragments. *J. Exp. Zool.* **164**, 317–24.

Abercrombie, M., J. E. M. Heaysman and S. M. Pegrum 1970. The locomotion of fibroblasts in culture. III. Movements of particles on the dorsal surface of the leading lamella. *Exp. Cell Res.* **62**, 389–98.

Abercrombie, M., J. E. M. Heaysman and S. M. Pegrum 1971. The locomotion of fibroblasts in culture. IV. Electron microscopy of the leading lamella. *Exp. Cell Res.* **67**, 359–67.

Abercrombie, M., J. E. M. Heaysman and S. M. Pegrum 1972. The locomotion of fibroblasts in culture. V. Surface marking with concanavalin A. *Exp. Cell Res.* **73**, 536–9.

Abercrombie, M., D. M. Lamont and E. M. Stephenson 1968. The monolayering in tissue culture of fibroblasts from different sources. *Proc. R. Soc. Lond.* B **170**, 349–60.

Adams, R. J. 1982. Organelle movement in axons depends on ATP. *Nature (Lond.)* **297**, 327–9.

Adams, R. J. and D. Bray 1983. Rapid transport of foreign particles microinjected into crab axons. *Nature (Lond.)* **303**, 718–20.

Albertini, D. F. and J. I. Clark 1975. Membrane–microtubule interactions: concanavalin A capping induced redistribution of cytoplasmic microtubules and colchicine binding proteins. *Proc. Nat. Acad. Sci. USA* **72**, 4976–80.

Alexander, R. McN. 1968. *Animal Mechanics*. London: Sidgwick and Jackson.

Allan, R. B. and P. C. Wilkinson 1978. A visual analysis of chemotactic and chemokinetic locomotion of human neutrophil leucocytes. Use of a new chemotaxis assay with *Candida albicans* as gradient source. *Exp. Cell Res.* **111**, 191–203.

Amos, W. B. 1975. Contraction and calcium binding in the vorticellid ciliates.

In *Molecules and cell movement*, S. Inoué and R. E. Stephens (eds), 411–36. New York: Raven.

Amos, W. B., L. M. Routledge, T. Weis-Fogh and F. F. Yew 1976. The spasmoneme and calcium-dependent contraction in connection with specific calcium binding proteins. *Symp. Soc. Exp. Biol.* **30**, 273–302.

Anderson, D. C., L. J. Wible, B. J. Hughes, C. W. Smith and B. R. Brinkley 1982. Cytoplasmic microtubules in polymorphonuclear leukocytes: effects of chemotactic stimulation and colchicine. *Cell* **31**, 719–29.

Armstrong, P. B. 1977. Cellular positional stability and intercellular invasion. *Bioscience* **27**, 803–9.

Armstrong, P. B. 1984. Invasiveness of non-malignant cells. In *Invasion. Experimental and clinical implications*, M. M. Mareel and K. C. Calman (eds), 126–67. Oxford: OUP.

Armstrong, P. B. 1985. The control of cell motility during embryogenesis. *Cancer Metastasis Rev.* (in press).

Armstrong, P. B. and J. M. Lackie 1975. Studies on intercellular invasion *in vitro* using rabbit peritoneal neutrophil granulocytes (PMNs). I. Rôle of contact inhibition of locomotion. *J. Cell Biol.* **65**, 439–62.

Bard, J. B. L. and E. D. Hay 1975. The behaviour of fibroblasts from the developing avian cornea. Morphology and movement *in situ* and *in vitro*. *J. Cell Biol.* **67**, 400–18.

Bardele, C. F. 1974. Transport of material in the suctorian tentacle. *Symp. Soc. Exp. Biol.* **28**, 191–208.

Bardsley, W. G. and J. D. Aplin 1983a. Kinetic analysis of cell spreading. I. Theory and modelling of curves. *J. Cell Sci.* **61**, 365–73.

Bardsley, W. G. and J. D. Aplin. 1983b. Kinetic analysis of cell spreading. II. Substratum adhesion requirements of amniotic epithelial (FL) cells. *J. Cell Sci.* **61**, 375–88.

Baum, S. G., M. Wittner, J. P. Nadler, S. B. Horwitz, J. E. Dennis, P. B. Schiff and H. B. Tanowitz 1981. Taxol, a microtubule stabilizing agent, blocks the replication of *Trypanosoma cruzi*. *Proc. Nat. Acad. Sci. USA* **75**, 4571–5.

Begg, D. A., R. Rodewald and L. I. Rebhun 1978. The visualization of actin filament polarity in thin sections: evidence for the uniform polarity of membrane associated filaments. *J. Cell Biol.* **79**, 846–52.

Bellairs, R., A. Curtis and G. Dunn (eds) 1982. *Cell behaviour*. Cambridge: CUP.

Bennett, V. and J. Davis 1982. Immunoreactive forms of human erythrocyte ankyrin are localized in mitotic structures in cultured cells and are associated with microtubules in brain. *Cold Spring Harbor Symp.* **46**, 647–58.

Berg, H. C. 1975. Bacterial behaviour. *Nature (Lond.)* **254**, 389–92.

Berg, H. C. 1983. *Random Walks In Biology*. Princeton, NJ: Princeton University Press.

Berg, H. C. and D. A. Brown 1972. Chemotaxis in *Escherichia coli* analysed by three-dimensional tracking. *Nature (Lond.)* **239**, 500–4.

Berg, H. C., D. B. Bromley and N. W. Charon 1978. Leptospiral motility. *Symp. Soc. Gen. Microbiol.* **28**, 285–95.

Berg, H. C., M. D. Manson and M. P. Conley 1982. Dynamics and energetics of flagellar rotation in bacteria. *Symp. Soc. Exp. Biol.* **35**, 1–31.

Bershadsky, A. D. and V. I. Gelfand 1981. ATP-dependent regulation of cytoplasmic microtubule disassembly. *Proc. Nat. Acad. Sci. USA* **78**, 3610–13.

Bessen, M., R. B. Fay and G. B. Witman 1980. Calcium control of waveform in isolated flagellar axonemes of *Chlamydomonas*. *J. Cell Biol.* **86**, 446–55.

Besterman, J. M. and R. B. Low 1983. Endocytosis: a review of mechanisms and plasma membrane dynamics. *Biochem. J.* **210**, 1–13.

Binder, L. I., W. L. Dentler and J. L. Rosenbaum 1975. Assembly of chick brain tubulin onto flagellar microtubules from *Chlamydomonas* and sea urchin sperm. *Proc. Nat. Acad. Sci. USA* **72**, 1122–6.

Birchmeier, W., T. A. Libermann, B. A. Imhof, and T. E. Kreis 1982. Intracellular and extracellular components involved in the formation of ventral surfaces of fibroblasts. *Cold Spring Harbor Symp.* **46**, 755–68.

Black, M. M. and R. J. Lasek 1980. Slow components of axonal transport: two cytoskeletal networks. *J. Cell Biol.* **86**, 616–23.

Blake, J. R. and M. A. Sleigh 1974. Mechanics of ciliary locomotion. *Biol. Rev.* **49**, 85–125.

Blakemore, R. P. 1982. Magnetotactic bacteria. *Ann. Rev. Microbiol.* **36**, 217–38.

Bloodgood, R. A. 1982. Dynamic properties of the flagellar surface. *Symp. Soc. Exp. Biol.* **35**, 353–80.

Blum, J. J., A. Hayes, G. A. Jamieson, and T. C. Vanaman 1980. Calmodulin confers calcium sensitivity on ciliary dynein ATPase. *J. Cell Biol.* **87**, 386–97.

Bongrand, P., C. Capo and R. Depieds 1982. Physics of cell adhesion. *Prog. Surf. Sci.* **12**, 217–85.

Borisy, G. G. and J. B. Olmsted 1972. Nucleated assembly of microtubules in porcine brain extracts. *Science* **177**, 1196–7.

Bouck, G. B. and A. A. Rogalski 1982. Surface properties of the euglenoid flagellum. *Symp. Soc. Exp. Biol.* **35**, 381–98.

Boyd, A., G. Mandel and M. I. Simon 1982. Integral membrane proteins required for bacterial motility and chemotaxis. *Symp. Soc. Exp. Biol.* **35**, 123–38.

Branton, D. 1982. Membrane cytoskeletal interactions in the human erythrocyte. *Cold Spring Harbor Symp.* **46**, 1–6.

Bray, D. 1973. Model for membrane movements in the neural growth cone. *Nature (Lond.)* **244**, 93–6.

Bray, D. 1982. Filopodial contraction and growth cone guidance. In *Cell behaviour*, R. Bellairs, A. Curtis and G. Dunn (eds), 299–317. Cambridge: CUP.

Bray, D. and C. Thomas 1975. The actin content of fibroblasts. *Biochem. J.* **147**, 221–8.

Bretscher, M. S. 1976. Directed lipid flow in cell membranes. *Nature (Lond.)* **260**, 21–3.

Brinkley, B. R., S. M. Cox, D. A. Pepper, L. Wible, S. L. Brenner and R. L. Pardue 1981. Tubulin assembly sites and the organisation of cytoplasmic microtubules in cultured mammalian cells. *J. Cell Biol.* **90**, 554–62.

Brokaw, C. J., D. J. L. Luck and B. Huang 1982. Analysis of the movement of *Chlamydomonas* flagella: the function of the radial-spoke system is revealed by comparison of wild-type and mutant flagella. *J. Cell Biol.* **92**, 722–32.

Brooks, R. F. and F. N. Richmond 1983. Microtubule-organising centres during the cell cycle of 3T3 cells. *J. Cell Sci.* **61**, 231–45.

Brown, A. F. 1982. Neutrophil granulocytes: adhesion and locomotion on collagen substrata and in collagen matrices. *J. Cell Sci.* **58** 455–67.

Brown, S. S. and J. A. Spudich 1981. Mechanism of action of cytochalasin: evidence that it binds to actin filament ends. *J. Cell Biol.* **88**, 487–91.

Brown, S. S., H. L. Malinoff and M. S. Wicha 1983. Connectin: cell surface protein that binds both laminin and actin. *Proc. Nat. Acad. Sci. USA.* **80**, 5927–30.

Burridge, K. 1981. Are stress fibres contractile? *Nature (Lond.)* **294**, 691–2.

Burridge, K. and J. R. Feramisco 1980. Microinjection and localization of a 130 K protein in living fibroblasts: a relationship to actin and fibronectin. *Cell* **19**, 587–92.

Canale-Parola, E. 1978. Motility and chemotaxis of spirochetes. *Ann. Rev. Microbiol.* **32**, 69–99.

Carlsson, L., L-E. Nystrom, I. Sundkvist, F. Markey and U. Lindberg 1977. Actin polymerizability is influenced by profilin, a low molecular weight protein in non-muscle cells. *J. Mol. Biol.* **115**, 465–83.

Carter, S. B. 1965. Principles of cell motility: the direction of cell movement and cancer invasion. *Nature (Lond.)* **208**, 1183–7.

Carter, S. B. 1967a. Haptotaxis and the mechanism of cell motility. *Nature (Lond.)* **213**, 256–60.

Carter, S. B. 1967b. Haptotactic islands. A method of confining single cells to study individual cell reactions and clone formation. *Exp. Cell Res.* **48**, 189–93.

Chen, W. -T. 1981. Surface changes during retraction-induced spreading of fibroblasts. *J. Cell Sci.* **49**, 1–13.

Chenoweth, D. E. and T. E. Hugli 1978. Demonstration of specific C5a receptor on intact human polymorphonuclear leucocytes. *Proc. Nat. Acad. Sci. USA* **75**, 3943–7.

Clark, T. G. and R. W. Merriam 1978. Actin in *Xenopus* oocytes. I. Polymerization and gelation *in vitro*. *J. Cell Biol.* **77**, 427–38.

Clarke, C. H. 1981. Motility and fruiting in the bacterium *Myxococcus xanthus*. In *Biology of the chemotactic response* J. M. Lackie and P. C. Wilkinson (eds), 155–71. Cambridge: CUP.

Cobbold, P. H. 1980. Cytoplasmic free calcium and amoeboid movement. *Nature (Lond.)* **285**, 441–6.

Cohen, C. M. and S. F. Foley 1980. Spectrin-dependent and -independent association of F-actin with the erythrocyte membrane. *J. Cell Biol.* **86**, 694–8.

Cohen, M. H. and A. Robertson 1971. Wave propagation in the early stages of aggregation of cellular slime molds. *J. Theor. Biol.* **31** 101–18.

Cooper, M. S. and R. E. Keller 1984. Perpendicular orientation and directional migration of amphibian neural cells in dc electrical fields. *Proc. Nat. Acad. Sci. USA* **81**, 160–4.

Cote, G. P. and L. B. Smillie 1981. Preparation and some properties of equine platelet tropomyosin. *J. Biol. Chem.* **256**, 11004–10.

Couchman, J. R. and D. A. Rees 1979. The behaviour of fibroblasts migrating from chick heart explants: changes in adhesion, locomotion and growth, and in the distribution of actomyosin and fibronectin. *J. Cell Sci.* **39**, 149–65.

Curtis, A. S. G. 1967. *The Cell Surface: its molecular role in morphogenesis.* London: Logos/Academic.

Curtis, A. S. G. 1984. Cell adhesion. In *Developmental control in animals and plants*, 2nd edn, C. F. Graham and P. F. Waring (eds), 98–119. Oxford: Blackwell Scientific.

Curtis, A. S. G. and J. D. Pitts (eds) 1980. *Cell adhesion and motility.* Cambridge: CUP. (Symp. Brit. Soc. Cell Biol. **3**.)

Dahlstrom, A., J. Haggendal, P. -O. Heiwall, P. -A. Larsson and N. R. Saunders

1974. Intra-axonal transport of neurotransmitters in mammalian neurons. *Symp. Soc. Exp. Biol.* **28**, 229–48.

Davis, E. M. 1980. Translocation of neural crest cells within a hydrated collagen lattice. *J. Embryol. Exp. Morphol.* **55**, 17–31.

Dennison, D. S. 1984. Phototropism. In *Advanced plant physiology*, M. B. Wilkins (ed.), 149–62. London: Pitman.

Devreotes, P. N. 1982. Chemotaxis. In *The Development of* Dictyostelium discoideum, W. F. Loomis (ed.), 117–68. New York: Academic.

Dierich, M. P., D. Wilhelmi and G. O. Till 1977. Essential role of surface bound chemoattractant in leucocyte migration. *Nature (Lond.)* **270**, 351–2.

Dunn, G. A. 1980. Mechanisms of fibroblast locomotion. In *Cell adhesion and motility*, A. S. G. Curtis and J. D. Pitts (eds), 409–23. Cambridge: CUP. (Symp. Brit. Soc. Cell Biol. 3.)

Dunn, G. A. 1981. Chemotaxis as a form of directed cell behaviour: some theoretical considerations. In *Biology of the chemotactic response*, J. M. Lackie and P. C. Wilkinson (eds), 1–26. Cambridge: CUP.

Dunn, G. A. 1982. Contact guidance of cultured tissue cell: a survey of potentially relevant properties of the substratum. In *Cell behaviour*, R. Bellairs, A. Curtis and G. Dunn (eds), 247–80. Cambridge: CUP.

Dunn, G. A. 1983. Characterising a kinesis response: time averaged measures of cell speed and directional persistence. In *Leukocyte locomotion and chemotaxis*, H. -U. Keller and G. O. Till (eds), 14–33. Basel: Birkhauser.

Dunn, G. A. and T. Ebendal 1978. Contact guidance on oriented collagen gels. *Exp. Cell Res.* **111**, 475–9.

Dunn, G. A. and J. P. Heath 1976. A new hypothesis of contact guidance in tissue cells. *Exp. Cell Res.* **101**, 1–14.

Ebashi, S. 1974. Regulatory mechanism of muscle contraction with special reference to the Ca–troponin–tropomyosin system. *Essays in Biochem.* **10**, 1–36.

Ebendal, H. and J. P. Heath 1977. Self-contact inhibition of movement in cultured chick heart fibroblasts. *Exp. Cell Res.* **110**, 469–73.

Ebisawa, K. 1983. $Ca^{2+}$-regulation not associated with phosphorylation of myosin light chain in aortic intima smooth muscle. *J. Biochem.* **93**, 935–7.

Edds, K. T. 1975. Motility in *Echinosphaerium nucleofilum*. I. An analysis of particle motions in the axopodia and a direct test of the involvement of the axoneme. *J. Cell Biol.* **66**, 145–55.

Edelman, G. M. 1984. Cell-adhesion molecules: a molecular basis for animal form. *Sci. Am.* **250**, 80–91.

Edelman, G. M., W. J. Gallin, A. DeLouvee, B. A. Cunningham and J. -P. Thiery 1983. Early epochal maps of two different cell adhesion molecules. *Proc. Nat. Acad. Sci. USA.* **80**, 4384–8.

Edwards, J. G. 1983. The biochemistry of cell adhesion. *Prog. Surf. Sci.* **13**, 125–96.

Englander, L. L. and H. L. Malech 1981. Abnormal movement of poly-morphonuclear neutrophils in the immotile cilia syndrome: cinemato-graphic analysis. *Exp. Cell Res.* **135**, 468–72.

Erickson, C. A. 1978. Analysis of the formation of parallel arrays by BHK cells *in vitro*. *Exp. Cell Res.* **115**, 303–15.

Erickson, C. A. 1985. Morphogenesis of the neural crest. In *Developmental biology: a comprehensive synthesis*, L. Browden (ed.). New York: Plenum.

Erickson, C. A. and E. A. Turley 1983. Substrata formed by combinations of extracellular matrix components alter neural crest cell motility *in vitro*. *J. Cell Sci.* **61**, 299–323.

Euteneuer, U. and J. R. McIntosh 1980. Polarity of midbody and phragmoplast microtubules. *J. Cell Biol.* **87**, 509–15.

Euteneuer, U. and J. R. McIntosh 1981. Structural polarity of kinetochore microtubules in PtK1 cells. *J. Cell Biol.* **89**, 338–45.

Euteneuer, U. and M. Schliwa 1984. Persistent, directional motility of cells and cytoplasmic fragments in the absence of microtubules. *Nature (Lond.)* **310**, 58–61.

Fletcher, M. P., B. E. Seligmann and J. I. Gallin 1982. Correlation of human neutrophil secretion, chemoattractant receptor mobilization and enhanced functional capacity. *J. Immunol.* **128**, 941–8.

Folkman, J. and A. Moscona 1978. Rôle of cell shape in growth control. *Nature (Lond.)* **273**, 345–8.

Ford-Hutchinson, A. W., M. A. Bray, M. V. Doig, M. E. Shipley and M. J. H. Smith 1980. Leukotriene B, a potent chemokinetic and aggregating substance released from polymorphonuclear leukocytes. *Nature (Lond.)* **286**, 264–5.

Forrester, J. V., J. M. Lackie and A. F. Brown 1983. Neutrophil behaviour in the presence of protease inhibitors. *J. Cell Sci.* **59**, 213–30.

Fujiwara, K. and T. D. Pollard 1976. Fluorescent antibody localization of myosin in the cytoplasm, cleavage furrow and mitotic spindle of human cells. *J. Cell Biol.* **71**, 848–75.

Fulton, C. and P. A. Simpson 1976. Selective synthesis and utilization of flagellar tubulin. The multi-tubulin hypothesis. *Cold Spring Harbor Conf. Cell Prolif.* **3**, 987–1006.

Gail, M. M. and C. W. Boone 1972. Cell-substrate adhesivity: a determinant of cell motility. *Exp. Cell Res.* **70**, 33–40.

Gallin, J. I. and P. G. Quie (eds) 1978. *Leucocyte chemotaxis: methods, physiology, and clinical implications*. New York: Raven.

Gallin, J. I., B. E. Seligmann and M. P. Fletcher 1982. Dynamics of human neutrophil receptors for the chemoattractant fMet-Leu-Phe. In *Leucocyte locomotion and chemotaxis*, H-U. Keller and G. O. Till (eds), 290–308. Basel: Birkhauser.

Geiger, B. 1979. A 130 K protein from chicken gizzard: its localization at the termini of microfilament bundles in cultured chicken cells. *Cell* **18**, 193–204.

Geiger, B., A. H. Dutton, K. T. Tokuyasu and S. J. Singer 1981. Immuno-electron microscopic studies of membrane–microfilament interactions. The distributions of α-actinin, tropomyosin and vinculin in intestinal epithelial brush border and in chicken gizzard smooth muscle cells. *J. Cell Biol.* **91**, 614–28.

Gerisch, G. and H. -U. Keller 1981. Chemotactic reorientation of granulocytes stimulated with micropipettes containing f-Met-Leu-Phe. *J. Cell Sci.* **52**, 1–10.

Gerisch, G., H. Fromm, A. Huesgen and U. Wick 1975. Control of cell-contact sites by cyclic-AMP pulses in differentiating *Dictyostelium* cells. *Nature (Lond.)* **255**, 547–9.

Gerisch, G., D. Huelser, D. Malchow and U. Wick 1975. Cell communication by periodic cyclic-AMP pulses. *Phil. Trans. R. Soc. Lond.* **B 272**, 181–92.

Gerisch, G., H. Krelle, S. Bozzaro, E. Eitle and R. Guggenheim 1980. Analysis of cell adhesion in *Dictyostelium* and *Polysphondylium* by the use of *Fab*. In *Cell adhesion and motility*, A. S. G. Curtis and J. D. Pitts (eds), 293–307. Cambridge: CUP.

Gershmann, H. and J. Drumm 1975. Mobility of normal and virus-transformed cells in cellular aggregates. *J. Cell Biol.* **67**, 419–35.

Gibbons, I. R., E. Fronk, B. H. Gibbons and K. Ogawa 1976. Multiple forms of dynein in sea urchin sperm flagella. *Cold Spring Harbor Conf. Cell Prolif.* **3**, 915–32.

Goldberg, D. J., D. A. Harris, B. W. Lubit and J. H. Schwartz 1980. Analysis of the mechanism of fast axonal transport by intracellular injection of potentially inhibitory macromolecules: evidence for a possible rôle of actin filaments. *Proc. Nat. Acad. Sci. USA* **77**, 7448–52.

Goldman, R. D., B. Chojnacki and M-J. Yerna 1979. Ultrastructure of microfilament bundles in Baby Hamster Kidney (BHK-21) cells. The use of tannic acid. *J. Cell Biol.* **80**, 759–66.

Goldman, R. D., T. Pollard and J. Rosenbaum (eds) 1976. Cell motility. *Cold Spring Harbor Conf. Cell Prolif.* **3** (3 vols).

Goldman, R. D., M. J. Yearna and J. A. Schloss 1976. Localisation and organisation of microfilaments and related proteins in normal and virus-transformed cells. *J. Supramol. Struct.* **5**, 155–88.

Goldman, R. D., E. Lazarides, R. Pollack and K. Weber 1975. The distribution of actin in non-muscle cells. The use of actin antibody in the localization of actin within the microfilament bundles of mouse 3T3 cells. *Exp. Cell Res.* **90**, 333–44.

Gooday, G. W. 1981. Chemotaxis in the eukaryotic microbes. In *Biology of the chemotactic response*, J. M. Lackie and P. C. Wilkinson (eds), 115–38. Cambridge: CUP.

Gray, J. and G. J. Hancock 1955. The propulsion of sea-urchin spermatozoa. *J. Exp. Biol.* **32**, 804–14.

Grell, K. G. 1973. *Protozoology*. Berlin: Springer.

Griffin, F. M., J. A. Griffin, J. E. Leider and S. C. Silverstein 1975. Studies on the mechanism of phagocytosis. I. Requirements for circumferential attachment of particle-bound ligands to specific receptors on the macrophage plasma membrane. *J. Exp. Med.* **142** 1263–82.

Griffith, L. M. and T. D. Pollard 1978. Evidence for actin filament–microtubule interaction mediated by microtubule-associated proteins. *J. Cell Biol.* **78**, 958–65.

Grinnell, F. A. 1978. Cellular adhesiveness and extracellular substrata. *Int. Rev. Cytol.* **58**, 65–144.

Grumet, M., S. Hoffman and G. M. Edelman 1984. Two antigenically related neuronal cell adhesion molecules of different specificities mediate neuron–neuron and neuron–glia adhesion. *Proc. Nat. Acad. Sci. USA* **81**, 267–71.

Guelstein, V. I., O. Y. Ivanova, L. B. Margolis, J. M. Vasiliev and I. M. Gelfand 1973. Contact inhibition of movement in the cultures of transformed cells. *Proc. Nat. Acad. Sci. USA* **70**, 2011–14.

Hader, D-P. 1979. Photomovement. In *Encyclopedia of plant physiology*, New Series, vol. 7, W. Haupt and M. E. Feinleib (eds), 268–309. Berlin: Springer.

Haimo, L. T., B. R. Telzer and J. L. Rosenbaum 1979. Dynein binds to and crossbridges cytoplasmic microtubules. *Proc. Nat. Acad. Sci. USA* **76**, 5759–63.

Halfen, L. N. and R. W. Castenholz 1971a. Gliding motility in the blue-green alga *Oscillatoria princeps*. *J. Phycol.* **7**, 133–45.

Halfen, L. N. and R. W. Castenholz 1971b. Energy expenditure for gliding motility in a blue-green alga. *J. Phycol.* **7**, 258–60.

Harris, A. K. 1973a. Behaviour of cultured cells on substrata of variable adhesiveness. *Exp. Cell Res.* **77**, 285–97.

Harris, A. K. 1973b. Location of cellular adhesions to solid substrata. *Dev. Biol.* **35**, 97–114.

Harris, A. K. 1973c. Cell surface movements related to cell locomotion. In *Locomotion of tissue cells*, R. Porter and D. W. Fitzsimons (eds), 1–26. Amsterdam: Elsevier. (CIBA Foundation Symp. **14**.)

Harris, A. K. 1976. Recycling of dissolved plasma membrane components as an explanation of the capping phenomenon. *Nature (Lond.)* **263**, 781–3.

Harris, A. K. 1982. Traction, and its relations to contraction in tissue cell locomotion. In *Cell behaviour*, R. Bellairs, A. Curtis and G. Dunn (eds), 109–34. Cambridge: CUP.

Harris, A. K. and G. A. Dunn 1972. Centripetal transport of attached particles on both surfaces of moving fibroblasts. *Exp. Cell Res.* **73**, 519–23.

Harris, A. K., P. Wild and D. Stopak 1980. Silicone rubber substrata: a new wrinkle in the study of cell locomotion. *Science* **208**, 177–9.

Hartshorne, D. J. 1982. Phosphorylation of myosin and the regulation of smooth muscle actomyosin. In *Cell and muscle motility*, vol. 2, R. M. Dowben and J. W. Shay (eds), 185–220. New York: Plenum.

Hartwig, J. H. and T. P. Stossel 1975. Isolation and properties of actin, myosin and a new actin-binding protein in rabbit alveolar macrophages. *J. Biol. Chem.* **250**, 5696–705.

Hartwig, J. H., J. Tyler and T. P. Stossel 1980. Actin-binding protein promotes the bipolar and perpendicular branching of actin filaments. *J. Cell Biol.* **87**, 841–8.

Hartwig, J. H., H. L. Yin and T. P. Stossel 1980. Contractile proteins and the mechanism of phagocytosis in macrophages. In *Mononuclear phagocytes: functional aspects*, R. van Furth (ed.), 971–96. The Hague: Nijhoff.

Haston, W. S. and J. M. Shields 1984. Contraction waves in lymphocyte locomotion. *J. Cell Sci.* **68**, 227–42.

Hathaway, D. R., C. R. Eaton and R. S. Adelstein 1981. Regulation of human platelet myosin light chain kinase by the catalytic subunit of cyclic AMP-dependent protein kinase. *Nature (Lond.)* **291**, 252–4.

Hazelbauer, G. L. 1981. The molecular biology of bacterial chemotaxis. In *Biology of the chemotactic response*, J. M. Lackie and P. C. Wilkinson (eds), 139–54. Cambridge: CUP.

Heath, J. P. 1981. Arcs: curved microfilament bundles beneath the dorsal surface of the leading lamellae of moving chick embryo fibroblasts. *Cell Biol. Int. Reports* **5**, 975–80.

Heath, J. P. 1983. Direct evidence for microfilament-mediated capping of surface receptors on crawling fibroblasts. *Nature (Lond.)* **302**, 532–4.

Heath, J. P. and G. A. Dunn 1978. Cell to substratum contacts of chick fibroblasts and their relation to the microfilament system. A correlated interference-reflexion and high-voltage electron-microscope study. *J. Cell Sci.* **29**, 197–212.

Heaysman, J. E. M. 1978. Contact inhibition of locomotion – a reappraisal. *Int. Rev. Cytol.* **55**, 49–66.

Heaysman, J. E. M. and S. M. Pegrum 1973a. Early contacts between fibroblasts. An ultrastructural study. *Exp. Cell Res.* **78**, 71–8.

Heaysman, J. E. M. and S. M. Pegrum 1973b. Early contacts between normal fibroblasts and mouse sarcoma cells – an ultrastructural study. *Exp. Cell Res.* **78**, 479–81.

Heaysman, J. E. M. and S. M. Pegrum 1982. Early cell contacts in culture. In *Cell behaviour*, R. Bellairs, A. Curtis and G. Dunn (eds), 49–76. Cambridge: CUP.

Heaysman, J. E. M. and L. Turin 1976. Interactions between living and zinc-fixed cells in culture. *Exp. Cell Res.* **101**, 419–22.

Heggeness, M. H., M. Simon and S. J. Singer 1978. Association of mito-chondria with microtubules in cultured cells. *Proc. Nat. Acad. Sci. USA* **75**, 3863–6.

Hellewell, S. B. and D. L. Taylor 1979. The contractile basis of amoeboid movement. VI. The solation-contraction coupling hypothesis. *J. Cell Biol.* **83**, 633–48.

Henrichsen, J. 1983. Twitching motility. *Ann. Rev. Microbiol.* **37**, 81–93.

Herman, I. M., N. J. Crisona and T. D. Pollard 1981. Relation between cell activity and the distribution of cytoplasmic actin and myosin. *J. Cell Biol.* **90**, 84–91.

Heslop, J. P. 1974. Fast transport along nerves. *Symp. Soc. Exp. Biol.* **28**, 209–28.

Hinnsen, H., J. D'Haese, J. V. Small and A. Sobieszek 1978. Mode of filament assembly of myosins from muscle and non-muscle cells. *J. Ultrastruct. Res.* **64**, 282–302.

Hirata, M., T. Mikawa, Y. Nonamura and S. Ebashi 1980. $Ca^{2+}$ regulation in vascular smooth muscle. II. $Ca^{2+}$ binding of aorta leiotonin. *J. Biochem.* **87**, 369–78.

Hisanaga, S. and H. Sakai 1983. Cytoplasmic dynein of the sea urchin egg. II. Purification, characterization and interactions with microtubules and Ca-calmodulin. *J. Biochem.* **93**, 87–98.

Hitchcock, S. E. 1977. Regulation of motility in non-muscle cells. *J. Cell Biol.* **74**, 1–15.

Hobson, A. C., R. A. Black and J. Adler 1982. Control of bacterial motility in chemotaxis. *Symp. Soc. Exp. Biol.* **35**, 105–23.

Holwill, M. E. J. 1974. Hydrodynamic aspects of ciliary and flagellar move-ment. In *Cilia and flagella*, M. A. Sleigh (ed.), 143–75. London: Academic.

Holwill, M. E. J. 1982. Dynamics of eukaryotic flagellar movement. *Symp. Soc. Exp. Biol.* **35**, 289–312.

Holwill, M. E. J. and M. A. Sleigh 1967. Propulsion of hispid flagella. *J. Exp. Biol.* **47**, 267–76.

Hubbe, M. A. 1981. Adhesion and detachment of biological cells *in vitro*. *Prog. Surf. Sci.* **11**, 65–138.

Hyafil, F., D. Morello, C. Babinet and F. Jacob 1980. A cell surface glycoprotein involved in the compaction of embryonal carcinoma cells and cleavage stage embryos. *Cell* **21**, 927–34.

Hyams, J. S. and H. Stebbings 1977. The distribution and function of micro-tubules in nutritive tubes. *Tissue and Cell* **9**, 539–47.

Inoué, S. and R. E. Stephens 1975. *Molecules and Cell Movement*. New York: Raven.

Inoué, S. and L. G. Tilney 1982. Acrosomal reaction of *Thyone* sperm. I. Changes in the sperm head visualized by high resolution video microscopy. *J. Cell Biol.* **93**, 812–10.

Isenberg, G., P. C. Rathke, N. Hulsman, W. W. Franke and K. E. Wolfarth-Botterman 1976. Cytoplasmic actomyosin fibrils in tissue culture cells. Direct proof of contractility by visualization of ATP-induced contraction in fibrils isolated by laser microbeam dissection. *Cell Tiss. Res.* **166** 427–44.

Izzard, C. S. and L. R. Lochner 1976. Cell-to-substrate contacts in living fibroblasts: an interference-reflexion study with an evaluation of the technique. *J. Cell Sci.* **21**, 129–59.

Izzard, C. S. and L. R. Lochner 1980. Formation of cell-to-substrate contacts during fibroblast motility: an interference-reflexion study. *J. Cell Sci.* **42**, 81–116.

Jacobs, M., H. Smith and E. W.Taylor 1974. Tubulin: nucleotide binding and enzymic activity. *J. Mol. Biol.* **89**, 455–68.

Jameson, L. and M. Caplow 1981. Modification of microtubule steady-state dynamics by phosphorylation of the microtubule-associated proteins. *Proc. Nat. Acad. Sci. USA* **78**, 3413–17.

Jennings, H. S. 1906. *Behaviour of the lower organisms.* New York: Columbia University. (Quoted in Naitoh & Eckert 1974.)

Jennings, M. A. and H. W. Florey 1970. Healing. In *General Pathology*, 4th edn, H. W. Florey (ed.), 480–548. London: Lloyd-Luke.

Job, D., E. H. Fischer and R. L. Margolis 1981. Rapid disassembly of cold-stable microtubules by calmodulin. *Proc. Nat. Acad. Sci. USA* **78**, 4679–82.

Jones, J. C. R. and J. B. Tucker 1981. Microtubule-organising centres and assembly of the double-spiral microtubule patterns in certain heliozoan axonemes. *J. Cell Sci.* **50**, 259–80.

Kaiser, D. 1979. Social gliding is correlated with the presence of pili in *Myxococcus xanthus. Proc. Nat. Acad. Sci. USA* **76**, 5952–6.

Kamiya, N. 1981. Physical and chemical basis of cytoplasmic streaming. *Ann. Rev. Plant Physiol.* **32**, 205–36.

Kamiya, R., H. Hotani and S. Asakura 1982. Polymorphic transition in bacterial flagella. *Symp. Soc. Exp. Biol.* **35**, 53–76.

Keller, H. -U. 1981. The relationship between leucocyte adhesion to solid substrata, locomotion, chemokinesis and chemotaxis. In *Biology of the chemotactic response*, J. M. Lackie and P. C. Wilkinson (eds), 27–52. Cambridge: CUP.

Keller, H. -U. 1983. Shape, motility and locomotion responses of neutrophil granulocytes. In *Leucocyte locomotion and chemotaxis*, H. -U. Keller and G. O. Till (eds), 54–72. Basel: Birkhauser.

Kerrick, W. G. L. and L. Y. W. Bourguignon 1984. Regulation of receptor capping in mouse lymphoma T cells by $Ca^{2+}$-activated myosin light chain kinase. *Proc. Nat. Acad. Sci USA* **81**, 165–9.

Kersey, Y. M. 1974. Correlation of polarity of actin filaments with protoplasmic streaming in Characean algae. *J. Cell Biol.* **63**, A330.

Kim, H., L. I. Binder and J. L. Rosenbaum 1979. The periodic association of MAP-2 with brain microtubules *in vitro. J. Cell Biol.* **80**, 266–76.

King, C. A. and K. Lee 1982. Effect of trifluoperazine and calcium ions on gregarine gliding. *Experentia* **38**, 1051–2.

Kirschner, M. W. 1980. Implications of treadmilling for the stability and polarity of actin and tubulin polymers *in vivo. J. Cell Biol.* **86**, 330–4.

Koch, G. L. E. 1980. Microfilament-membrane interactions in the mechanism of capping. *Symp. Brit. Soc. Cell Biol.* **3**, 410–44.

Koffer, A., W. B. Gratzer, G. D. Clarke and A. Hales 1983. Phase equilibria of cytoplasmic actin of cultured epithelial (BHK) cells. *J. Cell Sci.* **61**, 191–218.

Kolega, J. 1981. The movement of cell clusters *in vitro*: morphology and directionality. *J. Cell Sci.* **49**, 15–32.

Kolega, J., M. S. Shure, W. -T. Chen and N. D. Young 1982. Rapid cellular translocation is related to close contacts formed between various cultured cells and their substrata. *J. Cell Sci.* **54**, 23–34.

Komnick, H., W. Stockem and K. E. Wohlfarth-Bottermann 1973. Cell motility: mechanisms in protoplasmic streaming and amoeboid movement. *Int. Rev. Cytol.* **34**, 169–249.

Koshland, D. E. 1980. *Bacterial chemotaxis as a model behavioural system.* Distinguished Lecture Series of the Society of General Physiologists. New York: Raven.

Kreis, T. E. and W. Birchmeier 1980. Stress fiber sarcomeres of fibroblasts are contractile. *Cell* **22**, 555–61.

Kuczmarski, E. R. and J. L. Rosenbaum 1979. Chick brain actin and myosin. Isolation and characterization. *J. Cell Biol.* **80**, 341–55.

Kuczmarski, E. R. and J. A. Spudich 1980. Regulation of myosin self-assembly: phosphorylation of *Dictyostelium* heavy chain inhibits formation of thick filaments. *Proc. Nat. Acad. Sci. USA* **77**, 7292–6.


Lackie, J. M. 1980. The structure and organization of the cell surface. In *Membrane structure and function*, vol. 1, E. E. Bittar (ed.), 73–102. New York: Wiley.

Lackie, J. M. 1982. Aspects of the behaviour of neutrophil leucocytes. In *Cell behaviour*, R. Bellairs, A. Curtis and G. Dunn (eds), 319–48. Cambridge: CUP.

Lackie, J. M. and P. B. Armstrong 1975. Studies on intercellular invasion *in vitro* using rabbit peritoneal neutrophil granulocytes. II. Adhesive interactions between cells. *J. Cell Sci.* **19**, 645–52.

Lackie, J. M. and D. DeBono 1977. Interactions of neutrophil granulocytes (PMNs) and endothelium *in vitro*. *Microvasc. Res.* **13**, 107–12.

Lackie, J. M. and R. P. C. Smith 1980. Interactions of leucocytes and endothelium. In *Cell adhesion and motility*, A. S. G. Curtis and J. D. Pitts (eds), 235–72. Cambridge: CUP.

Lackie, J. M. and P. C. Wilkinson (eds) 1981. *Biology of the chemotactic response.* Cambridge: CUP. (Soc. Exp. Biol. Seminar Series **12**.)

Lackie, J. M. and P. C. Wilkinson 1984. Adhesion and locomotion of neutrophil leucocytes on 2-D substrata and in 3-D matrices. In *White cell mechanics: basic science and clinical aspects*, H. J. Meiselman, M. A. Lichtman and P. L. LaCelle (eds), 237–54. New York: Liss.

Lasek, R. J. and P. N. Hoffmann 1976. The neuronal cytoskeleton, axonal transport and axonal growth. *Cold Spring Harbor Conf. Cell Prolif.* **3**, 1021–50.

Lauffenburger, D. A. 1982. The influence of external concentration fluctuations on leucocyte chemotactic orientation. *Cell Biophys.* **4**, 177–209.

Letourneau, P. C. 1975. Cell-to-substratum adhesion and guidance of axonal elongation. *Dev. Biol.* **44**, 92–101.

Lin, D. C., K. D. Tobin, M. Grumet and S. Lin 1980. Cytochalasins inhibit nuclei-induced actin polymerization by blocking filament elongation. *J. Cell Biol.* **84**, 455–60.

Loomis, W. F. (ed.) 1982. *The development of* Dictyostelium discoideum. New York: Academic.

Lubbock, R. and W. B. Amos 1981. Removal of bound calcium from nematocyst contents causes discharge. *Nature (Lond.)* **200**, 500–1.

Luck, D. J. L., B. Huang and G. Piperno 1982. Genetic and biochemical analysis of the eukaryotic flagellum. *Symp. Soc. Exp. Biol.* **35**, 399–420.

Luduena, R. F. and D. O. Woodward 1973. Isolation and partial purification of α- and β-tubulin from outer doublets of sea urchin sperm and microtubules of chick-embryo brain. *Proc. Nat. Acad. Sci. USA* **70**, 3594–8.

Luna, E. J., V. M. Fowler, J. Swanson, D. Branton and D. L. Taylor 1981. A membrane cytoskeleton from *Dictyostelium discoideum*. I. Identification and partial characterization of an actin-binding activity. *J. Cell Biol.* **88**, 396–409.

Machemer, H. 1974. Ciliary activity and metachronism in Protozoa. In *Cilia and flagella*, M. A. Sleigh (ed.), 199–286. London: Academic.

Macnab, R. M. 1982. Sensory reception in bacteria. *Symp. Soc. Exp. Biol.* **35**, 77–104.

Mak, A. S., L. B. Smillie and G. R. Stewart 1980. A comparison of the amino acid sequences of rabbit skeletal muscle α- and β-tropomyosin. *J. Biol. Chem.* **255**, 3647–55.

Marcum, J. M., J. R. Dedman, B. R. Brinkley and A. R. Means 1978. Control of microtubule assembly–disassembly by calcium-dependent regulator protein. *Proc. Nat. Acad. Sci. USA* **75**, 3771–5.

Margolis, R. L. and L. Wilson 1978. Opposite end assembly and disassembly of microtubules at steady state *in vitro*. *Cell* **13**, 1–8.

Margolis, R. L., L. Wilson and B. I. Kiefer 1978. Mitotic mechanism based on intrinsic microtubule behaviour. *Nature (Lond.)* **272**, 450–2.

Maroudas, N. G. 1973. Chemical and mechanical requirements for fibroblast adhesion. *Nature (Lond.)* **244**, 353–5.

Matsudaira, P. T. and D. R. Burgess 1982. Organization of the cross-filaments in intestinal microvilli. *J. Cell Biol.* **92**, 657–64.

Matsuura, S., J. -I. Shioi, Y. Imae and S. Iida 1979. Characterization of the *Bacillus subtilis* motile system driven by an artificially created proton motive force. *J. Bacteriol.* **140**, 28–36.

McDonald, K., J. Pickett-Heaps, J. R. McIntosh and D. H. Tippit 1977. On the mechanism of anaphase spindle elongation in *Diatoma vulgare*. *J. Cell Biol.* **74**, 377–88.

McKeown, M. and R. A. Firtel 1982. Actin multigene family of *Dictyostelium*. *Cold Spring Harbor Symp.* **46**, 495–506.

Means, A. R. and J. Dedman 1980. Calmodulin – an intracellular calcium receptor. *Nature (Lond.)* **285**, 73–7.

Meiselman, H. J., M. A. Lichtman and P. L. LaCelle (eds) 1984. *White cell mechanics. Basic science and clinical aspects*. New York: Liss.

Meza, I., B. Huang and J. Bryan 1972. Chemical heterogeneity of proto-filaments forming the outer doublets from sea urchin flagella. *Exp. Cell Res.* **74**, 535–40.

Middleton, C. A. 1977. The effect of cell–cell contact on the spreading of pigmented retina epithelial cells in culture. *Exp. Cell Res.* **109**, 349–59.

Middleton, C. A. 1979. Cell-surface labelling reveals no evidence for membrane assembly and disassembly during fibroblast locomotion. *Nature (Lond.)* **282**, 203–5.

Middleton, C. A. 1982. Cell contacts and the locomotion of epithelial cells. In *Cell behaviour*, R. Bellairs, A. Curtis and G. Dunn (eds), 159–82. Cambridge: CUP.

Middleton, C. A. and J. A. Sharp 1984. *Cell locomotion* in vitro: *techniques and observations*. London: Croom Helm.

Mikawa, T., Y. Nonamura, M. Hirata, S. Ebashi and S. Kakiuchi 1978. Involvement of an acidic protein in regulation of smooth muscle contraction by the tropomyosin–leiotonin system. *J. Biochem.* **84**, 1633–6.

Mooseker, M. S., T. D. Pollard and K. Fujiwara 1978. Characterization and localization of myosin in the brush border of intestinal epithelial cells. *J. Cell Biol.* **79**, 444–53.

Mooseker, M. S., T. A. Graves, K. A. Wharton, N. Falco and C. L. Howe 1980. Regulation of microvillus structure: calcium-dependent solation and cross-linking of actin filaments in the microvilli of intestinal epithelial cells. *J. Cell Biol.* **87**, 809–22.

Mooseker, M. S., E. M. Bonder, B. G. Grimwade, C. L. Howe, T. C. S. Keller,

R. H. Wasserman and K. A. Wharton 1982. Regulation of contractility, cytoskeletal structure, and filament assembly in the brush border of intestinal epithelial cells. *Cold Spring Harbor Symp.* **46**, 855–70.

Murphy, D. B. and G. G. Borisy 1975. Association of high-molecular-weight proteins with microtubules and their rôle in microtubule assembly *in vitro*. *Proc. Nat. Acad. Sci. USA* **72**, 2696–700.

Murphy, D. B. and L. G. Tilney 1974. The rôle of microtubules in the movement of pigment granules in teleost melanophores. *J. Cell Biol.* **61**, 757–79.


Naccache, P. H., M. Volpi, H. J. Showell, E. L. Becker and R. I. Sha'afi 1979. Chemotactic factor-induced release of membrane calcium in rabbit neutrophils. *Science* **203**, 461–3.

Naitoh, Y. and R. Eckert 1974. Control of ciliary activity in Protozoa. In *Cilia and flagella*, M. A. Sleigh (ed.), 305–52. London: Academic.

Nath, J. and J. I. Gallin 1984. Modulation of tubulin tyrosinylation in human polymorphonuclear leukocytes (PMN). In *White cell mechanics: basic science and clinical aspects*, H. J. Meiselman, M. A. Lichtman and P. L. LaCelle (eds), 95–110. New York: Liss.

Neidel, J. E. and B. L. Dolmatch 1983. Cellular processing of the formyl peptide receptor. In *Leucocyte locomotion and chemotaxis*, H. -U. Keller and G. O. Till (eds), 309–22. Basel: Birkhauser.

Newell, P. C. 1978. Cellular communication during aggregation of *Dictyostelium*. The second Fleming Lecture. *J. Gen Microbiol.* **104**, 1–13.

Newell, P. C. 1981. Chemotaxis in the cellular slime moulds. In *Biology of the chemotactic response*, J. M. Lackie and P. C. Wilkinson (eds), 89–114. Cambridge: CUP.

Newell, P. C. and I. A. Mullens 1978. Cell-surface cAMP receptors in *Dictyostelium*. *Symp. Soc. Exp. Biol.* **32**, 161–202.

Nultsch, W. 1974. Movements. In *Algal physiology and biochemistry*, W. D. P. Stewart (ed.), 864–93. Oxford: Blackwell.


Ockleford, C. D. and J. B. Tucker 1973. Growth, breakdown, repair, and rapid contraction of microtubular axopodia in the heliozoan *Actinophrys sol.* *J. Ultrastruct. Res.* **44**, 369–87.

Oertel, D., S. J. Schein and C. Kung 1977. Separation of membrane currents using a *Paramoecium* mutant. *Nature (Lond.)* **268**, 120–4.

Ogihara, S., M. Ikebe, K. Takahashi and A. Tonamura 1983. Requirement of phosphorylation of *Physarum* myosin heavy chain for thick filament formation, actin activation of $Mg^{2+}$-ATPase activity, and $Ca^{2+}$ inhibitory superprecipitation. *J. Biochem.* **93**, 205–23.

Ohara, P. T. and R. C. Buck 1979. Contact guidance *in vitro*. A light, transmission and scanning electron microscope study. *Exp. Cell Res.* **121**, 235–49.

Omoto, C. K. and C. J. Brokaw 1982. Structure and behaviour of the sperm terminal filament. *J. Cell. Sci.* **58**, 385–409.

O'Neill, C. H., P. N. Riddle and P. W. Jordan 1979. The relation between surface area and anchorage dependence of growth in hamster and mouse fibroblasts. *Cell* **16**, 909–18.


Pallini, V., M. Bugnoli, C. Mencarelli and G. Scapigliati 1982. Biochemical properties of ciliary, flagellar and cytoplasmic dyneins. *Symp. Soc. Exp. Biol.* **35**, 339–52.

Parkinson, E. K. and J. G. Edwards 1978. Non-reciprocal contact inhibition of locomotion of chick embryonic choroid fibroblasts by pigmented retina epithelial cells. *J. Cell Sci.* **33**, 103–20.

Pate, J. L. and L. -Y. E. Chang 1979. Evidence that gliding motility in prokaryotic cells is driven by rotary assemblies in the cell envelopes. *Curr. Microbiol.* **2**, 59–64.

Pethica, B. A. 1980. Microbial and cell adhesion. In *Microbial adhesion to surfaces*, R. C. W. Berkeley, J. M. Lynch, J. Melling, P. R. Rutter and B. Vincent (eds), 19–45. Chichester: Ellis Horwood.

Pfeffer, W. 1884. Locomotorische Richtungsbewegungen durch chemische Reize. *Untersuchungen ans dem Botanischen Institut zu Tubingen* **1**, 363. (Quoted in Wilkinson 1982).

Piras, R. and M. M. Piras 1975. Changes in microtubule phosphorylation during cell cycle of HeLa cells. *Proc. Nat. Acad. Sci. USA* **72**, 1161–5.

Poff, K. L. and B. D. Whitaker 1979. Movement of slime molds. In *Encyclopedia of plant physiology*, New Series, vol. 7, W. Haupt and M. E. Feinleib (eds), 355–82. Berlin: Springer.

Pollard, T. D. and S. Ito 1970. Cytoplasmic filaments of *Amoeba proteus*. I. The rôle of filaments in consistency changes and movement. *J. Cell Biol.* **46**, 267–89.

Pollard, T. D. and E. D. Korn 1971. Filaments of *Amoeba proteus*. II. Binding of heavy meromyosin by thin filaments in motile cytoplasmic extracts. *J. Cell Biol.* **48**, 216–19.

Porter, R. and D. W. Fitzsimons (eds) 1973. *Locomotion of Tissue Cells*. Amsterdam: Elsevier. (CIBA Foundation Symp. **14** (New Series).)

Rashke, K. 1979. Movements of stomata. In *Encyclopedia of Plant Physiology*, New Series, vol. 7, W. Haupt and M. E. Feinleib (eds), 383–441. Berlin: Springer.

Rebhun, L. I. 1972. Polarized intracellular particle transport: saltatory movements and cytoplasmic streaming. *Int. Rev. Cytol.* **32**, 93–137.

Revel, J. P. 1974. Contacts and junctions between cells. *Symp. Soc. Exp. Biol.* **28**, 447–62.

Rich, A. and A. K. Harris 1981. Anomalous preferences of cultured macrophages for hydrophobic and roughened substrata. *J. Cell. Sci.* **50**, 1–7.

Rinaldi, R. and M. Opas 1976. Graphs of contracting glycerinated *Amoeba proteus*. *Nature (Lond.)* **260**, 525–6.

Roberts, K. 1974. Cytoplasmic microtubules and their functions. *Prog. Biophys. Mol. Biol.* **28**, 373–420.

Roberts, K. and J. S. Hyams (eds) 1979. *Microtubules*. London: Academic.

Roberts, T. M. and S. Ward 1982. Directed membrane flow on the pseudopods of *Caenorhabditis elegans* spermatozoa. *Cold Spring Harbor Symp.* **46**, 695–702.

Rohrschneider, L., M. Rosok and K. Shriver 1982. Mechanism of transformation by Rous sarcoma virus: events within adhesion plaques. *Cold Spring Harbor Symp.* **46**, 953–66.

Routledge, L. M. 1978. Calcium-binding proteins in the vorticellid spasmoneme. Extraction and characterization by gel electrophoresis. *J. Cell Biol.* **77**, 358–70.

Royer-Pokora, B., H. Beug, M. Claviez, H.-J. Winkhardt, R. R. Friis and T. Graf 1978. Transformation parameters in chicken fibroblasts transformed by AEV and MC29 avian leukemia viruses. *Cell* **13**, 751–60.

Salmon, E. D. and R. R. Segal 1980. Calcium-labile mitotic spindles isolated from sea urchin eggs (*Lytechinus variegatus*). *J. Cell Biol.* **86**, 355–65.

Satir, B. 1974. Membrane events during the secretory process. *Symp. Soc. Exp. Biol.* **28**, 399–418.

Satir, P. 1974. How cilia move. *Sci. Am.* **231**(4), 44–53.

Satir, P. 1982. Mechanisms and control of microtubule sliding in cilia. *Symp. Soc. Exp. Biol.* **35**, 179–202.

Satter, R. L. 1979. Leaf movements and tendril curling. In *Encyclopedia of plant physiology*, New Series, vol. 7, W. Haupt and M. E. Feinleib (eds), 442–84. Berlin: Springer.

Satter, R. L. and A. W. Galston 1981. Mechanisms of control of leaf movements. *Ann. Rev. Plant Physiol.* **32**, 83–110.

Schliwa, M. 1978. Microtubular apparatus of melanophores. Three-dimensional organisation. *J. Cell Biol.* **76**, 605–14.

Schliwa, M. 1979. Stereo high voltage electron microscopy of melanophores. Matrix transformations during pigment movements and the effects of cold and colchicine. *Exp. Cell Res.* **118**, 323–40.

Schliwa, M., K. B. Pryzwansky and U. Euteneuer 1982. Centrosome splitting in neutrophils: an unusual phenomenon related to cell activation and motility. *Cell* **31**, 705–17.

Schmitt, H., R. Josephs and E. Reisler 1977. A search for *in vivo* factors in regulation of microtubule assembly. *Nature (Lond.)* **265**, 653–5.

Scholey, J. M., K. A. Taylor and J. Kendrick-Jones 1980. Regulation of non-muscle myosin assembly by calmodulin-dependent light chain kinase. *Nature (Lond.)* **287**, 233–5.

Schollmeyer, J. E., L. T. Furcht, D. E. Goll, R. M. Robson and M. H. Stromer 1976. Localization of contractile proteins in smooth muscle cells and in normal and transformed fibroblasts. *Cold Spring Harbor Conf. Cell Prolif.* **3**, 361–88.

Schrevel, J., E. Caigneaux, D. Gros and M. Philippe 1983. The three cortical membranes of the gregarines. I. Ultrastructural organisation of *Gregarina blaberae. J. Cell Sci.* **61**, 151–79.

Schroeder, T. E. 1973. Actin in dividing cells: contractile ring filaments bind heavy meromyosin. *Proc. Nat. Acad. Sci. USA* **70**, 1688–92.

Schroeder, T. E. 1976. Actin in dividing cells: evidence for its rôle in cleavage but not mitosis. *Cold Spring Harbor Conf. Cell Prolif.* **3**, 265–78.

Schwab, M. E., R. Heumann and H. Thoenen 1982. Communication between target organs and nerve cells: retrograde axonal transport and site of action of nerve growth factor. *Cold Spring Harbor Symp.* **46**, 125–34.

Sefton, B. M., T. Hunter, E. A. Nigg, S. J. Singer and G. Walter 1982. Cytoskeletal targets for viral transforming proteins with tyrosine protein kinase activity. *Cold Spring Harbor Symp.* **46**, 939–52.

Segall, J. E., M. D. Manson and H. C. Berg 1982. Signal processing times in bacterial chemotaxis. *Nature (Lond.)* **296**, 855–7.

Sheetz, M. P. and J. A. Spudich 1983. Movement of myosin-coated fluorescent beads on actin cables *in vitro. Nature (Lond.)* **303**, 31–5.

Sheterline, P. 1983. *Mechanisms of cell motility: molecular aspects of contractility*. London: Academic.

Sheterline, P. and C. R. Hopkins 1981. Transmembrane linkage between surface glycoproteins and components of the cytoplasm in neutrophil leococytes. *J. Cell Biol.* **90**, 743–54.

Shields, J. M. and W. S. Haston 1985. Behaviour of neutrophil leucocytes in uniform concentrations of chemotactic factors: contraction waves, cell polarity and persistence. *J. Cell Sci.* **74**, 75–93.

Showell, H. J., R. J. Freer, S. H. Zigmond, E. Schiffman, S. Aswanikumar, B. Corcoran and E. L. Becker 1976. The structure-activity relations of synthetic peptides as chemotactic factors and inducers of lysosomal enzyme secretion for neutrophils. *J. Exp. Med.* **143**, 1154–69.

Silverman, M. and M. Simon 1974. Flagellar rotation and the mechanism of bacterial motility. *Nature (Lond.)* **249**, 73–4.

Sims, D. E. and J. A. Westfall 1982. Microfilament-associated adhering junctions (6 nm F-maculae adherentes) connect bovine pulmonary fibroblasts *in vivo*. *Eur. J. Cell Biol.* **28**, 145–50.

Sleigh, M. A. (ed.) 1974. *Cilia and flagella*. London: Academic.

Sleigh, M. A. 1976. Fluid propulsion by cilia and the physiology of ciliary systems. In *Perspectives in experimental biology*, vol. I, P. S. Davies (ed.), 125–34. Oxford: Pergamon.

Sleigh, M. A. and D. I. Barlow 1982. How are different ciliary beat patterns produced? *Symp. Soc. Exp. Biol.* **35**, 139–58.

Sloboda, R. D., S. A. Rudolph, J. L. Rosenbaum and P. Greengard 1975. Cyclic-AMP-dependent endogenous phosphorylation of a microtubule-associated protein. *Proc. Nat. Acad. Sci. USA* **72**, 177–81.

Smith, D. S., U. Jarlfors and R. Beranek 1970. The organization of synaptic axoplasm in the lamprey central nervous system. *J. Cell Biol.* **46**, 199–219.

Smith, R. P. C., J. M. Lackie and P. C. Wilkinson 1979. The effects of chemotactic factors on the adhesiveness of rabbit neutrophil granulocytes. *Exp. Cell Res.* **122**, 169–77.

Snyder, J. A. and J. R. McIntosh 1975. Initiation and growth of microtubules from mitotic centers in lysed mammalian cells. *J. Cell Biol.* **67**, 744–65.

Snyderman, R. 1983. Chemoattractant receptor affinity reflects its ability to transduce different biological responses. In *Leucocyte locomotion and chemotaxis*, H. -U. Keller and G. O. Till (eds), 323–36. Basel: Birkhauser.

Sobieszek, A. and J. V. Small 1976. Myosin-linked $Ca^{2+}$-regulation in vertebrate smooth muscle. *J. Mol. Biol.* **102**, 75–92.

Sobieszek, A. and J. V. Small 1977. Regulation of the actin–myosin interaction in vertebrate smooth muscle: activation *via* a myosin light-chain kinase and the effect of tropomyosin. *J. Mol. Biol.* **112**, 559–76.

Sobue, K., Y. Muramoto, M. Fujita and S. Kakiuchi 1981. Purification of a calmodulin-binding protein from chicken gizzard that interacts with F-actin. *Proc. Nat. Acad. Sci. USA* **78**, 5652–5.

Springer, M. S., M. F. Goy and J. Adler 1979. Protein methylation in behavioural control mechanisms and in signal transduction. *Nature (Lond.)* **280**, 279–84.

Spudich, J. L. and D. E. Koshland 1975. Quantitation of the sensory response in bacterial chemotaxis. *Proc. Nat. Acad. Sci. USA* **72**, 710–13.

Squire, J. M. 1981. *The structural basis of muscle contraction*. New York: Plenum.

Stadler, J. and W. W. Franke 1974. Characterization of the colchicine binding of membrane fractions from rat and mouse liver. *J. Cell Biol.* **60**, 297–303.

Stebbings, H. and C. E. Bennett 1976. The effect of colchicine on the sleeve element of microtubules. *Exp. Cell Res.* **100**, 419–23.

Stebbings, H. and J. S. Hyams 1979. *Cell motility*. London: Longman.

Steinberg, M. S. 1962a. On the mechanism of tissue reconstruction by dissociated cells. I. Population kinetics, differential adhesiveness, and the absence of directed migration. *Proc. Nat. Acad. Sci. USA* **48**, 1577–82.

Steinberg, M. S. 1962b. Mechanism of tissue reconstruction by dissociated cells. II. Time course of events. *Science* **137**, 762-3.

Steinberg, M. S. 1962c. On the mechanism of tissue reconstruction by dissociated cells. III. Free energy relations and the reorganisation of fused heteronomic tissue fragments. *Proc. Nat. Acad. Sci. USA* **48**, 1769–76.

Steinberg, M. S. 1970. Does differential adhesiveness govern self-assembly processes in histogenesis? Equilibrium configurations and the emergence of a hierarchy among populations of embryonic cells. *J. Exp. Zool.* **173**, 395–434.

Steinemann, C., M. Fenner, H. Binz and R. W. Parish 1984. Evidence that the invasive behaviour of mouse sarcoma cells is inhibited by blocking a 37 000 dalton plasma membrane glycoprotein with Fab. *Proc. Nat. Acad. Sci. USA* **81**, 3747–50.

Stendahl, O. I., J. H. Hartwig, E. A. Brotschi and T. P. Stossel 1980. Distribution of actin-binding protein and myosin in macrophages during spreading and phagocytosis. *J. Cell Biol.* **84**, 215–24.

Stephens, R. E. 1975. Structural chemistry of the axoneme: evidence for chemically and functionally unique tubulin dimers in outer fibers. In *Molecules and cell movement*, S. Inoué and R. E. Stephens (eds), 181–95. New York: Raven.

Storti, R. V., D. M. Coen and A. Rich 1976. Tissue-specific forms of actin in the developing chick. *Cell* **8**, 521–7.

Sugimoto, Y. 1981. Effect on the adhesion and locomotion of mouse fibroblasts by their interacting with differently charged substrates. *Exp. Cell Res.* **135**, 39–45.

Sugimoto, Y. and A. Hagiwara 1979. Cell locomotion on differently charged substrates. Effects of substrate charge on locomotive speed of fibroblastic cells. *Exp. Cell Res.* **120**, 245–52.

Summers, K. E. and I. R. Gibbons 1971. Adenosine triphosphate-induced sliding of tubules in trypsin-treated flagella of sea-urchin sperm. *Proc. Nat. Acad. Sci. USA* **68**, 3092–6.

Szent-Gyorgyi, A. G., E. M. Szentkiralyi and J. Kendrick-Jones 1973. The light chains of scallop myosin as regulatory subunits. *J. Mol. Biol.* **74**, 179–203.

Szollosi, D. 1970. Cortical cytoplasmic filaments of cleaving eggs: a structural element corresponding to the contractile ring. *J. Cell Biol.* **44**, 192–209.

Takahashi, K., C. Shingyoji and S. Kamimura 1982. Microtubule sliding in reactivated flagella. *Symp. Soc. Exp. Biol.* **35**, 159–78.

Tamm, S. L. and S. Tamm 1981. Ciliary reversal without rotation of axonemal structures in Ctenophore comb plates. *J. Cell Biol.* **89**, 495–509.

Taylor, D. L., Y. -L. Wang and J. M. Heiple 1980. Contractile basis of amoeboid movement. VII. The distribution of fluorescently labelled actin in living amoebas. *J. Cell Biol.* **86**, 590–8.

Taylor, D. L., J. S. Condeelis, P. L. Moore and R. D. Allen 1973. The contractile basis of amoeboid movement. I. The chemical control of motility in isolated cytoplasm. *J. Cell Biol.* **59**, 378–94.

Telzer, B. R. and L. T. Haimo 1981. Decoration of spindle microtubules with dynein: evidence for uniform polarity. *J. Cell Biol.* **89**, 373–8.

Thompson, D'A. W. 1942. *Growth and form*. Cambridge: CUP.

Tickle, C. 1982. Mechanisms of invasiveness: how cells behave when implanted into the developing chick wing. In *Cell behaviour*, R. Bellairs, A. Curtis and G. Dunn (eds), 529–54. Cambridge: CUP.

Tickle, C. and A. Crawley 1979. Infiltration and survival: the behaviour of normal, invasive cells implanted to the developing chick wing. *J. Cell Sci.* **40**, 257–70.

Tickle, C., A. Crawley and M. Goodman 1978a. Cell movement and the mechanism of invasiveness: a survey of behaviour of some normal and malignant cells implanted into the developing chick wing bud. *J. Cell Sci.* **31**, 293–322.

Tickle, C., A. Crawley and M. Goodman 1978b. Mechanisms of behaviour of epithelial tumours: ultrastructure of the interactions of carcinoma cells with embryonic mesenchyme and epithelium. *J. Cell Sci.* **33**, 133–55.

Tilney, L. G. 1975a. Actin filaments in the acrosomal reaction of *Limulus* sperm. Motion generated by alterations in the packing of the filaments. *J. Cell Biol.* **64**, 289–310.

Tilney, L. G. 1975b. The rôle of actin in non-muscle cell motility. In *Molecules and cell movement*, S. Inoué and R. E. Stephens (eds), 339–88. New York: Raven.

Tilney, L. G. 1976a. The polymerization of actin. II. How non-filamentous actin becomes non-randomly distributed in sperm: evidence for the association of this actin with membranes. *J. Cell Biol.* **69**, 51–72.

Tilney, L. G. 1976b. The polymerization of actin. III. Aggregates of non-filamentous actin and its associated proteins: a storage form of actin. *J. Cell Biol.* **69**, 73–89.

Tilney, L. G. 1978. Polymerization of actin. V. A new organelle, the actomere, that initiates the assembly of actin filaments in *Thyone* sperm. *J. Cell Biol.* **77**, 551–64.

Tilney, L. G. and S. Inoué 1982. Acrosomal reaction of *Thyone* sperm. II. The kinetics and possible mechanism of acrosomal process elongation. *J. Cell Biol.* **93**, 820-7.

Tilney, L. G. and N. Kallenbach 1979. Polymerization of actin. VI. The polarity of the actin filaments in the acrosomal process and how it might be determined. *J. Cell Biol.* **81**, 608–23.

Tilney, L. G., E. M. Bonder and D. J. DeRosier 1981. Actin filaments elongate from their membrane-associated ends. *J. Cell Biol.* **90**, 485–94.

Tilney, L. G., D. P. Kiehart, C. Sardet and M. Tilney 1978. Polymerization of actin. IV. Rôle of $Ca^{2+}$ and $H^+$ in the assembly of actin and in membrane fusion in the acrosomal reaction of echinoderm sperm. *J. Cell Biol.* **77**, 536–50.

Tomasek, J. J., E. D. Hay and K. Fujiwara 1982. Collagen modulates cell shape and cytoskeleton of embryonic corneal and fibroma fibroblasts: distribution of actin, α-actinin, and myosin. *Dev. Biol.* **92**, 107–22.

Tosney, K. W. and N. K. Wessells 1983. Neuronal motility: the ultrastructure of veils and microspikes correlates with their motile activities. *J. Cell Sci.* **61**, 389–411.

Traeger, L. and M. A. Goldstein 1983. Thin filaments are not of uniform length in rat skeletal muscle. *J. Cell Biol.* **96**, 100–3.

Trinick, J. and A. Elliott 1979. Electron microscope studies of thick filaments from vertebrate skeletal muscle. *J. Mol. Biol.* **131**, 133–6.

Trinkaus, J. P. 1984. *Cells into organs. The forces that shape the embryo*, 2nd edn. Englewood Cliffs, NJ: Prentice-Hall.

Tucker, J. B. 1972. Microtubule-arms and propulsion of food particles inside a large feeding organelle in the ciliate *Phascolodon vorticella*. *J. Cell Sci.* **10**, 883–903.

Tucker, J. B. 1977. Shape and pattern specification during microtubule bundle assembly. *Nature (Lond.)* **266**, 22–6.

Vallee, R. B., M. J. DiBartolomeis and H. Theurkauf 1981. A protein kinase bound to the projection part of MAP-2 (microtubule-associated protein 2). *J. Cell. Biol.* **90**, 568–76.

Van Houten, J. 1979. Membrane potential changes during chemokinesis in *Paramoecium*. *Science* **204**, 1101–3.

Van Oss, C. J., C. F. Gillman and A. W. Neumann 1975. *Phagocytic engulfment and cell adhesiveness.* New York: Marcel Dekker.

Vasiliev, J. M., I. M. Gelfand, L. V. Domnina, O. Y. Ivanova, S. G. Komm and L. V. Olshevskaja 1970. Effect of colcemid on the locomotory behaviour of fibroblasts. *J. Embryol. Exp. Morphol.* **24**, 625–40.

Vesely, P. and R. A. Weiss 1973. Cell locomotion and contact inhibition of normal and neoplastic rat cells. *Int. J. Cancer* **11**, 64–76.

Vicker, M. G., W. Schill and K. Drescher 1984. Chemoattraction and chemotaxis in *Dictyostelium discoideum*: Myxamoeba cannot read spatial gradients of cyclic adenosine monophosphate. *J. Cell Biol.* **98**, 2204–14.


Waddell, D. R. 1982. A predatory slime mould. *Nature (Lond.)* **298**, 464–5.

Wais-Steider, J. and P. Satir 1979. Effect of vanadate on gill cilia: switching mechanism in ciliary beat. *J. Supramol. Struct.* **11**, 339–47.

Wang, E. and R. D. Goldman 1974. Functions of cytoplasmic fibers in intracellular movements. *J. Cell Biol.* **63**, A726.

Wang, E. A. and D. E. Koshland 1980. Receptor structure in the bacterial sensing system. *Proc. Nat. Acad. Sci. USA* **77**, 7157–61.

Wang, E., J. A. Connolly, V. I. Kalnins and P. W. Choppin 1979. Relationship between movement and aggregation of centrioles in syncytia and formation of microtubule bundles. *Proc. Nat. Acad. Sci. USA* **76**, 5719–23.

Ward, S., Y. Argon and G. A. Nelson 1981. Sperm morphogenesis in wild-type and fertilization-defective mutants of *Caenorhabditis elegans*. *J. Cell Biol.* **91**, 26–44.

Weeds, A. 1982. Actin-binding proteins – regulators of cell architecture and motility. *Nature (Lond.)* **296**, 811–16.

Weinert, T. A. and P. Cappuccinelli 1982. A novel tubulin inactivating protein from *Dictyostelium discoideum*. In *Microtubules in microorganisms*, P. Cappuccinelli and N. R. Morris (eds), 129–42. New York: Dekker.

Weis-Fogh, T. and W. B. Amos 1972. Evidence for a new mechanism of cell motility. *Nature (Lond.)* **236**, 301–4.

Weisenberg, R. C., G. G. Borisy and E. W. Taylor 1968. The colchicine-binding protein of mammalian brain and its relation to microtubules. *Biochem.* **7**, 4466-79.

Weiss, P. 1958. Cell contact. *Int. Rev. Cytol.* **7**, 391–423.

Welsh, M. J., J. R. Dedman, B. R. Brinkley and A. R. Means 1978. Calcium-dependent regulator protein: localisation in mitotic apparatus of eukaryotic cells. *Proc. Nat. Acad. Sci. USA* **75**, 1867–71.

Weston, J. A., K. M. Yamada and K. L. Hendricks 1979. Characterization of factor(s) in culture supernatants affecting cell social behaviour. *J. Cell Physiol.* **100**, 445–55.

Wilkins, M. B. 1984. Gravitropism. In *Advanced plant physiology*, M. B. Wilkins (ed.), 163–85. London: Pitman.

Wilkinson, P. C. 1973. Recognition of protein structure in leucocyte chemotaxis. *Nature (Lond.)* **244**, 512–13.

Wilkinson, P. C. 1981. Peptide and protein chemotactic factors and their recognition by neutrophil leucocytes. In *Biology of the chemotactic response*, J. M. Lackie and P. C. Wilkinson (eds), 53–72. Cambridge: CUP.

Wilkinson, P. C. 1982. *Chemotaxis and inflammation*, 2nd edn. Edinburgh: Churchill Livingstone.

Wilkinson, P. C. and R. B. Allan 1978a. Binding of protein chemotactic factors to the surfaces of neutrophil leucocytes and its modification with lipid-specific bacterial toxins. *Mol. Cell. Biochem.* **20**, 25–40.

Wilkinson, P. C. and R. B. Allan 1978b. Chemotaxis of neutrophil leucocytes towards substratum-bound protein attractants. *Exp. Cell Res.* **117**, 403–12.

Wilkinson, P. C. and J. M. Lackie 1979. The adhesion, migration and chemotaxis of leucocytes in inflammation. In *Inflammatory reaction*, H. Z. Movat (ed.), 47–88. Berlin: Springer.

Wilkinson, P. C. and J. M. Lackie 1983. The influence of contact guidance on chemotaxis of human neutrophil leukocytes. *Exp. Cell Res.* **145**, 255–64.

Wilkinson, P. C. and I. C. McKay 1972. The molecular requirements for chemotactic attraction of leucocytes by proteins. Studies of proteins with synthetic side groups. *Eur. J. Immunol.* **2**, 570–7.

Wilkinson, P. C., J. M. Lackie and R. B. Allan 1982. Methods for measuring leucocyte locomotion. In *Cell Analysis*, vol. 1, N. Catsimpoolas (ed.), 145–94. New York: Plenum.

Wilkinson, P. C., J. M. Shields and W. S. Haston 1982. Contact guidance of human neutrophil leukocytes. *Exp. Cell Res.* **140**, 55–62.

Wilkinson, P. C., J. M. Lackie, J. V. Forrester and G. A. Dunn 1984. Chemokinetic accumulation of human neutrophils on immune-complex-coated substrata: analysis at a boundary. *J. Cell Biol.* **99**, 1761–8.

Williams, L. T., R. Snyderman, M. C. Pike and R. J. Lefkowitz 1977. Specific receptor sites for chemotactic peptides on human polymorphonuclear leucocytes. *Proc. Nat. Acad. Sci. USA* **74**, 1204–8.

Williamson, R. E. and C. C. Ashley 1982. Free $Ca^{2+}$ and cytoplasmic streaming in the alga *Chara. Nature (Lond.)* **296**, 647–51.

Wilson, L. 1970. Properties of colchicine binding protein from chick embryo brain. Interactions with Vinca alkaloids and podophyllotoxin. *Biochem.* **9**, 4999–5008.

Wiseman, L. L. and M. S. Steinberg 1973. The movement of single cells within solid tissue masses. *Exp. Cell Res.* **79**, 468-71.

Witman, G. B. and N. Minervini 1982. The mechanochemical cycle of dynein. *Symp. Soc. Exp. Biol.* **35**, 203–24.

Witman, G. B., J. Plummer and G. Sander 1978. *Chlamydomonas* flagellar mutants lacking radial spokes and central tubules. Structure, composition and function of specific axonemal components. *J. Cell Biol.* **76**, 729–47.

Yamada, K. M. 1983. Cell surface interactions with extracellular materials. *Ann. Rev. Biochem.* **52**, 761–99.

Yeh, P. -Z. and A. Gibor 1970. Growth patterns and motility of *Spirogyra* sp. and *Closteridium acerosum. J. Phycol.* **6**, 44–8.

Yin, H. L. and T. P. Stossel 1979. Control of cytoplasmic actin gel–sol transformation by gelsolin, a calcium-dependent regulatory protein. *Nature (Lond.)* **281**, 583–6.

Zigmond, S. H. 1974. Mechanisms of sensing gradients by polymorphonuclear leucocytes. *Nature (Lond.)* **249**, 450–2.

Zigmond, S. H. 1977. Ability of polymorphonuclear leukocytes to orient in gradients of chemotactic factors. *J. Cell Biol.* **75**, 606–16.

Zigmond, S. H. 1982. Polymorphonuclear leucocyte response to chemotactic gradients. In *Cell Behaviour*, R. Bellairs, A. Curtis and G. Dunn (eds), 183–202. Cambridge: CUP.

Zigmond, S. H. and J. G. Hirsch 1973. Leukocyte locomotion and chemotaxis. New methods for evaluation, and demonstration of a cell-derived chemotactic factor. *J. Exp. Med.* **137**, 387–410.

Zigmond, S. H. and S. J. Sullivan 1981. Receptor modulation and its consequences for the response to chemotactic peptides. In *Biology of the chemotactic response*, J. M. Lackie and P. C. Wilkinson (eds), 73–87. Cambridge: CUP.

Zigmond, S. H., H. I. Levitsky and B. J. Kreel 1981. Cell polarity: an examination of its behavioral expression and it consequences for polymorphonuclear leucocyte chemotaxis. *J. Cell Biol.* **89**, 585–92.

# INDEX

Numbers in bold type refer to pages on which text sections related to the topic commence. Page numbers annotated 'g' refer to glossary entries. Numbers in italic type refer to text figures.